Welfare, Law, and Globalization

Collected Papers in Theoretical Economics

Volume III

Welfare, Law, and Globalization

KAUSHIK BASU

OXFORD

UNIVERSITY PRESS

OXFORD
UNIVERSITY PRESS

YMCA Library Building, Jai Singh Road, New Delhi 110 001

Oxford University Press is a department of the University of Oxford.
It furthers the University's objective of excellence in research,
scholarship, and education by publishing worldwide in

Oxford   New York

Auckland   Cape Town   Dar es Salaam   Hong Kong   Karachi
Kuala Lumpur   Madrid   Melbourne   Mexico City   Nairobi
New Delhi   Shanghai   Taipei   Toronto

With offices in
Argentina   Austria   Brazil   Chile   Czech Republic   France   Greece
Guatemala   Hungary   Italy   Japan   Poland   Portugal   Singapore
South Korea   Switzerland   Thailand   Turkey   Ukraine   Vietnam

Oxford is a registered trademark of Oxford University Press
in the UK and in certain other countries

Published in India by Oxford University Press, New Delhi

ISBN-13: 978-019-568648-2
ISBN-10: 019-568648-9

Typeset in AGaramond 10.5/12.3 by Jojy Philip
Printed in India at De Unique , New Delhi 110 018
Published by Oxford University Press
YMCA Library Building, Jai Singh Road, New Delhi 110 001

Contents

Tables and Figures

Abbreviations

CF	Cardinal full comparable
CU	Cardinal unit comparable
DPT	First-difference preserving transformation
EEOC	Equal Employment Opportunity Commission
EPZ	Export Processing Zone
ESAF	Enhanced structural adjustment facility
GDP	Gross domestic product
GE	General equilibrium
GERC	General equilibrium reversal claim
GWP	Gibbardian waiver profile
ICC	International Criminal Court
ILO	International Labour Organization
IMF	International Monetary Fund
IT	Information technology
LSE	London School of Economics
OECD	Organization for Economic Co-operation and Development
OL	Ordinal level comparable
PAT	Positive affine transformation
PGRF	Poverty reduction and growth facility
PPP	Purchasing power parity

SAF   Structural adjustment facility

SCR   Social choice rule

SEK   Svensk Exportkredit

SEZ   Special Economic Zone

SWF   Social welfare function

SWFL   Social welfare functional

SWO   Social welfare ordering

SWR   Social welfare relation

TDA   Trade and Development Agency

TDP   Trade and Development Programme

UNDP   United Nations Development Programme

USAID   United States Agency for International Development

WTO   World Trade Organization

1 Introduction

1.1 BY WAY OF PREFACE

This collection of my papers brings together writings that reflect my first interests in economics and my last, or, to put it in more optimistic terms, my most recent. It was welfare economics and moral philosophy that enticed me into the discipline when I was a dilettante—reading economics, but with the idea of eventually abandoning it to become a lawyer. It was the aesthetic appeal of the abstract deductive reasoning that forms the core of welfare economics that made me decide to abandon law and instead do a PhD in economics from the London School of Economics (LSE). I plunged into the world of Kenneth Arrow, Paul Samuelson, Amartya Sen and also the early thinkers like David Hume and John Stuart Mill, never to look back again.

My initial interest arose while sitting on Amartya Sen's riveting lectures on social choice theory and welfare economics and listening to the other great teachers at LSE—Terence Gorman and Michio Morishima on economic theory, John Watkins and some visiting philosophers on ethics, the philosophy of science, and mathematical logic. I remember vividly my first encounter with Arrow's impossibility theorem concerning voting systems for aggregating individual preferences to arrive at society's preference. Subject to the ambiguity of what counts as 'single', I consider this the single most important result that has emerged out of the discipline of economics. There certainly are harder theorems that have been discovered, there are propositions that are mathematically more demanding, and there are theorems that have had more important policy implications and therefore have had a larger impact on our lives. What was amazing about Arrow's result was that it was so out of the blue. There have been some important theorems in economics, where, if you happened to know the relevant mathematics, half your task was done. But Arrow's theorem needed neither any special knowledge of mathematics (in fact that was one of the reasons why the initial, long proof was so hard to read) nor any prior training in economics or philosophy. What made this a remarkable achievement is that, in principle, anybody could have proved the theorem. It was just that the person had to have the capacity to think out of the box and the ability to conduct a very long proof of which each step was easy, but to conceive the whole thing—to be aware that there was light at the end of the tunnel and so it was worth continuing to establish each little step—was a staggering intellectual feat.

I was also interested in equilibrium analysis and maintain that, as far as books in economics go, the most beautiful of the last century was Gerard Debreu's *Theory of Value*. At just over 100 pages, the terse mathematical treatise is of course acknowledged to be a great work of science, but it is also an unparalleled artistic achievement. It has been said about some great paintings that not a stroke could be different without adversely affecting the creation. The same can be said about Debreu's masterpiece. It was perfectly constructed—spare and minimalist, with virtually not a word that could be considered redundant or out of place.

Unfortunately, general equilibrium theory was a full-time discipline. People devoted their lives to it. I never acquired the kind of mastery needed to seriously contribute to the field. I took courses on general equilibrium analysis and learned its basics, like all graduate students of economics in the mid-1970s, so that I could be a user of the theory. Briefly, as I was starting out on my PhD, I was tempted to work on an off-shoot of general equilibrium theory—namely, decentralized planning on which I knew the works of Edmond Malinvaud and Janos Kornai. My professors at the LSE saved me from what would have been a mistake, by deflecting me off that topic, for had I done a PhD on it, I would probably be the world's sole surviving practioner of decentralized planning theory.

Fortunately, my main interest was in welfare economics and I started doing research in the field—actually in a field that is best described as the intersection of welfare economics and choice theory. Not surprisingly, my first published papers were largely in this subdiscipline. At that stage, I have to confess, my research interest was driven by nothing else but aesthetics. I did not even pause to think about the practical utility of the discipline.

After I returned to India from London my interests and focus of research moved on to development economics, game theory, and industrial organization. The previous two volumes of my collected works are based on my papers in these fields. In more recent times I have got interested in law and economics and in globalization and international political economy and have tried to contribute to these fields. Unlike my earlier work, the mainspring of my research in this area is rooted in real-world policy concerns. I have been troubled by the way policy gets crafted and new laws take shape. The papers were often written in exasperation and in the hope of influencing policy.

What I realized only after some time is how deeply some of these more applied papers of recent times were rooted in my early theoretical research in welfare economics and choice theory. The present book therefore becomes an excellent opportunity for me to bring together these two sets of papers, written over a long stretch of time but tied together by a common methodology, and to make them available to students and fellow economists.

In bringing this project to fruition I am grateful to my editors at Oxford

University Press; to my research assistants of recent years at Cornell University, Hyejin Ku and Doug Mitaronda; and to my secretary, Johanna Schroeder.

1.2 WELFARE AND EQUILIBRIUM

As soon as Arrow's famous impossibility theorem was published and its nihilistic implications became clear, researchers started seeking routes to escape the seemingly water-tight grip of the theorem. It was gradually realized that the theorem was based on a rather narrow view of how individuals expressed their preferences—to wit, by writing down orderings over the alternatives from which choice had to be made. For purely voting situations, this was realistic, but human beings do carry in their heads more information about what they like and what they do not. People may not only prefer x to y and y to z but their preference for x over y may be much greater than their preference for y over z. On a ballot paper there may not be room to express these preferences over 'differences', but people do often express such preferences in other ways and, if allowing individuals to express greater information about their preferences is a way out of Arrow's impossibility result, then that may be a reason enough to study such preferences and build up a theory of social choice based on them.

Around the time that I was doing my PhD there was an active group of people writing papers on the subject of utility measurement and interpersonal comparisons. The works of Claude d'Aspremont, Charles Blackorby, Louis Gevers, Peter Hammond, Eric Maskin, Kevin Roberts, Amartya Sen, John Weymark, and many more stirred up a lot of excitement in the field. While working on that 'contemporary' work I also got interested in some early research on individual utility measurement. Much of this had happened in the 1930s, theories by O. Lange, H. Bernadelli, F. Zeuthen, and others, and there was the even earlier classic of Vilfredo Pareto. Marrying some of this early research with the work that was going on then, I was able to prove a theorem on conditions under which utilitarianism was the natural welfare function to use when aggregating individual preferences. The axiomatization of utilitarianism had already occurred in social choice theory—in fact, there were several alternative theorems on this. My paper looked at this largely from the point of view of information availability of individuals and the domain of their preference. The result was published in the *Journal of Mathematical Economics* in 1983 and is reprinted here as Chapter 2.

One question that had been asked in the 1930s but never satisfactorily answered is the following: Suppose an individual can compare not only alternatives, such as $x, y, z, w \ldots$, but also the gaps between the alternatives (i.e., if in terms of his preference the gap between x and y is greater than the gap between z and w), and the gaps between the gaps, and the gaps between the gaps between the gaps, and so on. Clearly, the measurability of utility of an individual who can make these gap comparisons is stronger than ordinal utility. But can we characterize the extent of measurability of utility of this person? I showed in an earlier paper, 'Determinateness of the Utility Function: Revisiting a Controversy of the Thirties' (*Review of Economic Studies*, 1982) that, provided that some weak conditions are satisfied, this individual's utility can be characterized as cardinal, that is, the individual's preferences can be characterized exactly by a family of utility functions that are positive affine transformations of one another. In other words, cardinal utility is what obtains in the limit of the process of being able to compare higher and higher order gaps. The paper appears as Chapter 3 in the present book.

I returned to some related problems of choice and utility after a long time in early 2000, when Tapan Mitra and I decided to revisit the age-old problem of aggregating infinite streams of returns. This can be thought of as a problem of individual choice or one of aggregating the preferences of different individuals or generations. If x_t denotes the utility earned in period t, then $x = (x_1, x_2, ...)$ is an infinite 'stream of returns'. If we are interested in individual decision making, then we can think of each of x_t as the same person's expected utility in period t. If we are interested in intergenerational problems we can think of x_t as generation t's utility or satisfaction. In this case the problem of aggregating a stream of returns is a problem of social choice. It has been known for a long time—from the works of Frank Ramsay, Tjalling Koopmans, Peter Diamond, and Lars-Gunnar Svensson— that if we wish to represent every stream of return by a real number (generated presumably by a welfare function) and then describe stream x as superior to y if and only if $w(x) > w(y)$, where w is the real-valued welfare function, then we tend to run into some ethical problems.

Mitra and I managed to show that this problem was actually more acute than had been realized. Consider two standard axioms of social choice: Pareto and Anonymity. If streams x and y are such that stream x does better than y in some periods and no where does worse, then clearly $w(x)$ should be greater than $w(y)$. This is the Pareto axiom. Next consider two streams that are identical in all but two periods, t and k, and in these two periods the streams get returns that are simply a transpose of each other, that is, $x_t = y_k$ and $x_k = y_t$. Then it must be the case that $w(x) = w(y)$. In other words, w must satisfy intergenerational equity. This is the Anonymity axiom. The theorem that we published in *Econometrica*, 2003, and is reprinted here as Chapter 4, shows that there is no social welfare function that satisfies the Pareto and Anonymity axioms.

This disturbingly negative result generated a lot of interest among researchers in the field and raised the inevitable question as to how we can weaken the axioms and regain possibility. Tapan Mitra and I put in a lot of effort into finding ways in which the Pareto axiom may be weakened or the domain of the exercise limited so as to recover a real-valued social welfare function. Such an exercise also inevitably meant encountering other new impossibility theorems. Chapter 5 reproduces the paper that charts the terrain of possibility and impossibility in aggregating infinite utility streams using a real-valued social welfare function.

These two relatively abstract papers have a close relation to some surprisingly practical problems. I actually confronted this abstract problem while doing some research on labour market regulations and how to protect certain fundamental rights of workers. This led to some critical questions which through a curious process linked to the problem I just described. I explain this more clearly in the next section.

The welfare economics of utility functions and achieved utilities has periodically come under criticism. I am not referring here to the exercise of showing that a particular constellation of ethical axioms is inconsistent, but a more fundamental, methodological criticism. Are the ingredients of standard welfare analysis right or are we barking up the wrong tree in viewing individual welfare in terms of achieved utilities? A particularly influential line of criticism was opened up by Amartya Sen when he, drawing on works of earlier philosophers, developed the capabilities and functionings approach to evaluating human well-being and selecting policies. This approach tries to integrate into a whole not just what human beings actually achieve but what they are capable of achieving. Given that human beings value not just the final functionings that they achieve but the freedoms that underlie these functionings and the rights that they have, this theory has the advantage of giving us a way to evaluate these wider preferences of human beings. This important approach is, however, not without its pitfalls and Chapter 6, which is a reprint of a paper of mine in *Social Choice and Welfare*, 1987, is a critical evaluation of the approach of functionings and capabilities.

I have mentioned in the previous section about my interest in equilibrium analysis when I was a student at LSE. One off-shoot of this interest that I pursued after my return to India was the study of non-Walrasian general equilibrium theory. There were researchers in New Delhi, such as V. K. Chetty and Dipankar Dasgupta, who were actively working on general equilibrium theory that arose from the works of Jacques Dreze, Edmond Malinvaud, Yves Younes, Jean-Michel Grandmont, and Jean-Pascal Benassy; I sat on a large number of lectures and seminars that were delivered on the subject. Questions arose about the welfare properties of equilibria that were non-Walrasian in the sense of markets being cleared not through the adjustment of prices but changes in the perceived rationings that individuals faced. There were lots of results floating around derived with an artillery of mathematics. To understand some of these results in my own way I started developing some diagrams to represent various non-Walrasian equilibria. The paper that arose, and was published in *Journal of Macroeconomics*, 1992, had no new results but a geometric method for representing and understanding existing work. I tried to develop the counterpart of the Edgeworth box for non-Walrasian equilibrium and welfare analysis. Some of these methods had also been developed by others, but I took the diagrammatic techniques further and, in any case, this paper, which appears here as Chapter 7, is meant to be of expository value. Its aim is to provide a useful teaching aid.

1.3 LAW, RIGHTS, AND WELL-BEING

The first six chapters of the book that follow this Introduction and constitute Part I provide a selection of my theoretical research in welfare economics. From here on I turn to papers that are much more applied and policy-oriented in their concern. (Or, as some theorists would say, it is downhill all the way from here.) Chapter 8 is an important marker in my own intellectual development. In this paper, I argue that law and economics should be done differently from the way it is typically practiced. In the standard view a new law is one which alters the payoff functions of individuals or changes the strategies open to individuals. It troubled me, however, about why that should be so. A new legislation is nothing but some words set down on paper. How could that alter anybody's set of strategies or payoff function? It was soon clear that those who took this conventional view of law and economics assumed, often unwittingly, that there were some agents of the state who mechanically enforced whatever was the law. If a new speeding law lowered the maximum allowable speed to 50 km per hour this would change the driver's payoff of driving at 60 km per hour (because when travelling at that speed he will now have to calculate the expected cost of being caught and fined) only if we assume that there is police who could catch him and magistrates who would then fine him.

But once we begin to think of the police and magistrates also as rational agents and not simply robots waiting to do whatever the law requires them to do, it becomes clear that the law cannot change the game that citizens play—that is, the payoff functions and strategy sets faced by all the agents (i.e., including the enforcers of the law) involved in the economy game. This seems convincing enough but what then was the role of the law? How would it ever alter human behaviour? I argued, and this is the central message of this chapter, that the law affects behaviour by doing nothing to the game that citizens play but by creating new beliefs and in particular, focal points. Much to my satisfaction I later found that one could interpret the works of David Hume on the state as taking a view that was very close to this. Of course, he never used the language of game theory but he had realized that in the end the state was nothing but a set of beliefs and expectation in the heads of citizens. I would later elaborate on this approach to law and economics in my book *Prelude to Political Economy: A Study of the Social and Political Foundations of Economics* (The MIT Press and Oxford University Press, 2000). Chapter 8 is a brief and, I hope, a useful summary of this approach.

Chapter 9, unlike the other chapters in this part of the book, is an abstract theoretical essay. It is motivated by an ongoing debate in social choice theory on individual rights and what they mean for aggregate welfare and, in particular, the Pareto axiom. In this paper I asked the question about how the terms of this debate would get altered if, when a person was granted a right, he or she was also given the right to waive or trade the right. The paper is a follow-up on a large literature in social choice theory that began with the classic paper of Amartya Sen highlighting a potential conflict between the principle of individual liberty that drew on the writings of John Stuart Mill and the Pareto axiom. The literature gave rise to a large body of writings and my paper followed a line of contributions that came from the philosopher Allan Gibbard, and economists Wulf Gaertner, Prasanta Pattanaik, and Kotaro Suzumura.

When pursuing my early interest in social choice theory and working on this particular abstract problem on individual rights, I did not know then that I would return to this question later while dealing with a practical problem on labour regulation. This happened with my work on sexual harassment in the workplace, which appears here as Chapter 10. In this paper I reviewed the sexual harassment law of the United States and then asked a question, which in the light of the Paretian economics that all of us economists are brought up on seemed to be deeply troubling. Suppose a firm puts up a large sign outside its employment office stating that anybody wanting to work there would be paid well, get good health benefits, and so on, but the employers would retain the right to sexually harass the workers. If this is made clear so that the harassment that occurs in this firm can be described as contractually sanctioned, then do we have any reason to have the state disallow such harassment? The Pareto principle says that when two rational persons agree to a contract that has no obvious negative fall-out on others, the state must not disallow this. Indeed, economic efficiency depends on such contracts being permitted.

Nevertheless, when we carry the same logic over to the case of sexual harassment, we are left with a feeling of discomfort. Is this discomfort valid, or is it just a manifestation of the urge that bureaucrats feel to intervene in other people's decisions even when there is little welfare ground for such intervention? The way this relates to the concerns of Chapter 9 is the following. Most of us would agree that workers should have the right not to be sexually harassed. But this does not automatically answer the question whether they have the right to waive this right. If they do, then a firm offering them high wages and wanting to have the freedom to harass them should be construed as permissible behaviour. The workers are simply being given the option of selling one of their rights. If they do not, then even contractual harassment should not be allowed. So the question boils down to: Do workers have the right to waive their rights? This is a very important practical question and arises in a variety of labour-related matters. In the early twentieth century, firms in the United States often required workers to give up their right to join trade unions as a precondition for joining the firm. This was called the 'yellow dog' contract. Yellow dog contracts were made illegal in 1921. In other words, not only did the workers have the right to join trade unions, they were not allowed to waive this right. And as with so much of our legislation, it has never been made clear whether there is a compelling reason why clauses such as the yellow dog ought to be allowed as one more right of the worker, namely, the right to waive a right, or not allowed because it erodes some other basic right (of course that would need to be demonstrated).

I have tried to develop a general principle about how to deal with rights to waive other rights and have now come to take the position that whether a person has the right to waive a right depends on whether the contract in question satisfies what I call the 'large numbers argument'. There are some contracts which when two people sign may not have any adverse impact on others, but when a large number of people sign such contracts it may begin to have a negative effect on others. The philosopher Derek Parfit had been concerned about morals which have the property that each act has a certain moral status whereas a set of such acts have a different moral status. This argument is difficult to formalize. One particular formalization requires us to visualize an infinite number of choices and to think of a welfare function which has the property that each of a set of choices raises welfare but the collection of an infinite number of such choices causes welfare to drop. The abstract setting then becomes very similar to the one used in Chapter 4. In a working paper that was the background to what I published on sexual harassment, I have raised questions about this link.[1]

This line of inquiry still has the potential to be taken further. It crops up in a variety of situations. Should a country have the freedom to ask its workers to give up the right to join trade unions in order to be able to work in an export processing zone? Should a poor worker have the right to waive the right to safe working conditions in order to work in a dangerous mine and earn the money he desperately needs? Our natural tendency is to jump to answer these questions relying on nothing but our intuition. I think this can be dangerous, entailing interventions where these should not happen and complacency where activism is required. I hope these chapters will provoke further research.

Some of these labour market questions arise sharply in the context of India where the tendency is to specify a variety of worker benefits and rights in advance by the government. Under what conditions is this justified and when should we rely instead on free contracting between the worker and the employer? This is the subject matter of Chapter 11. This chapter is also in a way a theoretical assessment of India's well-known Industrial Disputes Act, 1947.

Another less known but important labour law in India is the Child Labour (Prohibition and Regulation) Act, 1986. In Chapter 12, I illustrate how well-meaning laws can backfire. The act just mentioned imposes a fine on firms that are found employing children. Will this always cause child labour to decline? This chapter argues that the problem of child labour can actually be exacerbated by a law of this kind.

Finally, Chapter 13 deals with another familiar topic of law and economics— the problem of rent control of the kind found in New York and most Indian cities. It is shown how pathologies can arise in models of adverse selection and how adverse selection is only natural in rental markets since landlords have very little knowledge of the tenants' types. The aim again is to bring economic analysis to bear on the question of how to craft laws that promote fairness while enhancing welfare.

1.4 GLOBALIZATION AND REGULATION

The subject of international development has been a continuing interest of mine. Some of my papers on this subject were reprinted in *Collected Papers in Theoretical Economics, Volume 1: Development, Markets and Institutions.* My growing interest in law and economics, coupled with this earlier research in international development, led me to work on some issues of regulation and law related to globalization. These are reprinted in the third part of the present collection.

The law was once primarily a matter of the nation state. In this age of globalization, no longer is that true. Intercountry externalities, such as one nation's environmental practices affecting another nation's environment or the lowering of labour standards in one nation attracting capital away from another nation, and also our awareness of what is happening in distant lands means that, on an increasing range of subjects, we now have legislation across nations and also have global conventions and treaties for common rules. Even though the enforceability of such global regulation may be weak, they are not without bite, and this explains why so many nations resist signing global conventions, be it concerning the environment, labour standards, or rules of nuclear engagement. Understanding the economics of international law is extremely important for organizations such as the International Labour Organization and the World Trade Organization, which have to deal with international treaties and conventions all the time.

Chapter 14, which was delivered as a WIDER Annual Lecture in Helsinki, and Chapter 15, which was part of a conference presentation in Tokyo, directly address the subject of global labour laws and regulations. A motivating concern of these chapters is what has already been discussed above. We know from textbook economic theory that economic progress and efficiency critically hinge on people being able to sign contracts and rely on the fact that other signatories will fulfill their part of the terms. Hence, when we set down global labour standards, which typically take the form of placing restrictions on allowable contracts, are we not risking blighting the prospects of growth and efficiency of nations? This risk is genuine and larger than what occurs within an economy. Well-meaning activists in Washington, New York, and Geneva, often ask that workers worldwide over be paid a minimum wage or that workers be stopped from doing hazardous work or that children be stopped working forthwith. The risk is that some of these people, unaware of the reality that prevails in distant lands, will try to set conditions that will make the local conditions worse. In a poor nation like Ethiopia, where more than 40 per cent of children in the age group of ten to fourteen years do regular labour, if child labour is suddenly stopped by decree, it is likely that there will be starvation deaths and a rise in child prostitution. Enacting global regulations and rules is a particularly hard problem since the terrain of such regulation covers nations and regions with varied economies, institutions, and culture. Chapters 14 and 15 discuss the economics of such regulations and try to articulate some general principles about where it is right to intervene in the market and where not.

One special problem of our contemporary world is that it harbours more inequality than ever before in history and more than any single nation today. A simple back-of-the-envelope calculation that I did a few years ago showed that the world's ten richest people earn the same income as the entire population of Tanzania—some 35 million that year. Given that Tanzania itself has some very rich people, if we leave out the top 1 per cent of Tanzania, the gap between the world's richest and the poorest people is mindboggling. One reason for this great contemporary inequality is technological innovation and rising overall prosperity. In olden times even the richest monarchs could not afford luxuries that moderately rich people do today, simply because there was no technology for such luxuries. They could not go to distant Pacific Islands for holidays, certainly not with a reasonable assurance of returning and succumbed to illnesses that are today easily cured. Given that the poorest people today are, however, as badly off as the poorest people in history, the gap between the best-off and worst-off has risen.

Another reason why the world as a whole has so much more inequality than any country is because we have so few institutions of global governance. It is arguable that if any single country had the inequality that the world has, its government would not survive. Hence, the world manages to have such inequality by not having a global government.

However, some of the political tensions and insurgencies that we see today are probably manifestations of these huge disparities. Chapter 16 deals with what we can do about this. It takes the view that the rising inequality in the world and in so many individual nations is probably a concomitant of globalization. Hence, when an individual nation thinks of policies to control inequality, it has to be careful. It could, as this chapter demonstrates by constructing a model, make the problem of poverty worse in the nation, by causing cross-border movement of resources. In today's globalized world it is difficult for any single country to alone control inequality beyond a point. Inequality control has a global coordination aspect to it that has not been adequately recognized in the literature. It is argued in this chapter that there is today a need for a new international organization that helps nations coordinate their antipoverty and inequality control policies.

Chapter 17, written jointly with Hodaka Morita, and published recently in *European Economic Review*, moves away from the problem of labour to international credit and aid. It has been known for a long time that aid is not always motivated by concern for the recipient nation's well-being and is often an instrument to boost production in one's own country by creating international demand for its product. The paper is a game-theoretic study of what comes out of the brew of big multinationals wanting to sell goods and big banks wanting to lend money to poor nations to enable them to buy goods and generating usury profits for the banks. The chapter shows that if commercial lending is replaced by subsidized international credit, because of the strategic nature of interactions, the recipient nations can, in some situations, be worse off because more than the subsidy ends up flowing out of the nation and into the hands of the big corporations that sell products to this nation. This paper is meant to be a theoretical contribution to the old, discursive literature on 'tied aid'.

The closing chapter of this book is not theoretical, unlike the other chapters in this collection. But I include it in this collection of my primarily theoretical essays because it probably touched a chord and has received quite a lot of attention; it was published in a less-easily accessible journal—*Indicators: The Journal of Social Health*, 2002—and it brings to some kind of closure many issues of global governance discussed in papers included in this book.

The argument that underpins Chapter 18 is deceptively simple. Democracy has many features but an essential one is that people should have the right to vote and elect those who are to hold political office and can have an effect on people's lives. Note that one consequence of globalization is that there is greater interdependence in the lives of people who may be located physically far apart. This means that one nation can affect the lives of people in other nations in ways that were not conceivable earlier. By cutting-off trade or diverting capital flows, and by influencing the decisions of the International Monetary Fund or the World Trade Organization, one nation can often bring another nation to heel. Note also that there are great asymmetries in the ability of nations to exert this kind of influence on citizens of other nations. Cuba is unlikely to be able to do to US citizens what the US government can do to Cubans. It is entirely arguable that for ordinary Iraqis, more important than who gets elected as leader of Iraq is who gets elected as the president of the United States.

This growing ability of nations to affect lives of citizens of other nations (which is clearly a concomitant of globalization) and the fact that we do not have any form of global voting system, whereby the citizens of one nation can express their preference in the election of the leader of another nation, means that, as globalization occurs, there will be a natural tendency for global democracy to erode. There is, therefore, an urgent need in the world to bolster global democracy and, in its absence, create surrogates for global governance and improve the democratic features of international organizations. Otherwise, we will run increasingly into a politically destabilized world.

I have no clear solution to this problem but believe that in our eagerness to celebrate the positive fall-outs of globalization (and there are, no doubt, several), we are overlooking this major negative consequence of globalization. We ignore this at our own peril.

NOTE

1. *Sexual Harassment in the Workplace: An Economic Analysis with Implications for Worker Rights and Labor Standards Policy*, Department of Economics, Massachusetts Institute of Economics, Working Paper 02-11, 2002.

PART I
Welfare and Equilibrium

2 Cardinal Utility, Utilitarianism, and a Class of Invariance Axioms in Welfare Analysis

2.1 MOTIVATION

Does the ability to compare first-differences in utility imply cardinality? Attempts to answer this have been the source of some confusion. I have tried to argue elsewhere that this question has generated conflicting answers because of differences in implicit frameworks. In the 'traditional' framework (e.g., the one used by Samuelson [1947] in his *Foundations*) first-difference comparability does turn out to be equivalent to cardinality. On the other hand, in the 'modern' (choice theory) framework, these two concepts are distinct. The confusion has been persistent because while most people are aware of the two frameworks, they have considered the differences between these to be *stylistic* rather than *substantial*.

The existence of two frameworks for analysing concepts of intrapersonal utility measurement suggests that there must exist *alternative* frameworks for interpersonal comparisons as well. It is argued here that the recent literature on social welfare analysis (e.g., d'Aspremont and Gevers 1977; Hammond 1976; Maskin 1978; Roberts 1980; Sen 1977) has been developed (implicitly) within one framework—the modern one. This suggests that one could translate these definitions into a class of related ones in the 'traditional framework'. Such an exercise yields interesting insights. Some concepts which are distinct in the existing literature are logically indistinguishable in this alternative framework. A number of new informational characterizations of social welfare functionals are possible in this framework. Some results on utilitarianism and the conflict between utilitarianism

This paper was written while I was a visitor at CORE, Louvain-la-Neuve, and CEME, Brussels. I benefitted from discussions with Claude d'Aspremont and Amartya Sen and from seminars at Nuffield College, Oxford, and the Universities of Birmingham and Bonn.

First published in *Journal of Mathematical Economics*, Vol. 12 (1983).

and Rawlsianism are established as illustration. It is shown that the conflict between utilitarianism and Rawlsianism arises, in some sense, earlier than traditionally supposed; and also that we may think of utilitarianism as characterized by interpersonal *almost level* comparability and intrapersonal *difference* comparability.

I argue that there are good methodological reasons for adopting the traditional framework rather than the modern one. The traditional approach is shown to differ from the modern one in its use of a kind of 'unrestricted domain' assumption. Since the theoretical economist's formulation is typically undertaken a priori—without knowing what the actual decision problem is—the unrestricted domain assumption does appear attractive.

Instead of developing the results on cardinal utility and social choice theory separately, I establish an abstract result on affine functions which is then shown to lie at the nub of the two sets of problems. I state and prove the affine function theorem in the strongest form I am aware of. This result is shown to be the basis of a number of axiomatizations of cardinal utility (Lange 1933, 1934; Samuelson 1938 including Basu 1982). In all these papers and in many other attempts to axiomatize cardinal utility it is assumed that the image of the utility function is connected. For example, such an assumption is implicit in axiom A6 of Suppes and Winet (1955). The present theorem, by making use of a standard lemma in analysis (the principle of extension by continuity), shows that this is not necessary. The theorem is stated in Section 2.2 and proved in Section 2.6.

Broadly speaking, there are two classes of axiomatization of cardinal utility without introducing risk. One approach begins by assuming that real-valued utility functions exist and then imposes conditions on these to derive cardinal utility. Lange (1933), Samuelson (1938), Bernadelli (1934), and Basu (1980, 1982) are examples of this, though Samuelson does go beyond. The other approach begins *without* assuming the existence of utility functions. The 'primitives' are binary and quaternary relations. Conditions are imposed on these relations and then it is shown that (i) utility functions exist and (ii) utility is cardinal. Of course, steps (i) and (ii) are often diffused in the proofs. The best-known result in this approach is that of Suppes and Winet (1955). There are many other such attempts (see Fishburn 1970, chapter 6). In a sense the first approach corresponds to step (ii) of the second approach. Therefore, while our Theorem 1 is a direct generalization of the theorems in the first approach, it does throw some light on part of the second approach.

2.2 A THEOREM ON AFFINE FUNCTIONS

Let R be the set of real numbers.

DEFINITION. The mapping $f: B \to R$, where $B \subset R$, is a *positive affine transformation* (PAT) if $\exists a, b \in R$, with $b > 0$, such that $\forall t \in B$,

$$f(t) = a + bt.$$

DEFINITION. The mapping $f: B \to R$, where $B \subset R$, is a *first-difference preserving transformation* (DPT) if $\forall t_1, t_2, t_3, t_4 \in B$,

$$t_1 - t_2 \geq t_3 - t_4 \Leftrightarrow f(t_1) - f(t_2) \geq f(t_3) - f(t_4).$$

It is easy to see that if a mapping is a PAT, it must necessarily be a DPT: Suppose $f: B \rightarrow R$ is a PAT. Then there exists a positive real number b such that for all t_i, t_j in B, $f(t_i) - f(t_j) = b\,(t_i - t_j)$. Since b is positive, it is clear that f satisfies the first-difference preservation property. However, the reverse implication need not be true, that is, a function may be a DPT without being a PAT. This is demonstrated in the example in Section 2.3. Thus PAT and DPT are distinct concepts.

What is interesting, however, is that the two concepts become logically indistinguishable with a little structure on the domain B. For instance, *if the domain is unrestricted, that is $B = R$, then DPT and PAT are equivalent.* What is surprising is that for this equivalence result not only does B not have to coincide with R, but it need not even be a connected subset of R.

THEOREM 1. *If $f: B \rightarrow R$ and B is dense in a connected subset of R, then f is a DPT if and only if it is a PAT.*

By making use of the Principle of Extension by Continuity (Dunford and Schwartz 1967, p. 23) and the fact that points on a real line can be denoted exactly by their binary expansions, a simple and direct proof of this theorem is given in Section 2.6.

2.3 UTILITY THEORY

Let X be the set of alternatives. A *utility function, u,* is a real-valued function on X, that is, $u : X \rightarrow R$. The word *transformation* is reserved for a real-valued mapping defined on a subset of R. An individual is characterized not by a single utility function but by a whole class of permitted utility functions. Let $f: B \rightarrow R$ be a permitted transformation of a utility function, u, with B sufficiently large so that $u\,(X) \subset B$. Then $fu : X \rightarrow R$ is a permitted utility function. We could therefore think of an individual as an ordered pair $(u, \Omega \,|\, B)$, where u is a reference utility function and $\Omega \,|\, B$ is a collection of permitted transformations, f, defined on a domain $B \subset R$, with $u\,(X) \subset B$. Therefore, for each individual, $(u, \Omega \,|\, B)$, the set of permitted utility functions, $L(u, \Omega \,|\, B)$, is defined as follows:

$$L(u, \Omega \,|\, B) = \{ fu \,|\, f \in \Omega \,|\, B \}.$$

We assume that $\Omega \,|\, B$ always contains the identity transformation so that $u \in L\,(u, \Omega \,|\, B)$, that is, u is itself a permitted utility function.

If we assume $B = R$, that is, the transformations are defined on the entire domain, then the above method of characterizing an individual is equivalent to the traditional approach (e.g., the approach in Lange [1933] and Samuelson [1947]). This is the commonly used approach even today in many areas of economics. In the modern choice-theoretic approach, however, instead of beginning from an individual being an ordered pair $(u, \Omega \,|\, B)$ and then *deriving* the set of permitted utility functions $L\,(u, \Omega \,|\, B)$, we start *directly* by specifying a set of permitted utility functions, $L = \{u, \phi, \psi, \cdots\}$. This approach is, as a little reflection will show, equivalent to the above approach with the assumption that $u\,(X) = B$.

This is so because given ϕ, $u \in L$, we could always construct a transformation $f: u(X) \rightarrow R$ such that $f(u(x)) = \phi(x)$.

Hence the traditional and modern approaches could be thought of as special cases of the above characterization of an individual. The two approaches differ in their choice of the following *alternative* assumptions.

An individual is an ordered pair $(u, \varOmega \,|\, B)$, where:

Assumption N   (No restriction on transformation domain), i.e., $B = R$.

Assumption M   (Maximal restriction on transformation domain), i.e., $B = u(X)$.

Cardinal utility and first-difference comparable utility are defined as follows:

DEFINITION. Individual $(u, \varOmega \,|\, B)$ has cardinal utility if $\forall f \in \varOmega \,|\, B, \exists\, a, b \in R$, with $b > 0$, such that $\forall\, t \in B$,

$$f(t) = a + bt.$$

DEFINITION. Individual $(u, \varOmega \,|\, B)$ can compare first-differences of utility if $\forall f \in \varOmega \,|\, B, \forall\, t_1, t_2, t_3, t_4 \in B$,

$$t_1 - t_2 \geq t_3 - t_4 \Leftrightarrow f(t_1) - f(t_2) \geq f(t_3) - f(t_4).$$

In the light of these two definitions the conceptual difference between Assumptions N and M become clear. Given Assumption N, a transformation f is defined on R, even though in actual practice f is used to convert $u(X) \subset R$, and the properties of f (e.g., cardinality) are defined on the entire R [and not just on $u(X)$]. Hence it is a kind of unrestricted domain assumption: the transformation f is *equipped* to transform any element of R though it *actually* has to transform elements of $u(X)$. We shall therefore refer to the traditional framework, that is, one using Assumption N, as an *unrestricted domain framework*. The rationale for N is now evident and this clarifies why the unrestricted domain framework has been the more popular one in economics.

Cardinality and first-difference comparability appear to be distinct concepts and they have been treated as such in much of economics and psychometrics (see Seigel 1957). However, under a variety of situations the concepts become equivalent. The following result is an immediate corollary of Theorem 1.

COROLLARY 1.1 *Given Assumption N, or Assumption M with $u(X)$ being dense in a connected subset of R, the ability to compare first-differences of utility is equivalent to cardinality.*

This shows that in the traditional framework, first-difference comparability and cardinality are logically equivalent. But in the modern framework the concepts coincide only if certain conditions are satisfied. Corollary 1.1 is a stronger result than both the theorems in Basu (1982) and it shows that the results in Lange (1933) and Samuelson (1938) are applications of Theorem 1.

An interesting question which arises is that in the modern framework, that is, given Assumption M, can the restriction on $u(X)$ be relaxed even further while retaining the equivalence between first-difference comparability and cardinality? The answer to this is yes, as will be obvious from the example below. What is, however, not immediately clear is that what is the minimal structure on $u(X)$

which gives us the equivalence result. It may at first sight appear that the answer has something to do with *lacunae* in B (in the context of Theorem 1). A lacuna of B is a nondegenerate interval of R, disjoint with B and having a lower and an upper bound in B (Hildenbrand and Kirman 1976). But it is possible to show that the equivalence between PAT and DPT as in Theorem 1 does not hinge on the existence or nonexistence of lacunae in B.

EXAMPLES. Let $B^1 = [0,2] \cup [3,5]$ and $B^2 = \{0, 1, 5\}$. Both B^1 and B^2 have lacunae. On B^1 if a function f is a DPT it must be a PAT, but not so on B^2. Consider first $f: B^1 \to R$. Let f be a DPT. Since the closures of [0,2] and [3,5] are connected in R, by Theorem 1, the restrictions of f to [0,2] and [3,5] must be PATs. Thus $\exists$ a, b, c, $d \in R$ with b, d > 0:

$$\forall t \in [0,2], \quad f|[0,2](t) = a + bt,$$

$$\forall t \in [3,5], \quad f|[3,5](t) = c + dt.$$

If $b \neq d$ then $f(2) - f(0) \neq f(5) - f(3)$, which contradicts the fact that f is a DPT. Hence $b = d$. If $a \neq c$ then $f(3) - f(2) \neq f(4) - f(3)$, which contradicts the fact that f is a DPT. Hence $a = c$. Therefore f is a PAT on the entire domain of B^1.

Now consider $f: B^2 \to R$. Let $f(0) = 0, f(1) = 1, f(5) = 6$; such an f is a DPT but not a PAT.

Therefore, the absence or presence of lacunae does not in itself change anything. We move on, leaving the question of minimal restriction on B, which makes the concepts of DPT and PAT equivalent, an open one.

Theorem 1 throws light on another classical problem. In traditional discussions on the measurement of utility the term 'addible' utility often cropped up (this should not be confused with the modern concept of 'additive' utility). The precise meaning of addible utility was seldom stated, but a large number of economists (see, e.g., Alchian 1953; Majumdar 1958) seemed to suggest by this the simple idea that the addition of utilities from a group of alternatives $x_1, x_2, \cdots, x_n$ in X is 'meaningful'. That is, it is possible to say which among two groups of alternatives in X gives a greater utility. In other words, if in terms of one utility function, group 1 gives greater utility than group 2, then all permitted utility functions retain this ordering.

DEFINITION. Individual $(u, \Omega | B)$ has *addible* utility if $\forall f \in B, t_1, \cdots, t_n, s_1, \cdots, s_m \in B$,

$$\sum_{i=1}^{n} t_i \geq \sum_{i=1}^{m} s_i \leftrightarrow \sum_{i=1}^{n} f(t_i) \geq \sum_{i=1}^{m} f(s_i).$$

What this definition says is that if the sum of utilities that a person derives from n alternative bundles is greater than the sum derived from m alternative bundles, then all permissible transformations of the utility function retain this feature. It can be shown that given some domain restrictions, addibility is equivalent to what is often referred to as ratio-scale measurability (Roberts 1980) in the social choice literature.

DEFINITION. Individual $(u, \Omega | B)$ has a *ratio-scale measurable* utility if $\forall f \in B$, $\exists b > 0$, such that $\forall t \in B$,

$$f(t) = bt.$$

THEOREM 2. *Given Assumption N, or Assumption M with u (X) being dense in a connected subset of R. an individual (u, $\Omega \mid B$) has addible utility if and only if his utility is ratio-scale measurable.*

Given Theorem 1, the proof of Theorem 2 is straightforward and is relegated to Section 2.6.

2.4 TRANSFORMATION DOMAINS IN SOCIAL CHOICE

In the late 1970s there were attempts to enlarge the Arrowian framework of social choice theory to a utility-theoretic base. This has led to the opening up of an engaging area of research and a series of remarkable results (Arrow 1977; d'Aspremont and Gevers 1977; Deschamps and Gevers 1978; Hammond 1976; Maskin 1978; Roberts 1980; Sen 1977, 1981).

It has already been noted that there are two approaches to the problem of utility measurement: one making use of the unrestricted domain framework (Assumption N) and the maximal restriction approach (Assumption M). In the same spirit it is possible to conceive of two frameworks for social welfare analysis. I argue here that the existing literature (the references cited above) has, however, adopted the maximal restriction approach. There is no reason why that ought to be so. As argued above there is a certain desirable catholicity in the unrestricted domain framework. This idea is developed in this and the next sections in the context of social choice theory and this yields some surprising insights.

Let us first recall the standard framework using mainly Sen's (1977) notation. Let $X (\# X \geq 3)$ be the set of alternative social states and $H (\# H \equiv n < \infty)$ the set of individuals. Let $W_i(\cdot)$ be individual i's real-valued utility function defined on X. A *social welfare functional* (SWFL), F, specifies an ordering (i.e., a complete, reflexive and transitive binary relation), R, on X for each n-tuple $\{W_i(\cdot)\}_{i \in H}$. Hence

$$R = F(\{W_i(\cdot)\}).$$

The domain of F consists of all n-tuples $\{W_i(\cdot)\}$. Intrapersonal and interpersonal comparisons are introduced via alternative 'invariance requirements'. Consider three standard invariance axioms:

(CF) *Cardinal full comparable (or co-cardinal):* If $\{W_i(\cdot)\}$ and $\{W_i'(\cdot)\}$ are such that $\exists\ a,\ b \in R$, with $b > 0$, such that $\forall\ x \in X,\ \forall\ i \in H,\ W_i'(x) = a + bW_i(x)$, then $F(\{W_i(\cdot)\}) = F(\{W_i'(\cdot)\})$.

(CU) *Cardinal unit comparable:* If $\{W_i(\cdot)\}$ and $\{W_i'(\cdot)\}$ are such that $\exists\ a_1, \cdots, a_n, b \in R$, with $b > 0$, such that $\forall\ x \in X,\ \forall\ i \in H,\ W_i'(x) = a_i + b\ (W_i(x))$, then $F(\{W_i(\cdot)\}) = F(\{W_i'(\cdot)\})$.

(OL) *Ordinal level comparable (or co-ordinal):* If $\{W_i(\cdot)\}$ and $\{W_i'(\cdot)\}$ are such that $\exists$ a positive monotone transformation $\phi(\cdot)$, such that $\forall\ x \in X,\ \forall\ i \in H,\ W_i'(x) = \phi(W_i(x))$, then $F(\{W_i(\cdot)\}) = F(\{W_i'(\cdot)\})$.

Note that in these definitions, given two n-tuples $\{W_i(\cdot)\}$ and $\{W_i'(\cdot)\}$, we may conceive $\{W_i'(\cdot)\}$ as derived from $\{W_i(\cdot)\}$ by applying an n-tuple of transformations

$\{\phi_i\} = \{\phi_1, \cdots, \phi_n\}$ with ϕ_i being used to transform $W_i(\cdot)$, $\forall\, i \in H$. This is particularly transparent in Roberts' (1980) paper. Note, however, that in the above axioms each ϕ_i need operate only on $W_i(X)$. The properties of ϕ_i specified in these axioms are properties which it must satisfy on $W_i(X)$. This is not generally stated explicitly thereby corroborating the view that alternative domain restrictions of transformations have been typically thought of as differences in style rather than of substance.

As argued above, in the context of utility theory, there is good reason to suppose that each transformation should be potentially able to transform any element of R, despite the fact that it does not have to actually do so in a particular situation. And also, transformations should be thought of as satisfying a certain property (e.g., first-difference preservation) if they (i) satisfy the property over the domain of values that *actually occur* and (ii) have the *potential* for satisfying it over the entire domain (R). The above definitions merely require (i). The motivation for an approach which uses (i) and (ii) is similar in spirit to the motivation which had led economists to use the standard unrestricted domain assumption in social choice theory (i.e., to define a collective choice rule on *all possible* n-tuples of individual orderings). Such an approach would correspond to the unrestricted domain framework or the traditional approach to the problem of intrapersonal utility measurement.

2.5 UTILITARIANISM

In the unrestricted domain framework CF and CU would be defined as follows, using a bar (e.g., $\overline{\text{CF}}$) to distinguish the new definitions:

$(\overline{\text{CF}})$ If $\{W_i(\cdot)\}$ and $\{W_i'(\cdot)\}$ are such that $W_i'(\cdot) = \phi_i(W_i(\cdot))$, $\forall\, i \in H$, where $\{\phi_i\}$ is a transformation n-tuple defined on R such that $\exists\, a,\, b \in R$, with $b > 0$, such that $\forall\, t \in R$, $\forall\, i \in H$, $\phi_i(t) = a + bt$, then $F(\{W_i(\cdot)\}) = F(\{W_i'(\cdot)\})$.

$(\overline{\text{CU}})$ If $\{W_i(\cdot)\}$ and $\{W_i'(\cdot)\}$ are such that $W_i'(\cdot) = \phi_i(W_i(\cdot))$, $\forall\, i \in H$, where $\{\phi_i\}$ is a transformation n-tuple defined on R such that $\exists\, a_1, \cdots, a_n,\, b \in R$, with $b > 0$, such that $\forall\, t \in R$, $\forall\, i \in H$, $\phi_i(t) = a_i + bt$, then $F(\{W_i(\cdot)\}) = F(\{W_i'(\cdot)\})$.

It is not difficult to see that CF $\leftrightarrow$ $\overline{\text{CF}}$ and CU $\leftrightarrow$ $\overline{\text{CU}}$—the change of framework does not make a difference thus far. The next concept, however, is framework sensitive.

Let $\bar{a}$ be the invariance axiom given that interpersonal utilities are OL and that there exists one individual with difference comparable utility that is, he can compare his own first-differences in utility. (Of course, coupled with level comparability, the latter immediately implies that *everybody* can compare first-differences.)

$(\bar{a})$ If $\{W_i(\cdot)\}$ and $\{W_i'(\cdot)\}$ are such that $W_i'(\cdot) = \phi_i(W_i(\cdot))$, $\forall\, i \in H$, where $\{\phi_i\}$ is a transformation n-tuple defined on R such that

(1) $\exists\, k \in H$, $\forall\, t_1, t_2, t_3, t_4 \in R$, $t_1 - t_2 \geq t_3 - t_4 \leftrightarrow \phi_k(t_1) - \phi_k(t_2) \geq \phi_k(t_3) - \phi_k(t_4)$ and

(2) $\phi_i(\cdot) = \phi(\cdot)$, $\forall\, i \in H$ and $\phi(\cdot)$ is a positive monotone transformation, then $F(\{W_i(\cdot)\}) = F(\{W_i'(\cdot)\})$.

The counterpart of $\bar{\alpha}$ in the framework conventionally used in modern social choice theory, that is, in the maximal restriction framework (we may refer to such an axiom as α), is logically distinct from $\bar{\alpha}$.

What is surprising is that $\bar{\alpha}$ and CF are logically equivalent. In other words, in the unrestricted domain framework, the concept of cardinal full comparability is equivalent to ordinal level comparability coupled with intrapersonal difference comparability. This is precisely the assertion of the next theorem, the proof of which is given in Section 2.6.

THEOREM 3. *In the unrestricted domain framework, ordinal level comparability and the existence of an individual who can compare first-differences of his own utility is equivalent to cardinal full comparability, that is, $\bar{\alpha} \leftrightarrow CF$.*

Two of the most discussed SWFLs are utilitarianism and Rawlsianism. Consider two slight variants of these: An SWFL, F, is *utilitarian-type* if $\forall\, x, y \in X$, $\forall\, \{W_i(\cdot)\}$,

$$\left[\sum_{i=1}^{n} W_i(x) > \sum_{i=1}^{n} W_i(y) \right] \text{ implies } [xPy],$$

where P is the asymmetric part of $F(\{W_i(\cdot)\})$.

Now consider the leximin rule. Let $k(x)$ denote the kth worst-off person (ties being broken arbitrarily) in state x given an n-tuple $\{W_i(\cdot)\}$. A SWFL, F, is *leximin* if $\forall\, x, y \in X$, $\forall\, \{W_i(\cdot)\}$, xPy if and only if $\exists\, k \leq n$ such that $W_{k(x)}(x) > W_{k(y)}(y)$ and $\forall\, r < k$, $W_{r(x)}(x) > W_{r(y)}(y)$, where P is the asymmetric part of $F(\{W_i(\cdot)\})$.

Given the prominence of Rawlsianism and utilitarianism in modern social choice theory, it is but natural that a lot of effort has gone into isolating situations where these two ethical principles are in direct conflict. One of the most interesting findings in this direction is a result due to Deschamps and Gevers (1978). They showed that if an SWFL satisfies the properties of independence of irrelevant alternatives, the Pareto criterion, anonymity, a certain minimal equity axiom, and separability (for formal definitions, see Deschamps and Gevers 1978; Sen 1977)—we refer to this axiom set as the D–G axiom set—then if we assume cardinal full comparability, the leximin criterion and utilitarian-type criterion are in direct conflict because these are the only two SWFLs satisfying the D–G axiom set.

What Theorem 3 suggests is that if we adopt the unrestricted domain framework then this direct confrontation between utilitarianism and Rawlsianism arises as soon as we assume ordinal level comparability and the existence of an individual who can compare his own first-differences.

COROLLARY 3.1. *Given the D–G axiom set, if an SWFL satisfies α then it must be either utilitarian-type or leximin.*

This corollary is an immediate consequence of the D–G theorem and Theorem 3 above.

Let us now turn to the characterization of utilitariansim in the unrestricted domain framework. An SWFL, F, is *utilitarian* if $\forall\, x, y \in X$, $\forall\, \{W_i(\cdot)\}$,

$$\left[\sum_{i=1}^{n} W_i(x) \geq \sum_{i=1}^{n} W_i(y)\right] \text{ implies } [xRy],$$

where

$$R = F(\{W_i(\cdot)\}).$$

Note that one obvious implication of being able to make level comparisons between individuals i and j is that if $W_i(x) = W_i(y)$ and $W_j(z) = W_j(v)$, then given a permitted transformation n-tuple, $\{\phi_i\}$, $\phi_i(W_i(x)) - \phi_i(W_i(z)) = \phi_j(W_j(y)) - \phi_j(W_j(v))$. This part of level comparability may be referred to as almost level comparability. The invariance axiom $\bar{\beta}$ below is a formal statement of almost level comparability of interpersonal utilities plus intrapersonal first-difference comparability, with point (2) in the definition of $\bar{\beta}$ capturing the idea of almost level comparability discussed above.

($\bar{\beta}$)   If $\{W_i(\cdot)\}$ and $\{W_i'(\cdot)\}$ are such that $W_i'(\cdot) = \phi_i(W_i(\cdot))$, $\forall \, i \in H$, where $\{\phi_i\}$ is a transformation n-tuple defined on R such that

(1) $\forall \, k \in H$, $\forall \, t_1, t_2, t_3, t_4 \in R$, $t_1 - t_2 \geq t_3 - t_4 \Leftrightarrow \phi_k(t_1) - \phi_k(t_2) \geq \phi_k(t_3) - \phi_k(t_4)$ and

(2) $\forall \, i, j \in H$, $\forall \, t_1, t_2 \in R$, $\phi_i(t_1) - \phi_i(t_2) = \phi_j(t_1) - \phi_j(t_2)$, then $F(\{W_i(\cdot)\}) = F(\{W_i'(\cdot)\})$.

THEOREM 4. *In the unrestricted domain framework, almost ordinal level comparability and the ability of every individual to compare first-differences of his own utility is equivalent to cardinal unit comparability, that is, $\bar{\beta} \leftrightarrow \overline{CU}$.*

In a well-known paper, d'Aspremont and Gevers (1977) proved that if an SWFL satisfies the properties of independence of irrelevant alternatives, the Pareto criterion and anonymity (for formal definitions, see d'Aspremont and Gevers 1977; Sen 1977)—we refer to these as the D–G axiom set—and also CU, then it must be a utilitarian SWFL. This result, the fact that $\overline{CU} \leftrightarrow CU$, and Theorem 4 suggest the following characterization of utilitarianism.

COROLLARY 4.1. *If an SWFL satisfies the D–G axiom set and $\bar{\beta}$, it must be utilitarian.*

A similar exercise is possible for Rawlsianism difference principle, in particular, the leximin rule.

2.6 PROOFS

The proof of Theorem 1 is greatly facilitated by using a standard lemma in mathematics (see Dunford and Schwartz [1967] for a proof):

LEMMA (*Principle of Extension by Continuity*). *Let X and Y be metric spaces and let Y be complete. If $f : A \to Y$ is uniformly continuous on the dense subset A of X, then f has a unique continuous extension $g : X \to Y$. This unique extension is uniformly continuous on X.*

The following notation is useful. Given that $f: B \to R$, $\forall A \subseteq R$, $f(A) \equiv \{y \in R \mid y = f(t), t \in A \cap B\}$. Also $\forall A \subseteq R$, $\bar{A}$ denotes the closure of A in R.

PROOF OF THEOREM 1. Assume that $\bar{B}$ is a connected subset of R and that f is a DPT. If B is a singleton then the result is immediate. Hence assume $\# B > 1$.

As a first step note that f is a monotonically increasing function, that is

$$\forall t_1, t_2 \in B, \quad [t_2 > t_1] \Leftrightarrow [f(t_2) > f(t_1)]. \tag{2.1}$$

Let $t_1, t_2 \in B$. Then

$$t_2 > t_1 \Leftrightarrow t_2 - t_1 > t_1 - t_1$$

$$\Leftrightarrow f(t_2) - f(t_1) > f(t_1) - f(t_1), \text{ since } f \text{ is a DPT,}$$

$$\Leftrightarrow f(t_2) > f(t_1).$$

This establishes (2.1).

As a second step we prove that f is a uniformly continuous function. Let $\varepsilon > 0$. Choose $t_1, t_2 \in B$ such that $t_1 \neq t_2$ and $\mid f(t_1) - f(t_2) \mid < \varepsilon$. The existence of such t_1, t_2 may be proved as follows: Assume no such t_1, t_2 exist. If $[s_1, s_2]$ is a nondegenerate interval with $s_1, s_2 \in B$ it is easy to check that $\# ([s_1, s_2] \cap B) = \infty$. Hence, since no t_1, t_2 of the above description exists, f must be unbounded on $[s_1, s_2] \cap B$. Hence $\exists s_3 \in [s_1, s_2] \cap B$ such that $f(s_3) \notin [f(s_1), f(s_2)]$. This violates (2.1). Therefore, t_1, t_2 of the above description exists. Let $\mid t_1 - t_2 \mid = \delta$,

$$\forall s_1, s_2 \in B, \quad \mid s_1 - s_2 \mid < \delta \to \mid s_1 - s_2 \mid < \mid t_1 - t_2 \mid$$

$$\to \mid f(s_1) - f(s_2) \mid < \mid f(t_1) - f(t_2) \mid,$$

since f is a DPT,

$$\to \mid f(s_1) - f(s_2) \mid < \varepsilon.$$

Therefore f is uniformly continuous.

By the above lemma, $\exists g: \bar{B} \to R$ which is a unique continuous extension of f. The proof is completed by showing that g is a PAT on $\bar{B}$. This, in turn, is proved by taking an arbitrary bounded and nondegenerate interval in $\bar{B}$ and proving that the restriction of g on this is a PAT.

Let $[t_*, t^*]$, $t_* \neq t^*$, be such an arbitrary interval in $\bar{B}$. Let $s_* \equiv g(t_*)$, $s^* \equiv g(t^*)$, $\hat{t} \equiv \frac{1}{2}(t_* + t^*)$, and $\hat{s} \equiv \frac{1}{2}(s_* + s^*)$. We first prove

$$g([t_*, \hat{t}[) = [s_*, \hat{s}[\quad \text{and} \quad g([\hat{t}, t^*[) = [\hat{s}, s^*[, \tag{2.2}$$

that is, each half is mapped into its corresponding half in the range. It is easily checked that by virtue of (2.1), g is also a monotonically increasing function. This, and the continuity of g, implies that $g([t_*, t^*[) = [s_*, s^*[$ and $\exists$ a unique $s \in [s_*, s^*[$ such that $g([t_*, \hat{t}[) = [s_*, s[$ and $g([\hat{t}, t^*[) = [s, s^*[$.

To prove (2.2) we simply have to show that $s = \hat{s}$. Without loss of generality assume $s > \hat{s}$. This implies $s - s_* > s^* - s$. Hence $\exists t_1, t_2 \in [t_*, \hat{t}[\cap B$ such that $g(t_1) - g(t_2) > s^* - s$. Choose $t_3, t_4 \in [\hat{t}, t^*[\cap B$ such that $t_3 - t_4 \geq t_1 - t_2$. Since $g([\hat{t}, t^*[) = $

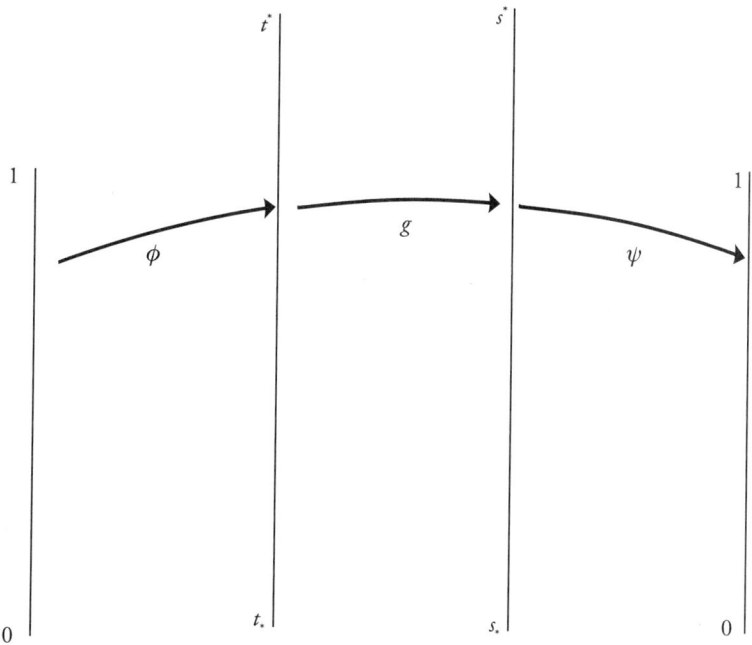

Figure 2.1: The Composite Mapping

$[s, s^*[$, hence $g(t_3) - g(t_4) < s^* - s$. Thus, $g(t_3) - g(t_4) < g(t_1) - g(t_2)$. Since $t_1, t_2, t_3,$ $t_4 \in B$, this violates the fact that f is a DPT. This establishes (2.2).

Let $\phi = [0, 1[\to [t_*, t^*[$ be a PAT with $\phi([0, 1[) = [t_*, t^*[$. Similarly define $\psi : [s_*,$ $s^*[\to [0, 1[$ such that ψ is a PAT with $\psi([s_*, s^*[) = [0, 1[$. Let $(g \mid I)$ be the restriction of g to $[t_*, t^*[$. Define a composite mapping $\Delta \equiv \psi (g \mid I) \phi$. Clearly $\Delta :$ $[0, 1[\to [0, 1[$. Figure 2.1 illustrates this mapping.

It will be shown that Δ is an identity mapping. Let $a \in [0, 1[$ and let $a_1 a_2 a_3 \cdots$ be the binary expansion of a and let $b_1 b_2 b_3 \cdots$ be the binary expansion of $\Delta(a)$. We first prove $a_1 = b_1$. Without loss of generality, suppose $a_1 = 0$. Thus $a \in [0, \frac{1}{2}[$. Hence $\phi(a) \in [t_*, \hat{t}[$. By (2.2), $(g \mid I) \phi(a) \in [s_*, \hat{s}[$. Hence $\psi(g \mid I) \phi(a) \in [0,$ $\frac{1}{2}[$. Thus $b_1 = 0$.

In the same way we could prove $a_2 = b_2$. Since $a_1 = 0$, we know that $\phi(a)$ lies in the first half of $[t_*, t^*[$, that is, $\phi(a) \in [t_*, \hat{t}[$. Now, since t_*, t^* were chosen arbitrarily above, we could set t_* as before and t^* as what was earlier $\hat{t}$ and by a similar proof show that whichever half of $[t_*, \hat{t}[\phi(a)$ happens to be in, $(g \mid I) \phi(a)$ would be in a corresponding half in the range. Hence $a_2 = b_2$, and by repeating, $a_n = b_n, \forall$ n. Hence Δ is an identity mapping and in particular it is a PAT. By construction, ϕ and ψ are PATs. Hence g is a PAT on $[t_*, t^*[$. It is an immediate consequence of monotonicity and continuity of g that $g(t^*) = s^*$. Hence g is a PAT on $[t_*, t^*]$. Therefore, g is a PAT on $\bar{B}$ and its restriction, f, must be a PAT.

It is obvious, as has been shown in Section 2.2, that if f is a PAT, it must be a DPT.

Q.E.D.

PROOF OF THEOREM 2. Given addibility, it is obvious that $f \in \Omega \mid B$ implies f is a DPT. Given Assumption N or Assumption M with $u(X)$ being a connected subset of R, Theorem 1 implies $\exists\, a,\, b \in R$, with $b > 0$, such that $\forall\, t \in B, f(t) = a + bt$. Since individual $(u,\, \Omega \mid B)$ has addible utility, $1 + 0 = 0$ implies $f(1) + f(0) = f(1)$. Hence $f(0) = 0$. Therefore $a = 0$. Thus the individual's utility is *ratio-scale measurable*. The reverse implication is obvious.   Q.E.D.

PROOF OF THEOREM 3. Obviously $\bar{\alpha} \rightarrow \overline{CF}$. Assume $\overline{CF}$ is valid, and $\{W_i(\cdot)\}$ and $\{W_i{}'(\cdot)\}$ satisfy conditions (1) and (2) in the definition of $\bar{\alpha}$. It immediately follows that $\forall\, t_1,\, t_2,\, t_3,\, t_4 \in R,\, t_1 - t_2 \geq t_3 - t_4 \leftrightarrow \phi(t_1) - \phi(t_2) \geq \phi(t_3) - \phi(t_4)$. By Theorem 1, $\exists\, a,\, b \in R$, with $b \geq 0$, such that $\forall\, t \in R, \phi(t) = a + bt$. Then by $\overline{CF}$, $F(\{W_i(\cdot)\}) = F(\{W_i{}'(\cdot)\})$. Hence $\bar{\alpha}$ holds.   Q.E.D.

PROOF OF THEOREM 4. Assume $\overline{CU}$ is valid and that $\{W_i(\cdot)\}$ and $\{W_i'(\cdot)\}$ satisfy conditions (1) and (2) in the definition of $\bar{\beta}$. By Theorem 1, $\forall\, k \in H, \exists\, a_k,\, b_k \in R$, with $b_k > 0$, such that $\forall\, t \in R, \phi_k(t) = a_k + b_k t$. Since $\phi_i(t_1) - \phi_i(t_2) = \phi_j(t_1) - \phi_j(t_2)$, hence $b_i(t_1 - t_2) = b_j(t_1 - t_2)$. Therefore $b_i = b_j = b, \forall\, i, j \in H$. Hence by $\overline{CU}$, $F(\{W_i(\cdot)\}) = F(\{W_i'(\cdot)\})$. Hence $\bar{\beta}$ holds.   Q.E.D.

REFERENCES

Alchian, A., 1953, 'The meaning of utility measurement', *American Economic Review*, 42.

Arrow, K. J., 1977, 'Extended sympathy and the possibility of social choice', *American Economic Review*, 67.

Basu, K., 1980, *Revealed Preference of Government*, Cambridge University Press, Cambridge.

—— 1982, Determinateness of the utility function: Revisiting a controversy of the thirties, *Review of Economic Studies*, 49, (reprinted here as Chapter 3).

Bernadelli, H., 1934, 'Notes on the determinateness of the utility function', *Review of Economic Studies*, 2.

d'Aspremont, C. and L. Gevers, 1977, 'Equity and the informational basis of collective choice', *Review of Economic Studies*, 44.

Deschamps, R. and L. Gevers, 1978, 'Leximin and utilitarian rules: A joint characterization', *Journal of Economic Theory*, 17.

Dunford, N. and J. T. Schwartz, 1967, *Linear Operators, Part I: General Theory*, Interscience Publishers, New York.

Fishburn, P. C., 1970, *Utility Theory for Decision Making*, Wiley, New York.

Hammond, P. J., 1976, 'Equity, Arrow's conditions and Rawls difference principle', *Econometrica*, 44.

Hildenbrand, W. and A. P. Kirman, 1976, *Introduction to Equilibrium Analysis*, North-Holland, Amsterdam.

Lange, O., 1933, 'The determinateness of the utility function', *Review of Economic Studies*, 1.

—— 1934, 'Notes on the determinateness of the utility function', *Review of Economic Studies*, 2.

Majumdar, T., 1958, *The Measurement of Utility*, Macmillan, New York.

Maskin, E., 1978, 'A theorem on utilitarianism', *Review of Economic Studies*, 45.

Roberts, K. W. S., 1980, 'Interpersonal comparability and social choice theory', *Review of Economic Studies*, 47.

Samuelson, P. A., 1938, 'The numerical representation of ordered classifications and the concept of utility', *Review of Economic Studies*, 6.

—— 1947, *Foundations of Economic Analysis*, Harvard University Press, Cambridge, MA.

Seigel, S., 1957, 'A method for obtaining an ordered metric scale', *Psychometrica*, 21.

Sen, A. K., 1977, 'On weights and measures: Informational constraints in social welfare analysis', *Econometrica*, 45.

—— 1981, 'Social choice theory', in K. J. Arrow and M. Intriligator (eds.), *Handbook of Mathematical Economics* (Vol. III), North-Holland, Amsterdam.

Suppes, P. and M. Winet, 1955, 'An axiomatization of utility based on the notion of utility differences', *Management Science*, 1.

3 Determinateness of the Utility Function
Revisiting a Controversy of the 1930s

3.1 INTRODUCTION: PARETO'S ASSUMPTIONS

In an important paper on utility theory, Lange (1933) alleged that a large number of Lausanne economists, particularly Pareto, Amoroso, and Pietri–Tonelli, had assumed that an individual could compare (a) levels of utility, and (b) first-differences of utility, and they had also assumed that utility was not cardinal. This, Lange claimed, was not consistent and he set out to show that (a) and (b) implied cardinal utility. A controversy ensued in which economists from Allen (1934) to Zeuthen (1936), including Alt (1936), Bernadelli (1934), Lange (1934), Phelps-Brown (1934), and Samuelson (1938), participated.

None of the existing proofs are, however, rigorous and, moreover, none of these authors specified their framework of analysis clearly. This is a major shortcoming since the validity of the Lange hypothesis hinges critically on the framework of analysis. In (what in retrospect may be referred to as) the framework of the 1930s or simply the traditional framework, the theorem is valid. It is more robust than generally supposed, because it is valid in the absence of differentiability and even continuity of the utility function. This is shown in the next section.

In the framework of modern social choice theory (see Sen 1970, chapters 7 and 7*) the Lange hypothesis is invalid. Section 3.3 tries to locate the critical difference in the two frameworks which leads to these divergent implications.

While it is true that Pareto did assume—though only on occasions (see Pareto, 1906, chapter 4, section 32, of his *Manual*)—that individuals could compare

This paper was presented at the Fourth World Congress of the Econometric Society at Aix-en-Province, September 1980. I benefited a lot from two related seminars I attended at the Indian Statistical Institute, New Delhi. I am grateful to Sukhamoy Chakravarty, Ranen Das, V.K. Chetty, and an anonymous referee for many useful comments.

First published in *Review of Economic Studies*, Vol. 49, (1982).

first-differences of utility, Lange's charge of inconsistency in Pareto's *Manual* is not completely tenable since it is not clear that Pareto was operating in the framework of the 1930s.

It is then argued in Section 3.3 that if, however, the utility functions are continuous and are defined on a connected topological space, then even in the modern framework the Lange hypothesis is valid. To be able to compare first-differences implies an ability to compare all higher-order differences!

The main theme of this paper is to examine and comment on the controversy of the 1930s in the light of modern developments.

3.2 LANGE'S CLAIM

Let X be the set of alternatives. A utility function, u, is a real-valued function on X, that is, $u: X \to R$, where R is the set of all real numbers. A *transformation* is a mapping from R to R. An individual is characterized by a reference utility function, u, and a set of permitted transformations, Ω. The individual's preference may be represented by u or any composite function fu, where $f \in \Omega$. Note that $fu: X \to R$.

Thus given u and Ω we have the set of permitted utility functions, $L(u, \Omega)$, defined as follows:[1]

$$L(u, \Omega) = \{ fu \mid f \in \Omega \}.$$

The agent is supposed to have those preferences which are implied by *all* the utility functions in $L(u, \Omega)$. For example, the welfare change from x to y is larger than the change from a to *b* if and only if

$$|\phi(x) - \phi(y)| > |\phi(a) - \phi(b)| \forall \phi \in L(u, \Omega).$$

The above characterization of an agent is a formal statement of the traditional approach. There are, of course, many alternative ways of describing an agent which are equivalent to the above description.

The essential difference between the modern analysis (Sen 1970, 1979) and the analysis of the 1930s is in the choice of 'primitives'. The modern analysis begins by assuming that an individual has a set L of utility functions. The traditional analysis begins by assuming the existence of u and Ω, and L is derived from these. The difference appears innocuous but a failure to appreciate this leads to confusions. For the time being we assume that an individual is simply an ordered pair (u, Ω).

Now we may introduce the two axioms of Lange. Axiom 1 asserts that individuals can compare utility levels. (Elements of R are denoted by u_i and this should not be confused with the utility function u.)

AXIOM 1. $\forall f \in \Omega, \forall u_1, u_2 \in R$

$$u_1 \geq u_2 \leftrightarrow f(u_1) \geq f(u_2).$$

In other words, only positive monotonic transformations are permitted. Clearly, the ordering on X generated by all the utility functions in $L(u, \Omega)$ will be the same if Axiom 1 is true ($\forall x, y \in X$, $u(x) \geq u(y)$ implies $f(u(x)) \geq f(u(y))$, $\forall f \in \Omega$).

While the entire Hicksian consumer theory is based on ordinal utility, that is, Axiom 1, we may in certain situations require a sharper conception of utility. For instance, one interpretation of the law of diminishing marginal utility requires first differences of utility to be comparable.[2] Axiom 2 states that only those transformations are permitted which retain the ordering over first-differences given by u.

AXIOM 2.　　$\forall f \in \Omega, \forall u_1, u_2, u_3, u_4 \in R$

$$u_1 - u_2 \geq u_3 - u_4 \Leftrightarrow f(u_1) - f(u_2) \geq f(u_3) - f(u_4).$$

An individual, (u, Ω), has cardinal utility if $\forall f \in \Omega$, $\exists$ scalars a and b ($b>0$) such that for all $u_i \in R, f(u_i) = a + bu_i$. In this framework, Lange's theorem is valid. That is conditions (a) and (b), interpreted as Axioms 1 and 2, do imply that utility is cardinal. While Lange's theorem is valid, neither Lange nor Samuelson proved it in its complete generality. Lange's main proof requires u to be differentiable. But Lange's (1933) geometric proof, while incomplete, is in this respect superior. Here he dispenses with differentiability. What he does use implicitly is the continuity of the utility function. But even this is dispensable.

Before proving Lange's theorem in the next section, it is worth observing a small result which many have overlooked. Axioms 1 and 2 are not two independent axioms at all. Whenever Axiom 2 is satisfied, Axiom 1 is automatically satisfied. This is easily established by assuming $u_3 = u_4$ in Axiom 2.

3.3 TWO FRAMEWORKS AND THEOREMS

In modern choice theory, particularly social choice, instead of characterizing an individual as (u, Ω) and deriving $L(u, \Omega)$, it is typical to begin by directly specifying a set L of permitted utility functions. An individual is thought of as simply a set L. Any preference which is endorsed by all the utility functions in L is a preference of the individual.

In this framework an individual who can compare first-differences in utility is characterized by Axiom 2*.

AXIOM 2*. Let $u \in L$. For all $\phi \in L$, for all $x, y, a, b \in X$.

$$u(x) - u(y) \geq u(a) - u(b) \Leftrightarrow \phi(x) - \phi(y) \geq \phi(a) - \phi(b).$$

An L satisfying Axiom 2* may be labelled as an individual whose utility is *quasi-cardinal* of degree one (Basu 1980). Cardinal utility, in this framework, is defined as a set L such that $\forall u, \phi \in L, \exists a, b$ ($b > 0$) such that $\forall x \in X, \phi(x) = a + bu(x)$.

In this framework, being able to compare first-differences of utility (i.e., Axiom 2*) does not imply cardinality. This is easily demonstrated with the example in Basu (1980, pp. 67–8). If this is the implicit framework of Pareto's utility analysis then Pareto's assumptions or first-differences being comparable and utility being noncardinal cannot be criticized as being inconsistent.

The critical difference between Axioms 2 and 2* which leads to these different implications is not difficult to isolate. Consider the function $\phi: X \to R$ in

Axiom 2*. Corresponding to this we could derive a transformation f such that $\forall$ $x \in X, f(u(x)) = \phi(x)$. We shall refer to f as the transformation associated with ϕ. Now Axiom 2* may be expressed in the following equivalent manner. For all x, y, $a, b \in X, u(x) - u(y) \geq u(a) - u(b) \Leftrightarrow f(u(x)) - f(u(y)) \geq f(u(a)) - f(u(b))$. Now compare this to Axiom 2. Function f here is required to satisfy the same property as is required of f in Axiom 2, but with one major difference: in Axiom 2, f has to satisfy the property on the entire domain of R whereas here it has to satisfy the property on $u(X) = \{y \in R \mid y = u(x), x \in X\}$, which is a subset of R. This is the crucial difference. In the modern framework (given an arbitrarily chosen reference utility function $u \in L$) for any $\phi \in L$, the associated transformation f operates only on the relevant real numbers, namely $u(X)$. And the property that the ordering over first-differences is unchanged by transformations of the utility function is restricted to first-differences of elements of $u(X)$. Thus the traditional framework could be thought of as the modern framework with the additional assumption of an unrestricted domain for the transformations.[3] It is this unrestricted domain assumption which makes the derivation of cardinal utility possible.

So in the modern framework the ability to compare first-differences of utility does not imply cardinality. What additional assumptions do we need to guarantee cardinality in this framework? One line of approach explored in chapter 6 of Basu (1980) yields the result that for cardinality it is necessary (and sufficient) that all higher-order differences be comparable. Here we pursue a different route that yields a more surprising result.

THEOREM 1. *If u(X) is a connected subset of R, then quasi-cardinality of degree one implies cardinality.*

PROOF.[4] Given that $u(X)$ is a connected subset of R, $\forall x, y \in X, \exists t \in X$ such that

$$u(x) - u(t) = u(t) - u(y). \tag{3.1}$$

Let $\phi \in L$. Axiom 2* implies

$$\phi(x) - \phi(t) = \phi(t) - \phi(y). \tag{3.2}$$

In other words $\forall x, y \in X, \exists t \in X$ such that

$$u(x) + u(y) = 2u(t), \tag{3.3}$$

$$\phi(x) + \phi(y) = 2\phi(t), \tag{3.4}$$

Now suppose,

$$\sum_{i=1}^{r} u(x_i) \geq \sum_{i=1}^{r} u(y_i), \tag{3.5}$$

where $r = 2^k$ (k is a positive integer) and $x_i, y_i \in X$, $\forall i$. Given Eq. (3.3), Eq. (3.5) implies that $\exists t_i, s_i \in X$ ($i = 1, \cdots, r'$) such that

$$\sum_{i=1}^{r'} 2u(t_i) \geq \sum_{i=1}^{r'} 2u(s_i), \tag{3.6}$$

where $r' = 2^{k-1}$. Similarly by Eq. (3.4),

$$\sum_{i=1}^{r} \phi\,(x_i) \geq \sum_{i=1}^{r} \phi\,(y_i) \tag{3.7}$$

is equivalent to

$$\sum_{i=1}^{r} 2\phi\,(t_i) \geq \sum_{i=1}^{r} 2\phi\,(s_i), \tag{3.8}$$

Thus Eq. (3.5) holds iff Eq. (3.6) holds, and Eq. (3.7) holds iff Eq. (3.8) holds.

Hence for any permutations $(x_1, \cdots, x_r)$ and $(y_1, \cdots, y_r)$ from X for which Eq. (3.5) holds, there exist permutations $(t_1, \cdots, t_r)$ and $(s_1, \cdots, s_r)$ from X such that [Eq. (3.8) iff Eq. (3.6)] implies Eq. (3.7) iff Eq. (3.5)]. It follows from Basu (1980, p. 82) that all $(k-1)$th order differences being comparable is equivalent to [Eq. (3.8) iff Eq. (3.6)] holding for all permutations $(t_1, \cdots, t_r)$ and $(s_1, \cdots, s_r)$ from X. Hence all $(k-1)$th order differences being comparable implies [Eq. (3.7) iff Eq. (3.5)] for all permutations $(x_1, \cdots, x_r)$ and $(y_1, \cdots, y_r)$ from X, which is equivalent to all kth order differences being comparable. Hence, given that first-order differences are comparable under quasi-cardinality of degree one, all kth order differences are comparable, $\forall k$. Thus by theorem 6*5 of Basu (1980), utility is cardinal.[5]

From the discussion above it is clear that Axiom 2 is equivalent to Axiom 2* with $u(X) = R$. Hence Lange's result follows directly from Theorem l.

THEOREM 2. *If an individual, (u, Ω), satisfies Axiom 2 then his utility must be cardinal.*

REMARK 1. In the light of Theorem 1, if X is a connected topological space and there exists a utility function L which is continuous, then Axiom 2* implies that utility is cardinal.

REMARK 2. It is easy to check that if utility is cardinal then not only first-differences but nth-order differences,[6] for all n, are comparable. Hence it is an immediate corollary of Theorem 1 that if the image of a utility function on X is an interval then to be able to compare first-differences implies an ability to compare all higher-order differences.

We have thus established Lange's result in a more general context, and have also established conditions under which the ability to compare first-differences of utility does imply cardinality, even within the modern framework.

NOTES

1. We assume that the identity function is an element of Ω. Thus $u \in L\,(u,\,\Omega)$, that is, u is a permitted utility function.
2. For an interpretation of the law of diminishing marginal utility that requires only ordinal utility and a restricted subset of first-differences of utility to be comparable, see Mayston (1976).
3. After writing $f = f(u_i)$, Lange, Samuelson, and others differentiate and consider the effect of arbitrary changes in the value of u_i on f. If f is defined on the domain $u(X)$

then of course these operations may not be permissible. Thus though $u(x)$, $\forall x \in X$, may take a limited number of values, the transformation f is implicitly assumed by Lange and others to be equipped to transform any $u_i \in R$ to $f(u_i) \in R$, though such an u_i may not actually occur.

4. I am indebted to an anonymous referee for the present version of the proof. This version makes use of an earlier result of mine: Theorem 6*5 in Basu (1980). A longer proof which is independent of earlier results is available with the author.

5. Theorem 6*5 in Basu (1980) asserts that if kth-order differences are comparable, $\forall k$, then utility is cardinal.

6. The meaning of nth-order difference is obvious by induction given that a second-order difference is a difference of differences.

REFERENCES

Allen, R. G. D., 1934, 'A Note on the determinateness of the utility function', *Review of Economic Studies*, 2.

Alt, F., 1936, 'On the measurability of utility', For English translation, see J. S. Chipman et al. (eds.), *Preferences Utility, and Demand*, Harcourt Brace Jovanovich, New York, 1971.

Basu, K., 1980, *Revealed Preference of Government*, Cambridge University Press, Cambridge.

Bernadelli, H., 1934, 'Notes on the determinateness of the utility function', *Review of Economic Studies*, 2.

Lange, O., 1933, 'The Determinateness of the Utility Function', *Review of Economic Studies*, 1.

——1934, 'Notes on the determinateness of the utility function', *Review of Economic Studies*, 2.

Mayston, D. J., 1976, 'On the Nature of Marginal Utility—A Neo-Marshallian Theory of Demand', *Economic Journal*, 86.

Pareto, V., 1906, *Manual of Political Economy*, Macmillan edition Milan, 1971.

Phelps-Brown, E. H., 1934, 'Notes on the determinateness of the utility function', *Review of Economic Studies*, 2.

Samuelson, P. A., 1938, 'The numerical representation of ordered classifications and the concept of utility', *Review of Economic Studies*, 6.

Sen A. K., 1970, *Collective Choice and Social Welfare*, Oliver and Boyd, Edinburg.

—— 1979, 'Interpersonal comparisons of welfare', in Boskin, M. (ed.), *Economics and Human Welfare*, Academic Press, Burlington, MA.

Zeuthen, F., 1936, 'On the determinateness of the utility function', *Review of Economic Studies*, 4.

4 Aggregating Infinite Utility Streams with Intergenerational Equity
The Impossibility of Being Paretian

with Tapan Mitra

4.1 INTRODUCTION

The subject of intergenerational equity in the context of aggregating infinite utility streams has been of enduring interest to economists. Ramsey (1928) had maintained that discounting one generation's utility or income vis-à-vis another's is 'ethically indefensible', and something that 'arises merely from the weakness of the imagination'. Since it is generally agreed that any such process of aggregation should satisfy the Pareto principle, Ramsey's observation raises the question of whether one can consistently evaluate infinite utility streams while respecting intergenerational equity and at least some form of the Pareto axiom.

In an important contribution to this literature, Svensson (1980) established the general possibility result (for a social welfare relation, SWR) that one can find an ordering that satisfies the axioms of Pareto and intergenerational equity. It is worth noting here that he obtains the (complete) ordering by nonconstructive methods; specifically, he defines a partial order satisfying the two axioms and then completes the order by appealing to Szpilrajn's lemma.

Svensson's paper builds on the earlier seminal contribution by Diamond (1965). In his paper, Diamond presents the celebrated theorem[1] that there does not exist any social welfare function (SWF, i.e., a function that aggregates an

The authors are grateful to Jorgen Weibull and two referees of *Econometrica* for comments and suggestions.

First published in *Econometrica*, (2003).

infinite stream into a real number) satisfying three axioms: Pareto, intergenerational equity, and continuity (in the sup-metric).

Neither contribution addresses the following question: Does there exist a SWF satisfying intergenerational equity and the Pareto axiom? Since continuity of the SWF in Diamond's exercise is a technical axiom (in contrast to the other two axioms), we think that this is an issue worth investigating.[2] The principal task of this paper is to present the general impossibility theorem that there is no SWF that satisfies the Pareto and equity axioms. In other words, we show (in Theorem 1) that any Paretian SWF is necessarily inequitable.

We can compare our result with Diamond's as follows. If we denote by Y (a subset of the reals) the set of admissible utility levels of each generation and by X the set of infinite streams of these utility levels, an SWF is a function from X to the reals. Our result implies that the continuity axiom is not needed for Diamond's impossibility theorem. Further, in our result, the utility space, Y, is unrestriced, while in Diamond's, it is taken to be the closed interval [0, 1]. Thus, apart from the continuity issue, the context in which our result is established is broader.

Our approach to the impossibility result is different from Diamond's. His approach is to show that if there is an SWF on X (when $Y = [0, 1]$) satisfying equity and the Pareto principle, then it cannot be continuous in the sup-metric. The proof of our result, roughly speaking, involves showing that in trying to represent any SWR respecting equity and the Pareto principle, one 'runs out of real numbers'. Thus, our approach has an affinity with the demonstration that lexicographic preferences do not have a real-valued representation.

Since the approaches are different, it is of some interest to enquire whether in Diamond's context (i.e., with $Y = [0, 1]$), one can dispense with the continuity axiom in his impossibility result *by following his own approach*. We do this in Theorem 2. We show that the extent of continuity needed for Diamond's technique to work is already implied by the Pareto axiom in the [0, 1] utility space case, making a separate continuity axiom superfluous.[3]

We establish (in Theorem 2) a stronger result than the one just described. We show that one does not need the full strength of the Pareto axiom; a weak version of it (and equity) suffices. This distinguishes it from Theorem 1, where the full strength of the Pareto axiom is utilized.

4.2 PARETIAN SWFS ARE INEQUITABLE

Let $\mathbb{R}$ be the set of real numbers, $\mathbb{N}$ the set of positive integers, and M the set of non-negative integers. Suppose $Y \subset \mathbb{R}$ is the set of all possible utilities that any generation can achieve. Then $X = Y^{\mathbb{N}}$ is the set of all possible utility streams. If $\{x\} \in X$, then $\{x\} = (x_1, x_2, \cdots)$, where, $\forall\, t \in \mathbb{N}$, $x_t \in Y$ represents the amount of utility that the tth generation (i.e., the generation of period t) earns. For all $y, z \in X$, we write $y \geq z$ if $y_i \geq z_i$, for all $i \in \mathbb{N}$; we write $y > z$ if $y \geq z$ and $y \neq z$; and we write $y \gg z$ if $y_i > z_i$, for all $i \in \mathbb{N}$.

If Y has only one element, then X is a singleton, and the problem of ranking or evaluating infinite utility streams is trivial. Thus, without further mention, the set Y will always be assumed to have at least two distinct elements.

An SWF is a mapping $W: X \rightarrow$ R. Consider now the axioms that we may want the SWF to satisfy. The first axiom is the standard Pareto condition.

PARETO AXIOM: *For all x, y $\in$ X, if x > y, then W (x) > W (y).*

The next axiom is the one that captures the notion of 'intergenerational equity'. We shall call it the 'anonymity axiom'.[4] It is equivalent to the notion of 'finite equitableness' (Svensson 1980) or 'finite anonymity' (Basu 1994).[5]

ANONYMITY AXIOM: *For all x, y $\in$ X, if there exist i, j $\in$ N such that $x_i = y_j$ and $x_j = y_i$, and for k $\in$ N − {i, j}, $x_k = y_k$, then W (x) = W (y).*

The principal result of this paper can now be stated as follows.

THEOREM 1: *There does not exist any SWF satisfying the Pareto and anonymity axioms.*

PROOF: Assume, on the contrary, that there is an SWF, $W: X \rightarrow \mathbb{R}$, that satisfies the Pareto and anonymity axioms. Since $Y \subset \mathbb{R}$ contains at least two distinct elements, there exist real numbers y^0 and y^1 in Y, with $y^1 > y^0$. Without loss of generality, and to ease the writing, we may suppose that $y^1 = 1$ and $y^0 = 0$.

Let Z denote the open interval $(0, 1)$, and let $r_1, r_2, r_3, \cdots$ be an enumeration of the rational numbers in Z. Let z be an arbitrary number in Z. We define the set: $E(z) = \{n \in \mathbb{N}: r_n < z\}$. Clearly, $E(z)$ is an infinite set. We now define a sequence $a(z) = (a(z)_1, a(z)_2, \cdots)$ as follows:

$$a(z)_n = \begin{cases} 1 & \text{if } n \in E(z), \\ 0 & \text{otherwise.} \end{cases}$$

Note that the sequence will have an infinite number of 1's and an infinite number of 0's. Define the sequence $b(z) = (b(z)_1, b(z)_2, \cdots)$ as the same as sequence $a(z)$ except that the first 0 appearing in the $a(z)$ sequence is replaced by 1. By the Pareto axiom, we have $W(b(z)) > W(a(z))$. We denote the closed interval $[W(a(z)), W(b(z))]$ by $I(z)$.

Now, let p, q be arbitrary real numbers in Z, with $q > p$. Clearly, we must have $E(p) \subset E(q)$, for if $n \in E(p)$, then $r_n < p$, and since $p < q$, we must have $r_n < q$, so that $n \in E(q)$. Further, there are an infinite number of rational numbers in the interval (p, q). Thus, comparing the sequence $a(p)$ with the sequence $a(q)$, we note that

(1) if $n \in \mathbb{N}$, and $a(p)_n = 1$, then $a(q)_n = 1$ and

(2) there are an infinite number of $n \in \mathbb{N}$ for which $a(p)_n = 0$ and $a(q)_n = 1$.

We now proceed to compare the sequence $a(q)$ with the sequence $b(p)$. Let $m \in \mathbb{N}$ be the index for which the sequence $a(p)$ differs from the sequence $b(p)$, that is, $a(p)_m = 0$ and $b(p)_m = 1$. There are two cases to consider: (i) $a(q)_m = 1$, (ii) $a(q)_m = 0$. In case (i), we clearly have $a(q) > b(p)$, and so

$$W(a(q)) > W(b(p)). \tag{4.1}$$

In case (ii), we proceed as follows. Let M be the smallest integer for which $a(p)_M$ = 0 while $a(q)_M$ = 1. By observation (2) above, such an M exists. Also, clearly M ≠ m, for $a(q)_M$ = 1 while $a(q)_m$ = 0. Since $b(p)$ differs from $a(p)$ in only the index m, we have $b(p)_M$ = 0. Now, define $b'(p)$ as follows: $b'(p)_m$ = 0, $b'(p)_M$ = 1, $b'(p)_n$ = $b(p)_n$ ∀ $n \in \mathbb{N}$ such that $n \neq m$, $n \neq M$. Since $b(p)_m$ = 1 and $b(p)_M$ = 0, the anonymity axiom implies that

$$W(b'(p)) = W(b(p)). \tag{4.2}$$

Comparing $b'(p)$ with $a(q)$, we note that $b'(p)_m$ = 0 = $a(q)_m$, $b'(p)_M$ = 1 = $a(q)_M$, and ∀ $n \in \mathbb{N}$ such that n ≠ m, $n \neq M$, we have $b'(p)_n$ = $b(p)_n$ = $a(p)_n$ ≦ $a(q)_n$ by observation (1). By observation (2), we must therefore have $a(q) > b'(p)$, so that by the Pareto axiom:

$$W(a(q)) > W(b'(p)). \tag{4.3}$$

Combining (4.2) and (4.3), we get

$$W(a(q)) > W(b(p)). \tag{4.4}$$

Thus, in both cases (i) and (ii), we have $W(a(q)) > W(b(p))$. This means that the interval $I(q)$ = $[W(a(q)), W(b(q))]$ is disjoint from the interval $[W(a(p)), W(b(p))]$, the latter interval lying entirely to the left of the former interval on the real line.

To summarize, we have shown that the intervals $I(z)$ associated with distinct values of $z \in (0, 1)$ are nonoverlapping. But this means that to each real number $z \in (0, 1)$, we can associate a distinct rational number [in the interval $I(z)$], contradicting the countability of the set of rational numbers. Q.E.D.

REMARKS: 1. The construction (used in the above proof) of an uncountable family of distinct nested sets $E(z)$, with each set containing an infinite number of positive integers, can be found in Sierpinski (1965, p. 82).

2. The general possibility theorem of Svensson (1980) on SWRs shows that there is no inherent conflict between the *concepts* of equity and the Pareto principle. Theorem 1 shows that a conflict arises when we try to obtain an evaluation of utility streams in terms of real numbers, while respecting these two properties.

3. The simplest example of our setup is one where the utility space, Y, is {0, 1}. One might interpret this as follows: there are precisely two states in which each generation might find itself, a good state and a bad state. The utility obtained by each generation is 1 in the good state and 0 in the bad state. Theorem 1 tells us that even in this simple setup, there is no SWF that respects anonymity and the Pareto axiom.

4. It is of interest to note that our general impossibility result does use the full strength of the Pareto axiom. A weaker version of the Pareto axiom may be written as follows.

WEAK PARETO AXIOM: *For all x, y $\in$ X, if there exists j $\in$ $\mathbb{N}$ such that $x_j > y_j$, and, for all k ≠ j, $x_k = y_k$, then W(x) > W(y). For all x, y $\in$ X, if x ≫ y, then W(x) > W(y).*

It can be shown (see theorem 3 in Basu and Mitra 2002) that when $Y \subset$ M (the set of non-negative integers), there is an SWF (on X) satisfying the anonymity and weak Pareto axioms.[6] In particular, when $Y = \{0, 1\}$, there is an SWF *on X*, respecting anonymity and weak Pareto axioms.

4.3 GENERALIZING THE DIAMOND–YAARI RESULT

Diamond (1965) had shown that (when the utility space Y is [0, 1]) there is no SWF satisfying anonymity, Pareto, and a continuity axiom (in the sup-metric on X). It follows, of course, from our Theorem 1 that the continuity axiom is redundant for his impossibility result.

The exercise in this section has two objectives. First, we show that the redundance of the continuity axiom for Diamond's impossibility result can be shown without any reference to our Theorem 1, but rather by following his own approach and exploiting more fully the particular structure of the utility space when Y is [0, 1]. Second, we establish that (apart from the continuity axiom) the full power of the Pareto axiom is also not needed for his impossibility result; a weaker version of the Pareto axiom (which we call the dominance axiom) suffices for this purpose.

DOMINANCE AXIOM: *For all x, y $\in$ X, if there exists j $\in$ $\mathbb{N}$ such that $x_j > y_j$, and, for all k $\neq$ j, $x_k = y_k$, then W(x) > W(y). For all x, y $\in$ X, if x $\gg$ y, then W(x) $\geq$ W(y).*

It is not as if we wish to recommend the use of such a weak form of the Pareto condition, but since we are going to prove an impossibility result, clearly it is better to use as weak an axiom as one can. Further, our proof indicates that it is precisely the dominance axiom that is needed to obtain the impossibility result.

THEOREM 2: *Assume Y = [0, 1]. There does not exist any SWF satisfying dominance axiom and the anonymity axiom.*

PROOF: To establish the theorem, assume that $Y = [0, 1]$ and that there exists an SWF, W: $X \rightarrow \mathbb{R}$, that satisfies the dominance and anonymity axioms.

Denote the vector $(1, 1, 1, \cdots)$ in X by e. Define the sequence $\bar{u}$ in X as follows:

$$\bar{u} = (1, 1, 0, 1/2, 1,0, 1/4, 2/4, 3/4, 1, \cdots). \tag{4.5}$$

This sequence is best understood as sequence u, defined in Eq. (4.7), with the first term changed from 0 to 1. For $s \in I \equiv (-0.5, 0.5)$, define

$$\bar{x}(s) = 0.5\bar{u} + 0.25(1 + s)e. \tag{4.6}$$

Then $(1/8)e \leq \bar{x}(s) \leq (7/8)e$, and so $\bar{x}(s) \in X$ for each $s \in I$.

Define the function, $f: I \rightarrow \mathbb{R}$ by $f(s) = W(\bar{x}(s))$. By the dominance axiom, f is monotonic nondecreasing in s on I. Thus f has only a countable number of points of discontinuity in I. Let $a \in I$ be a point of continuity of the function f.

Define the sequence u in X as follows:

$$u = (0,1, 0,1/2, 1,0, 1/4, 2/4, 3/4, 1, \cdots), \tag{4.7}$$

and then define

$$x(a) = 0.5u + 0.25(1 + a)e. \tag{4.8}$$

Clearly, $x(a) \in X$, and $\bar{x}_1(a) > x_1(a)$, while $x_j(a) = \bar{x}_j(a)$ for each $j \in \mathbb{N}$, with $j \neq 1$. By the dominance axiom, $W(\bar{x}(a)) > W(x(a))$. If we denote $[W(\bar{x}(a)) - W(x(a))]$ by θ; then $\theta > 0$.

Denote $\max(0.5 - a, 0.5 + a)$ by Δ; then $\Delta > 0$. Since f is continuous at a, given the θ defined above, there exists $\delta \in (0, \Delta)$ such that: $0 < |s - a| < \delta$ implies $|f(s) - f(a)| < \theta$. Note that for $0 < |s - a| < \delta$, we always have $s \in I$.

We define (following Diamond), for each $k \in \mathbb{N}$, a sequence u^k by starting with the sequence u, interchanging the initial 0 with the $(k + 1)^{\text{th}}$ 1 appearing in u, and then interchanging the sequence $(1/2^k, 2/2^k, \cdots, (2^k - 1)/2^k, 0)$ with $(0, 1/2^k, 2/2^k, \cdots, (2^k - 1)/2^k)$ so that

$$u^k = (1, 1, 0, 1/2, 1, \cdots, 0, 0, 1/2^k, 2/2^k, \cdots, (2^k - 1)/2^k, 0, \cdots). \tag{4.9}$$

Now, for each $k \in \mathbb{N}$, we use u^k to define $x^k(a)$ as follows:

$$x^k(a) = 0.5u^k + 0.25(1 + a)e. \tag{4.10}$$

Clearly, $x^k(a) \in X$ for each $k \in \mathbb{N}$. Comparing the expressions for $x(a)$ and $x^k(a)$ in Eqs. (4.8) and (4.10), and using the expressions for u and u^k in Eqs. (4.7) and (4.9) respectively, we see that the anonymity axiom yields

$$W(x^k(a)) = W(x(a)) \quad \text{for all } k \in \mathbb{N}. \tag{4.11}$$

Choose $K \in \mathbb{N}$ with $K \geq 2$ such that $(1/2^{K-2}) < \delta$, and define $S = [a - (1/2^{K-2})]$. We note that $0 < (a - S) < \delta$, and so $S \in I$, and

$$W(\bar{x}(S)) = f(S) > f(a) - \theta = W(\bar{x}(a)) - \theta. \tag{4.12}$$

We now compare the welfare levels associated with $x^K(a)$ and $\bar{x}(S)$ as follows. Note that $x^K(a) = 0.5u^K + 0.25(1 + a)e = 0.5\bar{u} + 0.25(1 + a)e - 0.5(\bar{u} - u^K) = \bar{x}(a) - 0.5(\bar{u} - u^K) \geq \bar{x}(a) - 0.5(1/2^K)e = 0.5\bar{u} + 0.25(1 + a)e - 0.5(1/2^K)e = 0.5\bar{u} + 0.25[1 + a - (1/2^{K-1})]e = 0.5\bar{u} + 0.25[1 + a - (1/2^{K-2})]e + 0.25(1/2^{K-1})e \gg 0.5\bar{u} + 0.25(1 + S)e = \bar{x}(S)$. Thus, by the dominance axiom, we have

$$W(x^K(a)) \geq W(\bar{x}(S)). \tag{4.13}$$

Using Eqs. (4.11)–(4.13), we obtain: $W(\bar{x}(a)) - \theta = W(x(a)) = W(x^K(a)) \geq W(\bar{x}(S)) > W(\bar{x}(a)) - \theta$, a contradiction, which establishes our result. Q.E.D.

REMARK: If we compare the dominance axiom with the weak Pareto axiom introduced in Section 4.2, we see that the last inequality in the statement of the former is a weak inequality, unlike in the statement of the latter. Hence, weak Pareto is stronger than dominance. It follows that when $Y = [0, 1]$, there is no SWF satisfying the weak Pareto and the anonymity axioms.

NOTES

1. Diamond attributes the result to Yaari.
2. In this connection, we may note that the line of research initiated by Koopmans (1960), and continued by Diamond (1965) and others, establishes that Paretian social welfare relations, continuous in suitably defined metrics, necessarily exhibit 'impatience' in the sense that the current generation receives more favourable treatment than future generations. Burness (1973) explores the impatience implications of continuously differentiable Paretian SWFs.
3. The underlying argument is that, if an SWF were to exist, the Pareto axiom would imply that it would have some monotonicity properties, and monotone functions on [0, 1] can be discontinuous at most at countably many points.
4. In informal discussions throughout the paper, the terms equity and anonymity are used interchangeably.
5. The anonymity axiom figures prominently in the social choice theory literature, where it is stated as follows: The social ordering is invariant to the information regarding individual orderings as to who holds which preference ordering. Thus, interchanging individual preference profiles does not change the social preference profile. (See May 1952; Sen 1977).
6. Our weak Pareto axiom is stronger than the standard form of the weak Pareto axiom used in social choice theory, where typically it is stated as follows: If every individual in a society is better off, then society is better off. See, for example, Sen (1977). The possibility result just mentioned would, of course, hold with the standard form of the weak Pareto axiom in our context.

REFERENCES

Basu, K., 1994, 'Group Rationality, utilitarianism and Escher's "waterfall" ', *Games and Economic Behavior*, 7, 1–9.

Basu, K. and T. Mitra, 2002, *Aggregating infinite utility streams with inter-generational equity*, MIT Department of Economics Working Paper No. 02–19.

Burness, H. S., 1973, 'Impatience and the preference for advancement in the timing of satisfactions', *Journal of Economic Theory*, 6, 495–507.

Diamond, P., 1965, 'The Evaluation of infinite utility streams', *Econometrica*, 33, 170–7.

Koopmans, T. C., 1960, 'Stationary ordinal utility and impatience', *Econometrica*, 28, 287–309.

May, K. O., 1952, 'A set of independent necessary and sufficient conditions for simple majority decision', *Econometrica*, 20, 680–4.

Ramsey, F. P., 1928, 'A mathematical theory of savings', *Economic Journal*, 38, 543–59.

Sen, A. K., 1977, 'On weights and measures: Informational constraints in social welfare analysis', *Econometrica*, 45, 1539–72.

Sierpinski, W., 1965, *Cardinal and Ordinal Numbers*, Polish Scientific, Warsaw.

Svensson, L.-G., 1980, 'Equity among generations', *Econometrica*, 48, 1251–6.

5 Possibility Theorems for Aggregating Infinite Utility Streams Equitably

with Tapan Mitra

5.1 INTRODUCTION

The need to aggregate and evaluate infinite streams of returns or utility arises in several areas of economics, ranging from intergenerational welfare theory to environmental economics. The subject of intergenerational equity in the context of aggregating infinite utility streams has been of enduring interest to economists, starting with the work of Ramsey (1928), who had maintained that discounting one generation's utility or income *vis-à-vis* another's was 'ethically indefensible', and something that 'arises merely from the weakness of the imagination'. His conjecture about the difficulty of aggregating infinite streams, while respecting intergenerational equity, turned out to be compelling, as a large number of impossibility theorems were proved subsequently by a number of authors, starting with the seminal works of Koopmans (1960) and Diamond (1965).

This problem has been confronted in the philosophy literature as well. Co wen and Parfit (1992), for instance, discussed the problem of aggregating the welfares of future generations, at length, and reached the conclusion that discounting the costs and benefits of future generations cannot be ethically justified. Hence, if we

The paper benefited greatly from a presentation and discussion at the IEA Roundtable Meeting in Hakone, Japan, 10–12 March, 2005, and from a detailed report by Yongsheng Xu. This is an area where, over the years, we have had conversations with and comments from a large number of economists and would like to thank, in particular, Geir Asheim, Claude d'Aspremont, Kuntal Banerjee, Marc Fleurbaey, Nick Kiefer, Wlodek Rabinowicz, Tomoichi Shinotsuka, Kotaro Suzumura, and Jorgen Weibull.

First published in John Roemer and Kotaro Suzumuru (eds), *Intergenerational Equity and Sustainability*, Palgrave (2006).

want to be morally correct, we must be 'against the discount rate'. The problem that they do not address and is germane to our paper, is the logical feasibility of what they recommend. If we do decide to go along with their advice and give equal importance to future returns, then how do we aggregate future streams of returns when these stretch into infinity? Simply adding up will often not work—it may not give us a real number and could lead to a violation of the Pareto principle.

Yet, it would be wrong to abandon the effort to search for a social welfare function (SWF) that aggregates infinite streams of returns and satisfies intergenerational anonymity and some form of the Pareto criterion. In reality, we encounter this problem all the time. In deciding whether to build a dam on a river, which will help irrigation and generate electricity but damage fauna and flora, we clearly face a problem of choosing among long streams of utility, stretching far into the future. Even if we believe that the world has a finite future, since we do not know its termination date, we effectively face an infinite decision problem.

Moreover, every time we analyse an infinitely repeated game, we are forced to confront an infinite decision problem. And, if we are to pass judgement on which among a set of possible outcomes is superior, we are compelled to contend with precisely the problem that is the concern of this paper.

In Diamond's celebrated paper (1965) he had shown that there is no SWF that aggregates infinite utility streams while satisfying the Pareto condition, a weak form of anonymity and a continuity property.[1] In a recent paper (Basu and Mitra 2003), we tried to show that the problem is more discouraging because the impossibility result survives even if we do not impose any continuity restriction on the SWF. Are we then completely into a *cul-de-sac*? This paper tries to answer this in the negative.

We can think of many routes to getting possible results. In an elegant paper, Svensson (1980) had shown that if, instead of seeking a (real-valued) SWF, we merely search for the ability to rank infinite streams of utilities, then it is possible to prove that the requirements of equity and the Pareto principle are compatible. He does this, however, with the use of Szpilrajn's theorem, which implies a nonconstructive proof. Related results have been obtained by Bossert, Sprumont, and Suzumura (2004) and Suzumura and Shinotsuka (2003).

Though we delve briefly into this, our main aim in this paper is to look for possibility theorems that satisfy *representability*, that is, the existence of real-valued SWFs. More precisely our aim is to delineate the frontier of possibility and impossibility results for the existence of real-valued SWFs. We consider, in particular, weakening the Pareto axiom and exploring domain restrictions.

It does seem that in reality the domain of values that individual utilities can take is often quite limited. The simple assumption that an individual's utility can be represented by any real number may be mathematically convenient but is unrealistic. Given the limits of human perception, it is much more realistic to suppose that individual utilities can take a finite number of values or, at most, a countably infinite number of values. Thus, exploring the implications of such domain restrictions certainly seems worthwhile.

Of course, domain restrictions by themselves will not yield possibility theorems, given the general impossibility theorem of Basu and Mitra (2003, theorem 1), which applies to all domains however restrictive they may be.[2] But, we will try to show that, as soon as we combine domain restrictions with weaker versions of the Pareto axiom, the scope for the use of SWFs expands considerably (Theorem 3).

Our investigation also reveals that the particular nature of the domain restriction may be quite important for such possibility results. Under domain restrictions of other types, even the weak Pareto axiom is seen to be incompatible with the requirement of an equitable SWF (Theorem 4). However, if the postulated version of Pareto is sufficiently weak, then it is possible to generate equitable and Paretian SWFs without any domain restrictions (Theorem 5).

It is true that the exercise that we undertake in this paper is abstract and theoretical but it is motivated by the practical concern for shedding light on what is feasible once we reject the standard (inequitable) method of aggregating streams by discounting the returns that accrue to future generations.

5.2 FORMAL SETTING AND BASIC RESULTS

Let $\mathbb{R}$ be the set of real numbers, $\mathbb{N}$ the set of positive integers, and M the set of non-negative integers. Suppose $Y \subset \mathbb{R}$ is the set of all possible utilities that any generation can achieve. Then $X = Y^{\mathbb{N}}$ is the set of all possible utility streams. If $\{x_t\} \in X$, then $\{x_t\} = (x_1, x_2, \cdots)$, where, $\forall t \in \mathbb{N}$, $x_t \in Y$ represents the amount of utility that the generation of period t earns. For all $y, z \in X$, we write $y \geq z$ if $y_i \geq z_i$, for all $i \in \mathbb{N}$; we write $y > z$ if $y \geq z$ and $y \neq z$; and we write $y \gg z$ if $y_i > z_i$, $\forall i \in \mathbb{N}$.

If Y has only one element, then X is a singleton, and the problem of ranking or evaluating infinite utility streams is trivial. Thus, without further mention, the set Y will always be assumed to have at least two distinct elements.

An SWF is a mapping $W: X \rightarrow \mathbb{R}$. Consider now the axioms that we may want the SWF to satisfy. The first axiom is the standard Pareto condition.

PARETO AXIOM: *For all $x, y \in X$, if $x > y$, then $W(x) > W(y)$.*

The next axiom is the one that captures the notion of 'intergenerational equity'. We shall call it the 'anonymity axiom'.[3] It is equivalent to the notion of 'finite equitableness' (Svensson 1980) or 'finite anonymity' (Basu 1994).[4]

ANONYMITY AXIOM: *For all $x, y \in X$, if there exist $i, j \in \mathbb{N}$ such that $x_i = y_j$ and $x_j = y_i$, and for every $k \in \mathbb{N} - \{i, j\}$, $x_k = y_k$, then $W(x) = W(y)$.*

We shall begin by stating the main impossibility theorem that was established in Basu and Mitra (2003, theorem 1). This will be the setting in which we can then ask the question of what is possible.

THEOREM 1: *There does not exist any SWF satisfying the Pareto and Anonymity axioms.*

It is the rather sparse requirement of this theorem that is at the root of the frustration that this field of inquiry has generated. Note, in particular, that the impossibility result does not depend on any continuity postulate of the SWF; and, it applies to all domains of the SWF.

Before exploring the routes out of this, it is useful to place the problem in perspective by recalling Svensson's (1980) important theorem. Let us suppose that we abandon the search for an SWF and instead look for a social welfare ordering[5] (SWO). We then have the result due to Svensson (1980) that there is an SWO which satisfies the (appropriate relational versions of the) Pareto and Anonymity axioms. For reasons of completeness we briefly review Svensson's result. We do this also because, the use of a variant of Szpilrajn's Theorem [due to Suzumura 1983, theorem A(5)] allows us to give a particularly easy proof of it. Furthermore, Svensson (1980) restricts his exercise to the case where Y is the closed interval $[0,1]$; we state the version of his result which applies to any utility space Y. His proof, as well as ours, applies to this more general setting.

Formally, an SWO is a binary relation, $\succeq$ on X, which is complete and transitive. We use $\succ$ and $\sim$ to denote, respectively, the asymmetric and symmetric parts of $\succeq$. The properties of Pareto and Anonymity for an SWO are easy to define. We shall call these axioms $\succeq$-Pareto and $\succeq$-Anonymity to distinguish them from the axioms applied to an SWF.

$\succeq$-PARETO AXIOM: *For all $x, y \in X$, $x > y$ implies $x \succ y$.*

$\succeq$-ANONYMITY AXIOM: *For all $x, y \in X$, if there exist $i, j \in \mathbb{N}$, such that $x_i = y_j$ and $x_j = y_i$ and for every $k \in \mathbb{N} - \{i, j\}$, $x_k = y_k$, then $x \sim y$.*

First, let us give a statement of Suzumura's result. Let Ω be a set of alternatives. If R is a binary relation on Ω and R^* an ordering on Ω, we shall say that R^* is an *ordering extension* of R if, for all $x, y \in \Omega$, xRy implies xR^*y. We say that R is *consistent* if, for all $t \in \mathbb{N}$, and for all $x^1, x^2, \cdots, x^t \in \Omega$, $[x^1Rx^2$ and not x^2Rx^1, and for all $k \in \{2, 3, \cdots, t-1\}$, $x^kRx^{k+1}]$ implies not x^tRx^1.

LEMMA 1 (SZPILRAJN'S COROLLARY [SUZUMURA 1983]): *A binary relation R on Ω has an ordering extension if and only if it is consistent.*

Before proving the next theorem it is useful to introduce some new notation. If $\sigma: \mathbb{N} \to \mathbb{N}$ is a permutation and there exists $t \in \mathbb{N}$, such that $\forall k > t, \sigma(k) = k$, then we shall call σ *a finite permutation*. Given a finite permutation, σ, we shall use $n(\sigma)$ to denote the smallest integer t which has the property that, $\forall k > t, \sigma(k) = k$. Given a finite permutation, σ, and $x \in X$, we shall use $x(\sigma)$ to denote $y \in X$, where y is obtained by permuting the elements of x using σ.

In contrast to Theorem 1, we now have the following theorem.

THEOREM 2 (SVENSSON 1980): *There exists a social welfare ordering satisfying the $\succeq$-Pareto and $\succeq$-Anonymity axioms.*

PROOF: Define two binary relations, P and I, on X, as follows. For all $x, y \in X$, if $x > y$ then xPy. And if there exists i, j such that $x_i = y_j$ and $x_j = y_i$, and $x_k = y_k$ for all $k \neq i, j$, then xIy. Now define the binary relation R as follows: $xRy \Leftrightarrow xPy$ or xIy.

To see that R is consistent, suppose $t \in N$ and $x^1, x^2, \cdots, x^t \in X$ such that

(A) x^1Rx^2 and not x^2Rx^1,

(B) x^kRx^{k+1}, for all $k \in \{2, 3, \cdots, t-1\}$.

We have to show that not x^tRx^1.

Note that (A) and (B) can be written equivalently as

(A') x^1Px^2 and
(B') x^kPx^{k+1}, or x^kIx^{k+1}, for all $k \in \{2, 3, \cdots, t-1\}$.

Note that (A') and $[x^2Px^3$ or $x^2Ix^3]$ imply that there exists a finite permutation, σ_3, such that:

(A'') $x^1Px^3 (\sigma_3)$.

Next note that (A'') and $[x^3Px^4$ or $x^3Ix^4]$ imply that there exists a finite permutation, σ_4, such that $x^1Px^4(\sigma_4)$. Continuing in the same way we get the result that there exists a finite permutation, σ_t, such that $x^1Px^t(\sigma_t)$. This implies not $[x^tPx^1$ or x^tIx^1]. Therefore, not x^tRx^1.

Hence, by Szpilrajn's corollary, R has an ordering extension $\succeq$. Clearly $\succeq$ satisfies the $\succeq$-Pareto axiom and the $\succeq$-Anonymity axiom. *Q.E.D.*

For a long time, researchers have conjectured that the impossibility of having an SWF satisfying Pareto and Anonymity axioms was a problem of *representability*, that is, of there not being 'enough real numbers' to do the job. Since Diamond's theorem (1965) showed that the requirements of Pareto, Anonymity, and continuity were inconsistent, the conjecture remained an open one. But in the light of Theorem 1 above we can state a corollary which (a) confirms the conjecture and (b) clarifies the relation between Theorems 1 and 2 in a way that is especially useful. Towards this end, define:

REPRESENTABILITY: An SWO, $\succeq$, is *representable* if there exists a mapping, $f: X \rightarrow \mathbb{R}$ such that, for all $x, y \in X$, $x \succeq y \Leftrightarrow f(x)$ $f(y)$.

In the light of Svensson's result, Theorem 1 can be restated as follows.

COROLLARY 1: *There does not exist an SWO satisfying the $\succeq$-Pareto axiom, the $\succeq$-Anonymity axiom, and representability.*

PROOF: If a representable SWO satisfies the $\succeq$-Pareto axiom and the $\succeq$-Anonymity axiom, the real-valued function, $f: X \rightarrow \mathbb{R}$ that represents the SWO, must satisfy the Pareto and Anonymity axioms. But we know from Theorem 1 that no such f exists. This establishes the result. *Q.E.D.*

Corollary 1 makes the nature of the impossibility clear. If we are looking for an equitable SWO (i.e., one satisfying the anonymity principle) to evaluate infinite streams of returns, we have to be prepared to weaken the Pareto axiom or to give up the representability requirement. There is a case for exploring both these avenues. In a recent paper, Bossert, Sprumont, and Suzumura (2004) have looked at the possibilities that emerge when one does not require representability.[6] In what follows, we explore what is possible by relaxing the Pareto axiom.

5.3 WEAKENING PARETO

It is arguable that for certain philosophical and even policy purposes we do not need the full power of the Pareto condition (even if we are committed Paretians) simply because all the possibilities that are technically allowed in our specification of the domain may not arise under any eventuality. Indeed for certain ethical discourses involving the comparison of the moral worth of individual actions and universalizable rules (see Basu 1994) it may be enough to be armed with some weaker forms of Paretianism.

One idea that may be of interest is to restrict the analysis to cases where one state is obtained from another through changes in a finite number of periods. For such cases it is enough to use the following weakening of Pareto that we shall call 'weak dominance'.

WEAK DOMINANCE AXIOM: For all $x, y \in X$, if for some $j \in \mathbb{N}$, $x_j > y_j$, while, for all $k \neq j$, $x_k = y_k$, then $W(x) > W(y)$.

Another version of Pareto—this one has been widely used in the literature (see Arrow 1963; Sen 1977—is the 'weak Pareto' axiom, as defined below.

WEAK PARETO AXIOM: *For all $x, y \in X$, if $x \gg y$, then $W(x) > W(y)$.*

A natural next step is to consider an axiom that combines the two above axioms. That is precisely what the next axiom does.

PARTIAL PARETO AXIOM: *The SWF, W, satisfies the weak dominance axiom and the weak Pareto axiom.*

The partial Pareto axiom demands that an SWF be positively sensitive to an increase in utility of a single generation, the utilities of other generations being unchanged (and therefore that it be positively sensitive to increases in utilities of any finite number of generations, the utilities of other generations being unchanged), and also that an SWF be positively sensitive to an increase in utilities of all generations. However, it need not be positively sensitive to an increase in utilities of an infinite number of generations, when the utilities of a (nonempty) set of generations is unchanged. This is the principal difference between the partial Pareto axiom and the Pareto axiom.

5.3.1 Possibility Results for Restricted Domains

Note that if we recognize that human perception or cognition is not endlessly fine, so that sufficiently small changes in well-being go unperceived, it seems reasonable to suppose that the set of feasible utilities will be a discrete set.[8] The same is true if the benefits are measured in money and there is a well-defined smallest unit, as is true for all currencies (Segerberg 1976). Thus, it seems worthwhile to explore whether, with $Y \subset M$ (which captures this very reasonable possibility), there is an SWF (on X) respecting Anonymity and one of the weaker versions of the Pareto axiom, introduced above.[10] It is interesting to note that the domain restriction allows us to establish the existence of an equitable SWF, which satisfies the strongest of these versions of Pareto, namely the partial Pareto axiom.

PROPOSITION 1: *Assume $Y \subset M$. There exists an SWF satisfying the partial Pareto and Anonymity axioms.*

PROOF: For each $x \in X$, let $E(x) = \{y \in X:$ there is some $N \in \mathbb{N}$, such that $y_k = x_k$ for all $k \in \mathbb{N}$, which are $N\}$. Let $\mathfrak{I}$ be the collection $\{E : E = E(x)$ for some $x \in X\}$. Then $\mathfrak{I}$ is a partition of X. That is, if E and F belong to $\mathfrak{I}$, then either $E = F$ or E is disjoint from F; further, $\cup_{E \in \mathfrak{I}} E = X$.

Define a function, $f: X \to M$ as follows. Given any $x \in X$, let $f(x) = \min\{x_1, x_2, \cdots\}$. Since $x_i \in M$ for all $i \in \mathbb{N}$, the set $\{x_1, x_2, \cdots\}$ is a nonempty subset of the set of non-negative integers and therefore has a smallest element (Munkres 1975, p. 32). Thus, f is well defined. By the axiom of choice, there is a function $g: \mathfrak{I} \to X$, such that $g(E) \in E$ for each $E \in \mathfrak{I}$.

Given any $x \in X$, we can denote for each $N \geq 1$, $(x_1, \cdots, x_N)$ by $x(N)$ and $(x_1 + \cdots + x_N)$ by $I(x(N))$. Next, given any x, y in $E \in \mathfrak{I}$, define $h(x, y) = \lim_{N \to \infty} [I(x(N)) - I(y(N))]$. Notice that h is well-defined, because given any x, y in $E \in \mathfrak{I}$, there is some $M \in \mathbb{N}$, such that $[I(x(N)) - I(y(N))]$ is a constant for all $N \geq M$. Given any x, y in $E \in \mathfrak{I}$, define $H(x, y) = 0.5\{h(x,y)/[1+|h(x,y)|]\}$. Then $H(x, y) \in (-0.5, 0.5)$.

We now define $W: X \to \mathbb{R}$ as follows. Given any $x \in X$, we associate with it its equivalence class, $E(x)$. Then, using the function g, we get $g(E(x)) \in E(x)$. Next, using the functions, h and H, we obtain $h(x, g(E(x)))$ and $H(x, g(E(x)))$. Finally, define $W(x) = f(x) + H(x, g(E(x)))$.

The Anonymity axiom can be verified as follows. If x, y are in X and there exist i, j in $\mathbb{N}$, such that $x_i = y_j$ and $x_j = y_i$, while $x_k = y_k$ for all $k \in \mathbb{N}$, such that $k \neq i, j$, then $E(x) = E(y)$. Furthermore, denoting this common set by E, we see that $h(x, g(E)) = h(y, g(E))$, and so $H(x, g(E)) = H(y, g(E))$. Further, the set $\{x_1, x_2, \cdots\}$ is the same as the set $\{y_1, y_2, \cdots\}$, so that $f(x) = f(y)$. Thus, we obtain $W(x) = W(y)$.

The partial Pareto axiom can be verified as follows. If x, y are in X and there exists $i \in \mathbb{N}$, such that $x_i > y_i$, while $x_k = y_k$ for all $k \in \mathbb{N}$, such that $k \neq i$, then $E(x) = E(y)$. Furthermore, denoting the common set by E, we see that $h(x, g(E)) > h(y, g(E))$. This implies that $H(x, g(E)) > H(y, g(E))$. Further, the smallest element of the set $\{x_1, x_2, \cdots\}$ is at least as large as the smallest element of the set $\{y_1, y_2, \cdots\}$, so that we have $f(x) \geq f(y)$. Thus, we obtain the desired inequality; $W(x) > W(y)$.

If $x, y \in X$, and $x \gg y$, then $E(x) \neq E(y)$. Thus, we will not be able to compare $H(x, g(E(X)))$ with $H(y, g(E(y)))$. However, we do know that $H(x, g(E(x))) > -0.5$, and $H(y, g(E(y))) < 0.5$. Further, since $x \gg y$, we have $f(x) \geq f(y) + 1$. Thus, we obtain

$$W(x) = f(x) + H(x, \ g(E)) > f(y) + 1 - 0.5 > f(y) + H(y, \ g(E)) = W(y). \qquad \text{Q.E.D.}$$

Proposition 1 has two shortcomings. First, it is a possibility result for an SWF, but we do not know how to construct the SWF whose existence is asserted, since our proof uses the axiom of choice.[10] The possible policy use of Proposition 1 is therefore limited. We should clarify, however, that though we give a proof using the axiom of choice and indeed know of no other proof, it is not the case that we have proved that the axiom of choice is necessary. Indeed, it remains a bit of an open conjecture as to whether the axiom of choice is necessary for the above proposition.

The second shortcoming can be seen by considering the setup, where $Y = \{0, 1\}$, so that we have the strongest possible domain restriction. Theorem 1 implies that there is no SWF respecting the Pareto and Anonymity axioms. And, Proposition 1 implies that there is an SWF satisfying the partial Pareto and Anonymity axioms. It follows that any SWF, W, so obtained, must violate the Pareto principle in a way that is particularly disturbing; that is, it must be the case that there exist alternatives $x, y \in X$ such that $x > y$, but $W(x) < W(y)$.

To see this, suppose on the contrary that there is an SWF, W, satisfying the Anonymity and partial Pareto axioms, and the following monotonicity condition:

CONDITION M (*Monotonicity*): For all $x, y \in X$, if $x > y$, then $W(x) > W(y)$.

We claim then that W must, in fact, satisfy the Pareto axiom. To see this, let $x, y \in X$ with $x > y$. There are three possibilities: (i) $x \gg y$; (ii) $x_i > y_i$ for $i \in F$, where F is a finite subset of $\mathbb{N}$ and $x_i = y_i$ for all $i \in \mathbb{N} \sim F$; (iii) $x_i > y_i$ for $i \in I$, where I is an infinite strict subset of $\mathbb{N}$, and $x_i = y_i$ for all $i \in \mathbb{N} \sim I$. In cases (i) and (ii), by the partial Pareto axiom, we must have $W(x) > W(y)$. In case (iii), let j be the smallest index in I, and define z as $z_j = y_j$ and $z_i = x_i$ for all $i \neq j$. Then, $z \in X$ and $z > y$, so that by Condition M, $W(z) \geq W(y)$. Also, comparing x and z, we see that they differ in only the jth index, and $x_j > y_j = z_j$, so that the partial Pareto axiom implies that $W(x) > W(z)$. Thus, $W(x) > W(y)$, and our claim is established. But, by Theorem 1, there is no SWF satisfying the Pareto and Anonymity axioms. Consequently, any SWF, W, satisfying the Anonymity and partial Pareto axioms, must violate Condition M.[11]

Both the shortcomings of Proposition 1 arise from the fact that we are trying to define an SWF that is sensitive to the utility of a single generation when the utilities of all other generations are unchanged. If we give up this sensitivity, and weaken our partial Pareto requirement to the weak Pareto one, we get a particularly satisfying possibility result on all domains X, when $Y \subset$ M.

THEOREM 3: *Assume $Y \subset$ M. Then the SWF, $W : X \to$ M, given by*

$$W(x) = \min \{x_1, x_2, \cdots\} \text{ for all } x \in X$$

satisfies the weak Pareto and Anonymity axioms. Further, it satisfies Condition M.

PROOF : The function, $W: X \to$ M,, given by $W(x) = min \{x_1, x_2, \cdots\}$ for all $x \in X$, is well defined (as already noted in the proof of Proposition 1). If $x, y \in X$ and $x \gg y$, then denoting an index, for which $\min \{x_1, x_2, \cdots\}$ is attained, by $k \in$ N, we have

$$W(y) = \min \{y_1, y_2, \cdots\} \leq y_k < x_k = \min \{x_1, x_2, \cdots\} = W(x),$$

so that the weak Pareto axiom is satisfied.

If $x, y \in X$ and there exist $i, j \in \mathbb{N}$, such that $x_i = y_j$ and $x_j = y_i$, while $x_k = y_k$ for all $k \in \mathbb{N}$, such that $k \neq i, j$, then the set $\{x_1, x_2, \cdots\}$ is the same as the set $\{y_1, y_2, \cdots\}$, so that $W(x) = W(y)$. Thus, the Anonymity axiom is satisfied.

Finally, if $x, y \in X$ and $x > y$, then denoting an index, for which $\min\{x_1, x_2, \cdots\}$ is attained, by $k \in \mathbb{N}$, we have

$$W(y) = \min \{y_1, y_2, \cdots\} \leqq y_k \leqq x_k = \min\{x_1, x_2, \cdots\} = W(x),$$

so that Condition M is satisfied. $\qquad$ Q.E.D.

The SWF in Theorem 3 can be explicitly written down, and this makes the possibility result especially useful for policy purposes.

5.3.2 Weakening Domain Restrictions

The above possibility results are obtained by weakening the Pareto axiom (to partial Pareto or to weak Pareto) and also considering a discrete domain. How would a change in the latter affect the results? It is especially useful to ask this question in the context where $Y = [0,1]$, since this is the standard framework used by Diamond (1965), Koopmans (1960), Svensson (1980), and others.

As it turns out, we run again into impossibility results, which means that with $Y = [0, 1]$, the weakening of Pareto to partial Pareto or to weak Pareto does not help to reverse the impossibility result of Theorem 1. To establish the first of these impossibility results, which follows directly from the result of Basu and Mitra (2003, theorem 2), it is useful to introduce a new axiom, the interest in which is purely constructive, so as to be able to explain the next result clearly.

DOMINANCE AXIOM: *For all $x, y \in X$, if there exists $j \in \mathbb{N}$ such that $x_j > y_j$, and, for all $k \neq j$, $x_k = y_k$, then $W(x) > W(y)$. For all $x, y \in X$, if $x \gg y$, then $W(x) \geq W(y)$.*

Note that the last inequality in the statement of this axiom is a weak inequality, unlike in the definition of the partial Pareto axiom. Hence, partial Pareto is stronger than dominance (which in turn is stronger than weak dominance).[12]

PROPOSITION 2: *Assume $Y \supset [0, 1]$. There is no SWF satisfying the partial Pareto and Anonymity axioms.*

PROOF: By theorem 2 of Basu and Mitra (2003), we know that there is no SWF satisfying the dominance and Anonymity axioms. The result is proved by noting that the partial Pareto axiom is stronger than the dominance axiom. $\quad$ Q.E.D.

When we weaken the partial Pareto axiom (of Proposition 2) to weak Pareto, the impossibility result persists, but it is a more subtle result, because the sensitivity of the SWF to a change in a single generation's utility (when the utilities of all other generations are unchanged) is not being imposed. The proof of it is, likewise, more intricate, combining the methods used by Basu and Mitra (2003, theorem 2) and by Fleurbaey and Michel (2003).

THEOREM 4: *Assume $Y \supset [0, 1]$. There is no SWF satisfying the weak Pareto axiom and the Anonymity axiom.*

PROOF: To establish the theorem, assume that there exists an SWF, $W: X \to \mathbb{R}$, which satisfies the weak Pareto and Anonymity axioms.

Denote the vector $(1, 1, 1, \cdots)$ in X by e. Define the sequences x and y in X as follows:

$$x = \left(\frac{1}{4}, \frac{2}{4}, \frac{3}{4}, \cdots, \frac{1}{4^k}, \cdots, \frac{4^k-1}{4^k}, \cdots \right), \tag{5.1}$$

$$ y = \left(\frac{1}{4} + \frac{1}{16}, \frac{2}{4} + \frac{1}{16}, \frac{3}{4} + \frac{1}{16}, \cdots, \frac{1}{4^k} + \frac{1}{4^{k+1}}, \cdots, \frac{4^k - 1}{4^k} + \frac{1}{4^{k+1}}, \cdots \right). \quad (5.2) $$

For $s \in I = (-0.5, 0.5)$, define

$$ y(s) = 0.5\,y + 0.25\,(1 + s)e. \quad (5.3) $$

Then $(1/8)e \le y(s) \le (7/8)e$, and so $y(s) \in X$ for each $s \in I$.

Define the function, $f: I \rightarrow \mathbb{R}$ by $f(s) = W(y(s))$. By the weak Pareto axiom, f is monotonic increasing in s on I. Thus f has only a countable number of points of discontinuity in I. Let $a \in I$ be a point of continuity of the function f.

Define the sequence $x(a)$ as follows:

$$ x(a) = 0.5x + 0.25\,(1 + a)e. \quad (5.4) $$

Clearly, $x(a) \in X$ and $y(a) \gg x(a)$. By the weak Pareto axiom, $W(y(a)) > W(x(a))$. We denote $[W(y(a)) - W(x(a))]$ by θ; then $\theta > 0$.

Denote max $(0.5 - a, 0.5 + a)$ by Δ, then $\Delta > 0$. Since f is continuous at a, given the θ defined above, there exists $\delta \in (0, \Delta)$, such that: $0 < |s - a| < \delta$ implies $|f(s) - f(a)| < \theta$. Note that for $0 < |s - a| < \delta$, we always have $s \in I$.

For $p \in \mathbb{N}$, let $r(p)$ denote the first nonzero remainder of the successive divisions of p by 4, and $q(p)$ the number of divisions with a zero remainder. [For example, $r(52) = 1$ and $q(52) = 1$.]

Define (following Fleurbaey and Michel 2003, p. 796), for each $k \in \mathbb{N}$, a sequence x^k as follows:

$$ x^k = \left(\frac{1}{4} + \frac{1}{16}, \frac{2}{4} + \frac{1}{16}, \frac{3}{4} + \frac{1}{16}, \cdots, \frac{1}{4^k} + \frac{1}{4^{k+1}}, \cdots, \frac{4^k - 1}{4^k} + \frac{1}{4^{k+1}}, \right. $$

$$ \frac{1}{4^{k+1}}, \cdots, \frac{4p}{4^{k+1}}, \cdots, \frac{4p}{4^{k+1}}, \frac{4p+2}{4^{k+1}}, \frac{4p+3}{4^{k+1}}, \cdots, \frac{4^{k+1} - 1}{4^{k+1}}, \frac{1}{4^{k+2}}, $$

$$ \left. \frac{2}{4^{k+2}}, \cdots, \frac{4^{k+2} - 1}{4^{k+2}}, \cdots \right), \quad (5.5) $$

where p runs from 1 to $4^k - 1$, and the term $[4p/(4^{k+1})]$ is repeated $q(4p)$ times if $r(4p) = 1$, and $q(4p) + 1$ times otherwise. Now, for each $k \in \mathbb{N}$, we use x^k to define $x^k(a)$ as follows:

$$ x^k(a) = 0.5\,x^k + 0.25(1 + a)e. \quad (5.6) $$

Clearly, $x^k(a) \in X$ for each $k \in \mathbb{N}$. Comparing the expressions for $x(a)$ and $x^1(a)$ in Eqs. (5.4) and (5.6), respectively, we see that $x^1(a)$ is obtained from $x(a)$ by a finite permutation, and that for all $k > 1$, $x^k(a)$ is obtained from $x^{k-1}(a)$ by a finite permutation. Thus, for every $k \in \mathbb{N}$, $x^k(a)$ is obtained from $x(a)$ by a finite permutation, and the Anonymity axiom yields

$$ W(x^k(a)) = W(x(a)) \qquad \forall\, k \in \mathbb{N} \quad (5.7) $$

Choose $K \in \mathbb{N}$ with $K \ge 2$ such that $(1/4^{K-2}) < \delta$, and define $S = [a - (1/4^{K-2})]$. We note that $0 < (a - S) < \delta$, and so $S \in I$, and

$$W(y(S)) = f(S) > f(a) - \theta = W(y(a)) - \theta. \qquad (5.8)$$

We now compare the welfare levels associated with $x^K(a)$ and $y(S)$ as follows. Notice that

$$x^K(a) = 0.5\,x^K + 0.25(1+a)e = 0.5\,y + 0.25(1+a)e - 0.5(y - x^K)$$

$$= y(a) - 0.5(y - x^K)$$

$$\geq y(a) - 0.5(1/4^K)e$$

$$= 0.5\,y + 0.25(1+a)e - 0.5(1/4^K)e$$

$$> 0.5y + 0.25\,[1 + a - (1/4^{K-1})]e$$

$$= 0.5\ y + 0.25\,[1+a - (1/4^{K-2})]e + 0.25(3/4^{K-1})e$$

$$\gg 0.5y + 0.25(1+S)e = y(S).$$

Thus, by the weak Pareto axiom, we have

$$W(x^K(a)) > W(y(S)) \qquad (5.9)$$

Using Eqs. (5.7)–(5.9), we obtain

$$W(y(a)) - \theta = W(x(a))$$

$$= W(x^K(a))$$

$$> W(y(S))$$

$$> W(y(a)) - \theta,$$

a contradiction that establishes our result. $\qquad$ Q.E.D.

It is worth noting that, with the domain restriction $Y \subset M$, weakening the Pareto axiom to the weak Pareto axiom led to a reversal of the impossibility result of Theorem 1 to the possibility result of Theorem 3. When $Y = [0,1]$, a similar weakening of the Pareto axiom (to the weak Pareto axiom) does not produce such a reversal.

This suggests that to recover possibility when $Y = [0,1]$, we need to go to a weaker form of Pareto. In fact, weak dominance is not weaker than weak Pareto, but we can establish the existence of an equitable SWF, which satisfies weak dominance. In fact, this possibility result holds with no domain restriction. Our proof employs the idea, already used in the proof of Proposition 1, of partitioning X into sets such that the members of each set differ from each other in only a finite number of indices. The proof of the possibility result then crucially hinges on (i) the use of the axiom of choice and (ii) the fact that weak dominance never requires one to compare the welfare of members in two different sets of the partition.

THEOREM 5: *There exists an SWF satisfying the weak dominance and anonymity axioms.*

PROOF: For each $x \in X$, let $E(x) = \{y \in X:$ there is some $N \in \mathbb{N}$, such that $y_k = x_k$ for all $k \in \mathbb{N}$, which are $\geq N\}$. Let $\mathfrak{I}$ be the collection $\{E : E = E(x)$ for some $x \in X\}$.

Then, $\Im$ is a partition of X. By the axiom of choice, there is a function, $g : \Im \to X$, such that $g(E) \in E$ for each $E \in \Im$.

Given any x, y in $E \in \Im$, define $h(x, y) = \lim_{N \to \infty} [I(x(N)) - I(y(N))]$. We now define $W : X \to \mathbb{R}$ as follows. Given any $x \in X$, we associate with it its equivalence class, $E(x)$. Then, using g, we get $g(E(x)) \in E(x)$, and, using h, we obtain $h(x, g(E(x)))$. Now, define $W(x) = h(x, g(E(x)))$. The Anonymity axiom and the weak dominance axioms are easily verified. *Q.E.D.*

REMARKS: (i) The weak dominance axiom compares utility streams which differ for only one generation. One could define a concept of finite dominance, which allows for comparisons between utility streams, in which one utility stream always has at least as much utility for each generation as the other and the utility streams differ for at most a *finite* number of generations.

FINITE DOMINANCE: If $x, y \in X$, and $x > y$, and there is $N \in \mathbb{N}$ such that $x_k = y_k$ for all $k > N$, then $W(x) > W(y)$.

Clearly, W satisfies finite dominance if and only if it satisfies weak dominance. In view of this, Theorem 5 is *equivalent* to the statement obtained by replacing 'weak dominance' by 'finite dominance'.

(ii) The possibility result of Theorem 5 can be contrasted with the impossibility result of Diamond (1965). When $Y = [0, 1]$, Diamond's result shows that there is no social welfare order, continuous in the sup metric, which satisfies the relational versions of the Anonymity and Pareto axioms. However, the proof of Diamond's impossibility result can be used to infer that there is no social welfare order, continuous in the sup metric, which satisfies the relational versions of Anonymity and weak dominance. Thus, continuity in the sup metric (in conjunction with Anonymity) is a stronger restriction than *representability* of a social welfare order in this context.

(iii) It can be checked that any SWF, W, satisfying the weak dominance and Anonymity conditions must violate Condition M, that is, there must exist x, $y \in X$, such that $x > y$ and $W(x) < W(y)$.

5.4 CONCLUDING REMARKS

We wanted to demarcate the boundary between what is possible and what is not and the set of results established in this paper tries to do that *vis-à-vis* variations of the Pareto axiom and the domain restriction for utilities. In setting out to write this paper we had wanted to display the positive side of this field, namely, the possibility theorems. We have done so. But now, at paper's end, we find that in the process we have also highlighted the robustness of the impossibility theorems of the literature. This is probably a reminder that we have no option but to play the hand that we are dealt.

Our investigation of domain restrictions for possibility theorems of equitable social welfare functions is, of course, not complete. If we restrict our attention to the Weak Pareto axiom, we have the possibility theorem (Theorem 3) when $Y = M$ and the impossibility theorem (Theorem 4) when $Y = [0, 1]$. One might be interested to know what results would hold if the domain restrictions place Y somewhere 'in between' these two cases. For example, Y could be the set of rationals in $[0, 1]$. Neither the method used to establish the possibility theorem when $Y = M$, nor the method used to prove the impossibility result when $Y = [0, 1]$, applies to this case.[13] We might hope that future research in this area will develop new methods capable of dealing with a wider class of domain restrictions.

NOTES

1. The continuity property postulated by Diamond is with respect to the sup-metric on $X = [0,1]^N$.
2. Of course, the case in which the period utility space is a singleton, and so the domain of the SWF is also a singleton, is ruled out in the framework of Basu and Mitra (2003, theorem 1).
3. In informal discussions throughout the paper, the terms 'equity' and 'anonymity' are used interchangeably.
4. The Anonymity axiom figures prominently in the social choice theory literature, where it is stated as follows: the social ordering is invariant to the information regarding individual orderings as to who holds which preference ordering. Thus, interchanging individual preference profiles does not change the social preference profile. For discussions of this axiom and its acceptability see May (1952) and Sen (1970, 1977).
5. An ordering is a binary relation which is complete and transitive.
6. In this connection, see also the papers by Suzumura and Shinotsuka (2003), and Xu (2005).
7. In fact, in some of the literature, what we are calling 'Weak Pareto' is often called 'Pareto', with the suffix 'strong' added to what we have called simply the 'Pareto axiom'.
8. The idea of setting a limit to the fineness of human perception has been used in a different context by Armstrong (1939) to argue that it is unreasonable to suppose that indifference is a transitive relation. For a discussion of this issue in individual choice theory, see Majumdar (1962).
9. While our choice of Y as a subset of the set of non-negative integers is motivated by the imprecision of human perception, the mathematical technique used to obtain our possibility result applies also to the case where $Y = \{(1/n): n \in \mathbb{N}\}$, where clearly human perception has to be considered to be sufficiently refined.
10. The use of the axiom of choice in proving *impossibility results* is, perhaps, less objectionable.
11. A weak version of Pareto, which requires that Condition M, together with what we have called the weak Pareto axiom, be satisfied, is quite appealing, and has been proposed and examined by Diamond (1965).
12. It is also worth noting that between dominance and weak Pareto axioms, neither is stronger than the other. They are in fact noncomparable in terms of strength. The same is true between weak dominance and weak Pareto axioms.
13. Of course, since the streams considered in the proof of Theorem 4 consist entirely of rational entries, imposing a continuity (in the sup-metric) axiom on the SWF will provide an impossibility result. But, using such a continuity axiom goes against the spirit of our chapter.

REFERENCES

Armstrong, W. E., 1939, 'The determinateness of the utility function', *Economic Journal*, 49, 453–67.

Arrow, K. J., 1963, *Social Choice and Individual Values* (2nd edition), John Wiley, New York.

Basu, K., 1994, 'Group rationality, utilitarianism and Escher's "waterfall" ', *Games and Economic Behavior*, 7, 1–9.

Basu, K. and T. Mitra, 2003, 'Aggregating infinite utility streams with intergenerational equity: The impossibility of being Paretian', *Econometrica*, 79, 1557–63.

Bossert, W., Y. Sprumont, and K. Suzumura, 2004, 'The possibility of ordering infinite utility streams', Universite de Montreal (Mimeo).

Cowen, T. and D. Parfit, 1992, 'Against the social discount rate', in P. Laslett and J. S. Fishkin (eds.), *Justice between Age-Groups and Generations*, Yale University Press, New Haven.

Diamond, P., 1965, 'The evaluation of infinite utility streams', *Econometrica*, 33, 170–7.

Fleurbaey, M. and P. Michel, 2003, 'Intertemporal equity and the extension of the Ramsey principle', *Journal of Mathematical Economics*, 39, 777–802.

Koopmans, T. C., 1960, 'Stationary ordinal utility and impatience', *Econometrica*, 28, 287–309.

Majumdar, T., 1962, *The Measurement of Utility* (2nd edition), Macmillan, London.

May, K. O., 1952, 'A set of independent necessary and sufficient conditions for simple majority decision', *Econometrica*, 20, 680–4.

Munkres, J., 1975, *Topology*, Prentice Hall, London.

Ramsey, F. P., 1928, 'A mathematical theory of savings', *Economic Journal*, 38, 543–59.

Segerberg, K., 1976, 'A neglected family of aggregation problems in ethics', *Nous*, 10, 221–44.

Sen, A. K., 1970, *Collective Choice and Social Welfare*, Holden-Day, San Fransisco.

—— 1977, 'On weights and measures: Informational constraints in social welfare analysis', *Econometrica*, 45, 1539–72.

Suzumura, K., 1983, *Rational Choice, Collective Decisions and Social Welfare*, Cambridge University Press, Cambridge.

Suzumura, K. and T. Shinotsuka, 2003, 'On the possibility of continuous, Paretian and egalitarian evaluation of infinite utility streams', Institute of Economic Research, Hitotsubashi University (Mimeo).

Svensson, L.-G., 1980, 'Equity among generations', *Econometrica*, 48, 1251–6.

Xu, Y., 2005, 'Pareto principle and intergenerational equity: Immediate impatience, universal indifference and impossibility', Palgrave Macmillan, Basingstoke, Hampshire, UK.

6   Achievements, Capabilities, and the Concept of Well-being
A Review of Commodities and Capabilities by Amartya Sen

6.1 INTRODUCTION

In modelling human welfare and choice, Amartya Sen has always been careful to make room for the imprecisions and inexactitudes which abound in life. Thus his books and papers in welfare economics[1] would almost invariably mention 'partial comparabilities', 'incompletenesses', and 'quasi-rankings'. However, in his early works, the kinds of imprecision he considered were amenable to very precise formulations and the inexactitudes with which he was concerned could be written about using an exacting language.

Sen's most recent book also parts with the pristine world of textbook economics but it is of a different genre from Sen's earlier works. The subject matter now is too diverse, too complex to be amenable to the relentless precision of social choice theory. The book examines a society peopled with characters whose entire welfare can no longer be captured in a unique real number, characters who have distinct notions of well-being, happiness, and desire-fulfillment. The aim is to evaluate such societies and the conditions of people and groups within such societies. Sen draws on different disciplines, uses several examples, and develops some empirical illustrations in order to persuade us that though this new approach to evaluation does not yet have an algebra, let alone systematic statistics, it is a direction we must pursue. The book under review, Sen reminds us towards the end, is 'no more than a beginning'.

I have benefited from discussions with Sanjay Subramanyam.
First published in *Social Choice and Welfare*, Vol 4 (1987).

The purpose of the present chapter is threefold: to present the essence of Sen's argument, to criticize it, and to suggest directions for developing it further.

6.2 SEN'S ARGUMENT

There are (at least) three sources of information for evaluating the welfare or standard of living of individuals or groups in society. These are:

(i) Market data (people's incomes and consumptions, their expenditure patterns, etc.),

(ii) direct surveys (i.e., information from directly asking people how well-off they are), and

(iii) social data (life expectancy, morbidity, malnutrition, literacy, etc.).

In economics our concern has been almost exclusively with (i) —with some exceptions such as the use of (ii) by the 'Leyden school' pioneered by Van Praag. While admitting that each of these three kinds of data have a role to play, Sen sets out to make a special case for the use of social data—or what he refers to as 'nonmarket data'. Of course, outside traditional economics, there is plenty of evidence of nonmarket data being used to criticize or praise social states. Newspaper editorials lament the poor quality of life in some societies because of the lack of freedom of speech; we often express dismay concerning some society for its high crime rate; demographers praise nations where infant mortality is low. However, such comments and discussions usually occur in fragments and somewhat in the shadow of the economist's practice of evaluation based on national incomes or utilities for which the economist can claim to have a consistent methodology and even a philosophy. A large part of this monograph is an effort to develop a philosophical base and systematic method for the use of nonmarket data in evaluating societies.

One reason why such data is scant is, according to Sen, that there is no cogent methodological foundation for the use of such information. So this monograph is supposed to provide a fillip not only to this method but also to the collection of the kind of data required for its practice. On this ultimate position, that is the usefulness of nonmarket data, Sen is extremely persuasive. From a reading of this book and from my earlier interest in analysing living standards in South Asia, I am convinced that nonmarket data ought to be used and more than currently is the practice among economists. However concerning some of the specific methods and underlying philosophy I have disagreements. Before going into these, it is useful to briefly recapitulate Sen's argument.

The central character in this study is the concept of 'well-being'. In traditional economic analysis one may jump to equate well-being with happiness. But that would be incorrect. A person who is poor and has some crippling ailment may nevertheless be happy because he has learnt to curb his desires, but clearly we would not describe him as having a high level of well-being or (what is related but nevertheless distinct) a high standard of living. Similarly well-being must not be equated with utility. This is so for many reasons. In modern economic theory, utility is often treated as a mere calibration system which reflects choice. If a person chooses x over y, then we say (by definition) that the utility he gets from x *is* greater than what he gets from y. Under this interpretation, utility cannot be equated with well-being because the motivations underlying a person's choice could be diverse and need not be consonant with his perceptions of his own well-being. Moreover, a person's preference may depend on his current condition. A woman who is exploited may choose (and prefer) to live in a society where such exploitation exists and it is quite possible that the same woman if she were based in a society where women were not exploited would prefer to live in such a society. As far as the notion of well-being is concerned, barring some fairly exceptional characterizations, we would not like to assert that whether i achieves a higher well-being in state x or in state y depends on which state i happens to be in currently. But, as the above example shows, a person's choice or even preference may depend on the state he happens to be in.

For a more specific definition of well-being it is essential to understand the concept of 'functioning'. Sen defines functioning as 'an achievement of a person: what he or she manages to do or to be'. This idea is captured in commonplace observations like: I had so far ignored the pain in my elbow but I'd better see a doctor now because it is beginning to affect my functioning. This remark also illustrates, what is important for Sen's story, that the functionings a person achieves depend partly on his choice. In the above example, I have the choice of not seeing my doctor and leaving my arm's functioning impaired.[2] Sen then goes on to define a person's 'being' as the vector of functionings he achieves and his 'well-being' as a valuation of his being. It should be clear that Sen does not attempt to give any specific way of doing this valuation. His aim is to persuade us that the valuation should be done on a particular domain, namely that of functionings. Sen warns us, on several occasions, that these are complex terms. And so they are. In fact one of my complaints is that Sen himself treats these concepts as simpler and more measurable than they actually are.

Data for this kind of valuation is not easily available. In the mean time one possibility is to use whatever meagre data are available for some preliminary exercises. This is what is done in two appendices in Sen's book. Appendix A compares living standards across countries by looking not only at per capita incomes but mortality, literacy, and life expectancy. A more literate people has the 'capability' of more diverse functionings than a less literate people. Similarly, a longer life gives us a greater capacity for the activities of life. This justifies looking at these extra-economic variables. In Appendix B, Sen examines the well-being of females relative to males within the family in India. Both these empirical studies are of great interest in themselves and I shall return to some of these issues later. However, I make one comment here.

While Sen is right in asserting that a society's well-being cannot be captured by simply looking at income levels, we must not make the mistake of thinking that income measures are totally devoid of registering improvements which lie beyond the realm of economics. Suppose that a direct measure of income shows that Kerala is behind, say, West Bengal but Kerala's health provisions are much better. In principle, an economist's computation of a Keralite's income (as opposed to the much-maligned accountant's calculation) should include the amount a Bengali would be willing to pay to achieve a Keralite's health provisions.[3] Though such computations are not done in practice, and even if they were, they would not capture the entirety of the concept of well-being, it is worth noting that an economist's measure is, atleast in principle, not totally oblivious of quality-of-life questions.

6.3 THE MEASURABILITY OF FUNCTIONINGS

There are two related questions concerning the measurability of functionings and capabilities. First, can a person's entire functioning status be characterized as an n-tuple? The use of the term 'a vector of functionings' suggests a kind of measurability which in reality may be elusive. Also it is not clear that the level of each functioning can be calibrated on a linear scale. As in the case of utility, the measurement of functioning is a complex question (e.g., should they be measured cardinally or ordinally?) and there will have to be research into this before Sen's approach becomes fully usable.

The second problem is best raised in the context of Appendix A. Here Sen studies intercountry data based on certain kinds of capabilities, namely those related to the length of life and education. He observes that the exercise is illustrative. A full-fledged analysis will have to look at a more complete list of capabilities. But (i) what constitutes a complete list of capabilities and (ii) how can we be sure that items in the list do not overlap? Consider problem (ii). Infant mortality and life expectancy at birth are distinct features of society but they also have a great deal of overlap. Thus to praise a society for having cut its infant mortality rate and for having raised its life expectancy at birth may involve more praise than ought to be heaped. On the other hand, to praise society for only one of these may be to praise it less than what it deserves. How do we separate out what seem to be inherent overlaps between concepts? Failure to do so would result in either double counting or under counting some innate quality of a society.[4] I have no answers to this problem but pose it as an open one.

6.4 THE INCOMPLETENESS OF RANKINGS

In this section I raise some issues concerning the incompleteness of rankings. First, I shall try to establish the need to distinguish between different kinds of incompleteness. Sen argues that two persons' subjective rankings, even if they are inconsistent with each other, may both be consistent with a unique objective ranking. He tries to establish this with the help of the following example.

Suppose A, B, and C are three possible personal states. Person 1 ranks A above B and person 2 ranks B above A. (I assume in addition that both rank C at the bottom.) Now suppose the objective ranking, denoted by R, is as follows:

$$R = \{(A, C), (B, C), (A, A), (B, B), (C, C)\}.$$

(x, y) is an element of R means x is at least as good as y on the basis of the objective ranking. Sen argues that since R is silent over the conflicting pair, $\{A, B\}$, R is consistent with both 1's and 2's orderings. If by consistent we mean the relation of 'subsuming' then of course this is so. But clearly Sen uses the term consistent here in the more primitive sense of compatibility. In this sense the consistency of R with 1's and 2's preferences depends on how we interpret the incompleteness over the pair $\{A, B\}$. There are at least two possible interpretations of this. The absence of (A, B) and (B, A) from R could be taken to mean either that A and B are unrankable or that the ranking of A and B are (as yet) unknown. Under the second interpretation it is right to say that the individual rankings are consistent with the objective one (meaning that, at the present level of information, there is no conflict between any of the subjective rankings and the objective one). Under the first interpretation, however, it is reasonable to argue that the individual rankings are inconsistent with the objective one because what the latter considers *unrankable* is ranked by the individuals. More important than its bearing on Sen's chapter 5 is the fact that this argument highlights how incompleteness can have different interpretations with different implications for welfare analysis.

Second, in dealing with partial information we need to distinguish between the axioms of completeness and exactness. Sen has emphasized that in conducting welfare analysis, whether across persons or nations, the conventional practice of searching for (complete) orderings may be a futile one. From this, he goes on to argue against the completeness axiom. It can, however, be shown that conventional methods of rankings not only assume completeness but also, implicitly, exactness. That is, if beginning from a fuzzy binary relation we are to arrive at an ordering, we have to impose the axioms of not only reflexivity, transitivity, and completeness out also exactness. Viewed in this manner it becomes clear that both completeness and exactness may be candidates for criticism based on precisely the kinds of insights that (though well-known today) originated in Sen's writings; and I have argued in Basu (1986) that for certain kinds of rankings, for example, that of societies based on income inequality, exactness may be the more untenable assumption.

6.5 OPPORTUNITIES AND ACHIEVEMENTS

It is argued in this monograph that in evaluating societies we should not be concerned only with the functionings that each individual achieves but also with the set of functionings available to each individual from which he makes his choice, that is, his 'capability set'. This raises a host of very interesting technical questions on how to rank sets based on a primitive ranking of elements. This problem has received considerable attention of late and it can boast its share of elegant 'impossibility' theorems.[5] Sen provides an excellent and characteristically lucid discussion of this difficulty (in chapter 7 of his work) but his main interest in this book is to defend the conceptual basis of a method which compares opportunities rather than achievements; and therefore it is this conceptual issue that this section focuses on.

A concern with opportunities rather than actual choices is not new. Though the

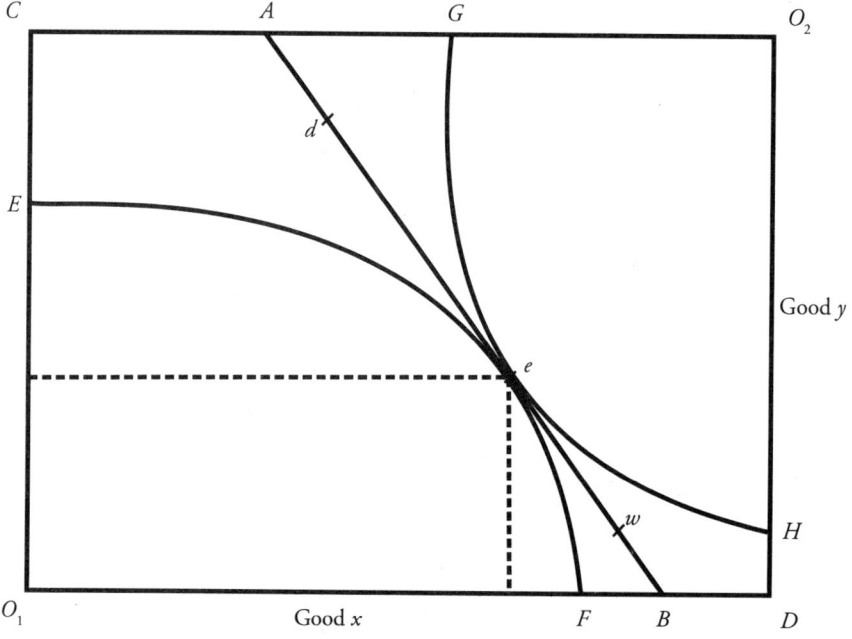

Figure 6.1: Feasibility Set in a Walras Economy.

traditional literature assumes a different space ('commodity' space, in contrast to Sen's 'functioning' space) the cores of the two sets of arguments are analogous. My comment in this section is sufficiently general for the space to be unimportant. I therefore choose the domain for ease of exposition.

One important problem of focussing attention on opportunity sets is that there is something illusory about the way 'opportunity' is defined in economics. This is best explained with an example. Let Figure 6.1 be a usual Edgeworth box of a 2-person 2-good exchange economy.

The figure depicts a general equilibrium. The endowments of persons 1 and 2 are at w. They face prices depicted by the line AB, and given their indifference curves, equilibrium occurs at e. In this economy person 1's choice is point e but his opportunity set in given by the area $CABO_1$. Similar comments apply to person 2, whose opportunity set is $ABDO_2$. But are these individuals really free to choose any point within their opportunity sets? The answer is no; because what is actually open to one person depends on what the other person actually chooses. Thus in Figure 6.1 it is not possible for 1 to choose d and for 2 to choose w. Similarly, if 2 chooses point e, 1 cannot choose point d. 1's belief that he can choose d is illusory. Once 2 has chosen e, the only choices actually open to 1 are those in the rectangle between e and O_1 (shown by broken lines).

Given that opportunity sets as traditionally defined include illusory choices, how much significance can we attach to the conventional opportunity sets? Also once this problem is appreciated, it becomes clear that opportunities can be increased vastly without changing anything. Consider the closed set bounded by $DFeECO_2$. From this remove all points on the curve FeE, except e. Let us call the set that remains Z_2. If instead of restricting 2's opportunity set to $ABDO_2$ we allow him to choose from Z_2, the equilibrium would remain unchanged. If we were evaluating this society on the basis of opportunities open to individuals, this exercise of making 2's opportunity set Z_2 would make this society appear better, but clearly our evaluation of this society should not hinge on such a ploy.

If we are to evaluate societies on the basis of opportunities open to each individual, we shall have to develop a definition of opportunity which is different from the traditional one. This definition will have to take into account the interdependence between one person's choice and another's opportunity set; and since a person's choice depends on his opportunity set, ultimately we shall have to deal with the interdependence of opportunity sets.

My own inclination would be to go along with Sen and evaluate well-being on the basis of functionings, but be content with achievements, instead of capabilities. One of Sen's principle reasons for wanting to evaluate capability sets is because this captures the extent of freedom open to a person. What I am suggesting here would therefore entail foregoing such a measure of freedom. However to base evaluation on achievement is not to ignore freedom altogether. There are two reasons why freedom is important: (i) for what it allows us to achieve and (ii) for its own sake. By ignoring capability sets we miss out on (ii) but not on (i). For reasons of interdependence, aspect (ii) of freedom is difficult to measure. Also, it seems to me that (i) is the more important aspect of freedom. Consequently, this is a loss which appears to me to be worth incurring.

6.6 INTRAFAMILY DISCRIMINATION

Appendix B of this book illustrates well the originality of Sen's approach to evaluation. The theme of this appendix is the alleged bias against females within the Indian household. If we were trying to study this in terms of the standard utility-based approach of welfare economics we would run into difficulties. This is for two reasons. First, as some anthropologists have pointed out, in much of South Asia people conceive welfare in terms of *groups* or *families*, and so the concept of *personal* welfare may not be meaningful or relevant. Second, women themselves are usually a party to the decisions which lead to bias against women. They may do this out of altruism or empathy, that is, their welfares may depend on the welfares of the male members of the households. What the well-being approach allows us to do is to assert without contradiction that the well-being of women relative to men is low even if the utility level of women relative to men is not.

This is the strength of Sen's approach. But this same appendix also demonstrates how difficult it is to translate this abstract method to empirical measurement. How does one establish the relative well-being of females within the Indian family? There are many ways of going about answering this. One way is to look at mortality and malnutrition statistics and this is what Sen does in the appendix. He notes that the sex ratio, that is, the ratio of female to male population, is not only less than one but has declined secularly since the beginning of this century. As far as malnutrition within the family goes, Sen himself has done several studies and he reports on these and others. But there are so many competing methods of doing this that one has to be cautious in generalising the results of any particular study.[6] I comment here on some specific problems.

One important problem for such studies is the difficulty of collecting information on people's age in Asia and Africa. Though in Sen's own studies great care is taken in collecting age data, this is, in general, a serious problem. And in case there is a systematic bias in age-reporting across sexual boundaries (e.g., if women's perceived age increases slower than men's perceived age) then this would introduce biases in a weight-for-age study of sex bias within the family. To try to avert this problem by doing a weight-for-height study raises other difficulties. For instance, a boy suffering from *chronic* malnutrition would be shorter than he would have been otherwise. Hence on a weight-for-height standard his malnutrition would appear to be less than it actually is.

6.7 CONCLUSION

Sen's book raises a variety of new and important issues concerning human well-being and its measurement. However it also leaves plenty of open questions and uncovered ground. Its argument for shifting our attention from commodities to functionings is powerful and I hope this will find increasing favour among welfare economists and other social scientists. Regarding its argument for shifting attention from achievements to capabilities, I am more skeptical.

Some of the concepts developed in this book can have important applications to questions beyond that of measurement, questions which Sen does not touch on in this book. For instance, the concept of 'exploitation' is difficult to describe in a utility-based framework. This is because people who are chronically exploited learn to adjust to their predicament and may achieve a reasonable level of utility and may not even strive for change. A definition of exploitation based on utility or choice would fail to identify such people. The concept of well-being can be useful in identifying the chronically exploited. This is just one of many possible directions that can be pursued from here.

NOTES

1. For instance, Sen (1970).
2. There is a serious question here as to whether what appears to be a choice cannot be reduced to an inherent characteristic of a person. This may be done by asking why a person chooses differently. If this can be traced to some inherent feature of the person then questions can be raised about the legitimacy of treating any functioning as chosen. This relates to the well-known problem of determinism and also to an interesting work on achieving equality by Roemer (1986).
3. There are some intricate index number questions involved here but it is harmless to gloss over these in the present context.
4. The problem is especially important to sort out if we are considering the use of formal measures like the weighted average of different characteristics.
5. See, for example, Barbera and Pattanaik (1984), Kannai and Peleg (1984).

6. Indeed a recent survey by Harris (1986) of empirical studies shows that the verdict on whether there is intra-family discrimination is mixed.

REFERENCES

Barbera, S. and P. K. Pattanaik, 1984, 'Extending an order on a set to the power set: some remarks on Kannai and Peleg's approach', *Journal of Economic Theory*, 32, 185–191.

Basu, K. 1987, 'Axioms for a fuzzy measure of inequality', *Mathematical Social Sciences*, Vol. 14.

Harris, B. 1986, 'Intra-family distribution of hunger in South Asia' (Mimeo).

Kannai, Y. and B. Peleg, 1984, 'A note on the extension of an order on a set to the power set', *Journal of Economic Theory*, 32, 172–5.

Roemer, J., 1986, 'Equality of resources implies equality of welfare', Quarterly *Journal of Economics*, Vol. 101, 751–86.

Sen, A. K., 1970, *Collective choice and social welfare*, Holden-Day, San Fransisco. (Republished by North-Holland in 1979.)

7 A Geometry for Non-Walrasian General Equilibrium Theory

7.1 INTRODUCTION

The emergence of non-Walrasian general equilibrium theory and its rapid advance through the 1970s have played a critical role in providing a common analytical platform for micro- and macroeconomic theory. Despite this important role, the dissemination of non-Walrasian ideas has been quite slow. The reason for this is the terseness of the subject which renders its main theorems inaccessible, not only to the laity but also to much of the economics profession.

The idea of a Walrasian equilibrium can be transmitted to the virtual layman with the help of the Edgeworth–Bowley box. It would be useful to have a systematic geometry for non-Walrasian equilibrium analysis as well. Some geometric developments have indeed occurred in this area, for example, see Bohm and Muller (1977), Dreze and Muller (1980), Grandmont (1977a).[1] The aim of the present chapter is to push this line of inquiry towards completion. By drawing together the scattered developments in the literature, this chapter provides a geometry which can capture the problems and theorems of fix-price general equilibrium theory of the kind developed in Dreze (1975), Dreze and Muller (1980), Grandmont (1977b), and Hahn (1978).[2] An important focus of the chapter is on the 'coupons equilibrium' and its variants. But, because a purpose of the paper is pedagogic, it first goes through discussions of existence, uniqueness, and optimality before taking on the subject of coupons more extensively.

Throughout the chapter I consider an economy with two individuals and three goods—one of which is money. An Edgeworth box for such an economy is indeed really a box, in the sense of being three dimensional, as shown in Figure 7.1.

I have benefited from the comments of Tariq Banuri, John Greedy, Dipanker Dasgupta, Janos Kornai, and an anonymous referee.

First published in *Journal of Macroeconomics*, (1992).

However, for a clear geometric analysis we need to work in two dimensions, so our first step is to translate our domain of analysis from three to two dimensions.

Note that, because in a fix-price model the initial endowments of the two agents, a and b, and prices are fixed, we can focus attention entirely on the budget hyperplane inside the Edgeworth box. This is shown as ABCDE in Figure 7.1. The strategy is to take out the slab ABCDE, lay it out flat, and do the analysis on it.

7.2 PRELIMINARY GEOMETRIC CONCEPTS

Throughout the chapter I consider three goods, labelled 0, 1, and 2, of which good 0 is supposed to be money. The prices of the three goods are given by $(p_0, p_1, p_2) \equiv p > 0$. There are two individuals labelled a and b. The endowment of individual i is given by $(w^i_0, w^i_1, w^i_2) = w^i$. We shall use w to denote (w^a, w^b).

The first problem that we confront is this. If we slice the Edgeworth box of this economy with a budget hyperplane, the area of the plane which lies within the box is our relevant domain. I shall call such an area the *PE-domain* since it is fixed as soon as we specify prices and endowments. What does a PE-domain look like?

It could look like ABCDE in Figure 7.1 but it could also look very different. It all depends on where the plane slices through the Edgeworth box. If it slices close to an origin it could be a triangle like *FGH* in Figure 7.1. But it could also be a parallelogram or a polygon.

Fortunately, there is an easy way of characterizing PE-domains in general. To do so, let us follow the convention of describing two triangles on a two-dimensional Euclidean surface as *parallel* if, for each side on each triangle, the opposite side on the other is parallel. That is, from each triangle you get the other by turning it around 180° and miniaturizing or expanding it if necessary. Figure 7.2 gives two examples of parallel triangles.

It is easy to see that a PE-domain takes those shapes that can be generated by

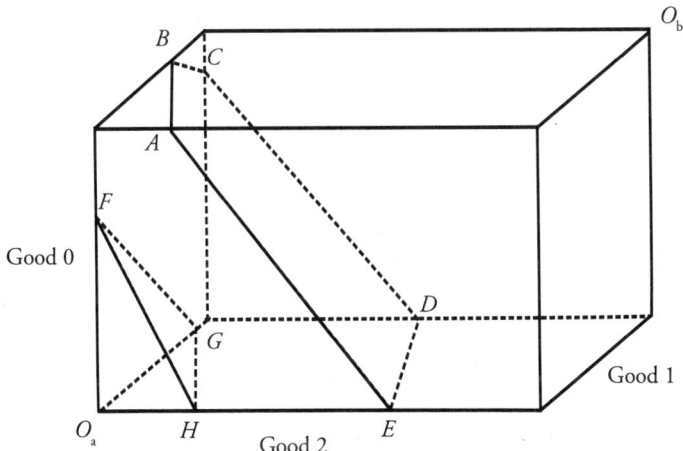

Figure 7.1: Edgeworth Box and PE-domains

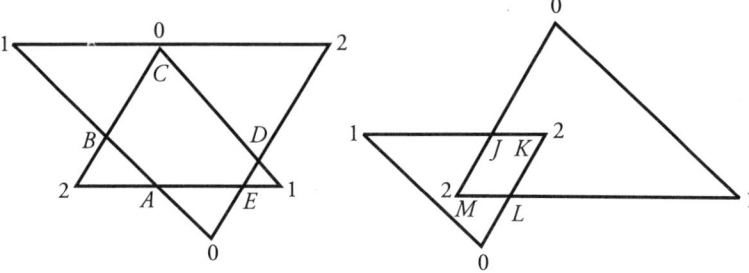

Figure 7.2: Budget Triangles

the intersection of two parallel triangles. Thus a PE-domain could be any of the following two areas in Figure 7.2: *ABCDE* and *JKLM*. One can sharpen ones intuition of why this should be so by cutting through loaves of bread in different directions and then contemplating the cross-sections.

To understand this characterization more formally it is useful to begin by considering a 2-by-2 Edgeworth box. Given an endowment and a price vector we get a PE-domain like *AB* in Figure 7.3. But there is another way of viewing *AB*. Given the price and endowment, we can first draw agent *a*'s budget set,[3] *CB*. Similarly, we can draw *b*'s budget set, AD. The PE-domain is nothing but the intersection of these two budget sets.

My characterization in the three-dimensional case uses exactly this principle. The two parallel triangles are the budget sets of the two players. That is, ABCDE in Figure 7.1 could be derived by extending this to the axes of player *a* to form *a*'s budget set, doing the same for b, and then taking the intersection of the two budget sets.

We shall follow some conventions throughout the chapter. Player *a*'s budget set will be drawn vertically with the top corner representing good 0 (i.e., money) and the right-hand bottom corner, good 1. This is shown in Figure 7.4. The budget triangle set of player *b* will therefore be upside down. The PE-domain will be

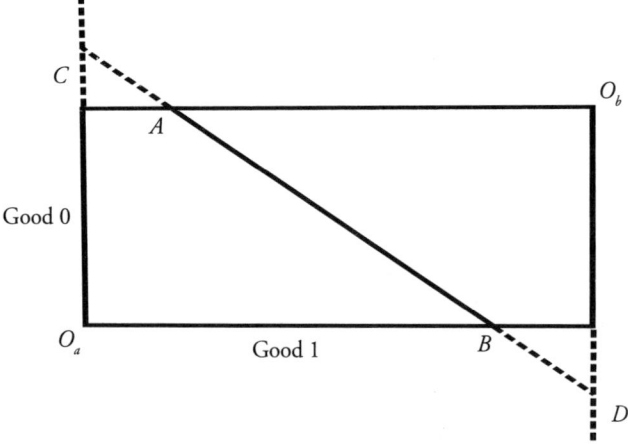

Figure 7.3: Budget Set

marked by an unbroken line as shown in Figure 7.4. The endowment point in the PE-domain will be marked w.

Now any point in an agent's budget set is a vector of commodities. Consider x and y in Figure 7.4. Which one of these two points contains a higher amount of, say, good 1 for person a? To answer this we have to draw straight lines through x and y which are parallel to the line 02. Whichever of x and y *is* on the line closer to point 1 has a larger amount of good 1. In the figure, y has more of good 1 for person a. Clearly for person b, x has more of good 1. Hence the broken lines through x and y could be described as *iso-good* 1 *lines*.

The next question is concerning indifference curves in the PE-domain. I shall make the usual assumptions on a consumer's preference relation [i.e., it must (i) be an ordering, (ii) subsume vector-dominance, (iii) be strictly convex, and (iv) be continuous]. In the three-dimensional Edgeworth box, indifference surfaces are saucer-shaped. The intersection of these with the PE-domain gives the indifference curves in the PE-domain. These will be strictly convex[4] and have a bliss point. We shall denote a's and b's bliss points by B_a and B_b. In the context of the full Edgeworth box, these are, of course, *constrained* bliss points, representing each player's highest-utility point on his budget set. Thus in Figure 7.4, B_a is the point which gives the greatest utility among all points in the triangle 012.

Pareto optima can be defined in many different ways in fix-price general equilibrium theory. We could think of a *constrained* Pareto optimum as a point on the PE-domain such that no other point on the PE-domain would constitute a Pareto improvement. In the terminology of Dreze and Muller (1980) such a point would be Pareto optimal at the 'second level'. The collection of all constrained Pareto optima will be referred to as the *constrained contract curve*. The line joining

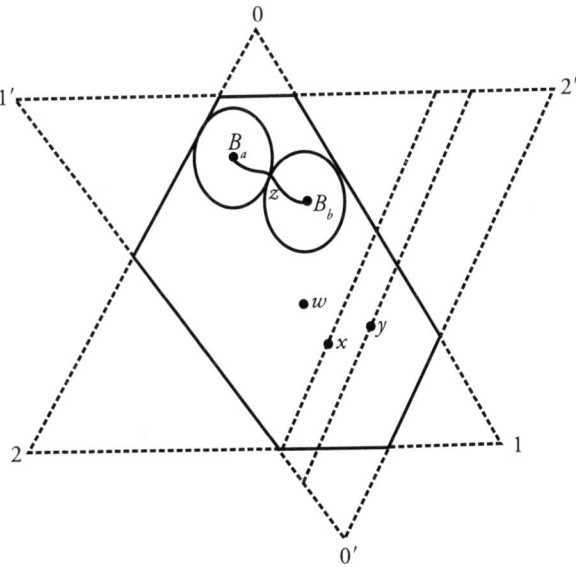

Figure 7.4: Constrained Bliss Point and Constrained Contract Curve

B_a and B_b and going through all points of tangency between indifference curves constitutes our contract curve as shown in Figure 7.4.

7.3 RATIONS AND CONSUMER CHOICE

Rationing in standard non-Walrasian economics takes the form of upper limits on net supply and net demand of an agent or a consumer, imposed separately on each good. I shall go along with this here and assume that a consumer can never be forced to trade more than he wants but he can be forced to trade less. This is known as the 'no forced trading' assumption in the literature.[5] We shall also assume, as is common in this literature, that money is never rationed.

A ration on net trade in commodity 1 is represented by a straight line parallel to the line 02 in Figure 7.4 drawn in the PE-domain. A consumer faced with such a ration is free to choose any consumption point on the PE-domain on the side of the ration line on which *w is* located. This follows from the no-forced-trade assumption. Hence whenever we draw a ration line, if this line does not pass through *w*, there is no need to indicate which is the feasible side of the ration line for it must always be the side on which *w* happens to be. If the ration line passes through *w*, it will be necessary to indicate which side is feasible.

Consider person *a* in Figure 7.5. If he faces a ration on good 1 represented by line AB, he would choose to consume at *E*. Note that this ration is a restriction on the net *supply* of good 1. If a ration on good 1 is imposed at A″B″ (i.e., a ration on demand) it is easy to see that the agent would choose to consume at B_a. It is, therefore, clear that by using a ration only on good 1, consumer *a* can be made to choose any point on the line $B_a E'$, which is derived by marking off the tangencies of lines parallel to 02 and the agent's indifference curves.

Moreover, the consumer cannot be directed to any other point by rationing *only* good 1. The line $B_a E'$ will be referred to in this paper as agent *a*'s 1-*rationed locus*.

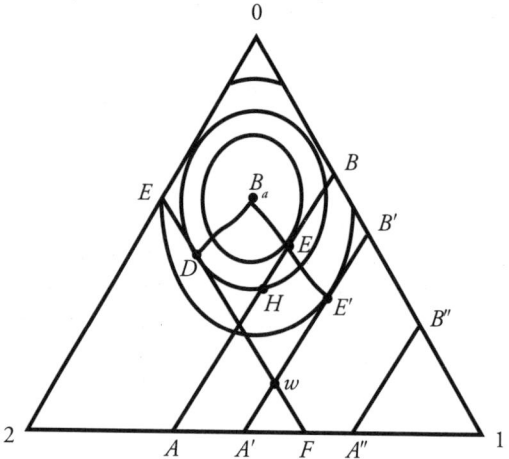

Figure 7.5: Rationed Locus

We could use a similar exercise and derive agent's 2-*rationed locus*, that is, the locus of points which the agent could be made to choose by rationing good 2 only. This is shown as line $B_a D$.

It is interesting to note here that a point like E (where an indifference curve is tangential to line AB) could represent a lower consumption of good 2 than point B_a. That is, a supply restriction on good 1 could enhance a's supply of good 2.[6] This is known as the 'spillover effect' (Benassy 1982) or 'forced substitution' (Kornai 1982), though Kornai speaks about this in the context of excess demand instead of supply.[7]

Before we can do a general equilibrium analysis we must go a step further and locate all points which an agent can be made to choose by using quantity rations on only goods 1 *and* 2. In Figure 7.5 this is represented by all points in the area $B_a E' wD$, that is, all points enclosed by the 1-rationed locus, 2-rationed locus, and lines through w which are parallel to 01 and 02.

This is easy to see. Consider a point like H. Suppose first there is a ration on good 1 represented by line AB. Now place a ration on good 2 by a line which goes through H. Faced with these two rations the agent will find it optimal to choose point H. This follows from the convexity of preferences. On line AB point E is the best. If he is prevented from reaching E, he would move as close to E as possible. This is given by point H.

The collection of all points which an agent can be made to choose by using quantity rations on goods 1 and 2 will be referred to as the 1–2 *rationed zone*. As already pointed out, in Figure 7.5, this is given by $B_a DwE'$.

7.4 GENERAL EQUILIBRIUM

Now we are in a position to give a full geometric representation of general equilibrium with quantity rationing *a la* Dreze (1975). I do this in two steps. A point, e, in the PE-domain and quantity rations on goods 1 and 2 on agents a and b constitute a *quasi-equilibrium* if both agents choose to consume at e when faced with the given quantity rations. A quasi-equilibrium is an *equilibrium* (or *Dreze equilibrium*) if there does not exist any good for which one agent is demand rationed and the other is supply rationed.[8, 9]

The existence of a quasi-equilibrium in general is easy to see. The original endowment point, to, with complete quantity restrictions (in the sense of zero trade being allowed) is clearly a quasi-equilibrium. This is not a very interesting equilibrium, and the purpose of Dreze's additional restriction is to rule out the no-trade point from qualifying as an equilibrium wherever possible.

The formal proof of the existence of an equilibrium is quite terse. One advantage of the diagrammatic method developed above is that one can 'see' and get an intuititive idea of this theorem. For this, it is useful to first identify the set of all quasi-equilibria, and then search for the equilibrium within this set.

In Figure 7.6, draw lines parallel to 01 and 02 through the initial endowment point w. This partitions the PE-domain into four zones which I shall refer to as the north, east, south, and west zones. Now the constrained bliss points of the

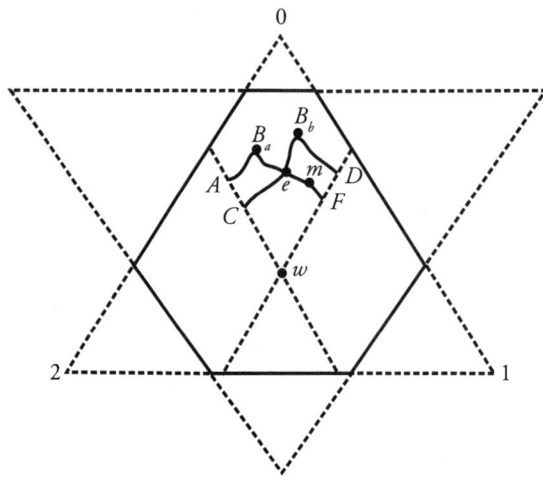

Figure 7.6: Quasi Equilibria and Equilibria: Case (i)

two agents could lie in the (i) same zone, (ii) adjacent zones, or (iii) opposite zones.

It is convenient to analyse the three cases separately. Consider (i) first, which is illustrated in Figure 7.6. Let $B_a A$ and $B_a F$ ($B_b C$ and $B_b D$) be agent a's (agent b's) 2-rationed locus and 1-rationed locus. From the discussion in the previous section it is obvious that for any point in the area $wCeF$, it is possible to impose quantity rations on the two agents such that the constrained optimum of both agents happen to occur at that point. Hence all points in $wCeF$ are quasi-equilibria.

Now consider any point in this area apart from e. Take, for instance, point m. Note that both agents have to be quantity constrained for good 1 and while agent a's supply is restricted, for agent b it is demand that is restricted. (Remember that agent b's budget set has to be viewed from the opposite side than that of a's.) Hence m cannot be an equilibrium. It is easy to check that point e is an equilibrium.

Using a similar analysis the other cases are simple to figure out. Hence I simply illustrate these in Figure 7.7 in two self-evident diagrams. In the left-hand panel,

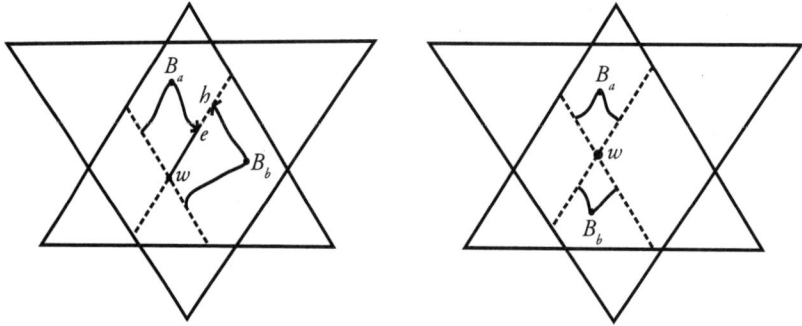

Figure 7.7: Quasi Equilibria and Equilibria: Cases (ii) and (iii)

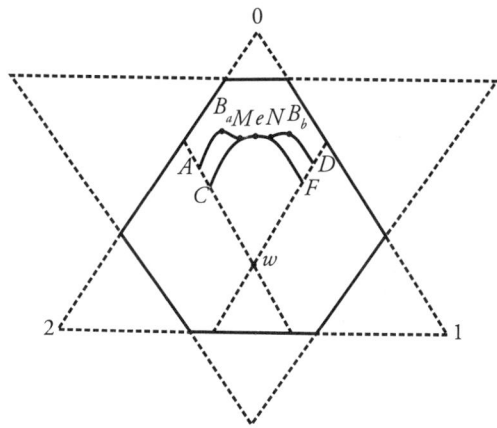

Figure 7.8: Multiple Dreze Equilibria

the line segment *ew* represents quasi-equilibria and *e* is an equilibrium. In the right-hand panel there is a unique quasi-equilibrium and equilibrium. This is the zero trade point, *w*.

In the above analysis I have omitted the case discussed in note 5, where the imposition of a quantity restriction on one commodity changes an agent from a net supplier (demander) to a net demander (supplier). This is because the complication is easy to handle and leaves the existence proposition unchanged.

From the three cases illustrated in Figures 7.6 and 7.7 it appears that the Dreze equilibrium is unique. However, the appearance is false. I shall now illustrate using the geometric technique developed thus far that an economy may have several Dreze equilibria; and this is so even when individual preferences are *strictly* convex.

In Figure 7.8 let player *b*'s 2-rationed locus be the line $B_b C$. It is possible for this to overlap with player *a*'s 1-rationed locus, which is represented by $B_a F$. Let *MN* be the segment over which the two loci overlap. It may now be checked that every point on MN is an equilibrium.

Consider a point *e* on the MN segment and draw two lines through it—one parallel to 01 and the other parallel to 02. Let the former represent a ration on player *b* and the latter a ration on player a. It is clear that with these constraints both players would choose to consume at point *e*. Hence *e* is a quasi-equilibrium. Since this quasi-equilibrium is sustained by rationing a's supply of good 1 and *b*'s supply of good 2, point *e* is also an equilibrium. Since *e* was an arbitrary point on MN, this establishes the non-uniqueness of the Dreze equilibrium. It is easy to check that the set of Dreze equilibria may be neither unique nor a continuum and may be finite.

7.5 OPTIMALITY

This section on optimality is kept brief because it overlaps with Bohm and Muller (1977). It is not omitted altogether for the sake of completeness and because the next section requires a familiarity with the matter covered here.

A *Pareto optimal* point is one such that there is no other point in the Edgeworth box at which both agents are better off. A *constrained Pareto optimal* point, on the other hand, is defined with respect to the PE-domain instead of the Edgeworth box. A formal definition is given in Section 7.2.

Consider a fix-price economy with endowment w and price vector p. Suppose further that p is not a Walras equilibrium price. Consider now the (Dreze) equilibrium of this economy. Is this Pareto optimal? The answer is: It may be so but only in very special cases. This is illustrated in Dreze and Muller's paper, in particular, their figure 7.1.

I shall concentrate here on the relation between equilibria and constrained Pareto optima. It will first be shown that an equilibrium of the kind discussed in the previous section need not in general be a constrained Pareto optimum.

Consider, for instance, case (ii) of the previous section. An equilibrium for this case is illustrated in the left panel of Figure 7.7. From the definition of an *i*-rationed locus it follows that *b*'s indifference curve through *h* and *a*'s indifference curve through *e* will both be tangential to the same line as shown in the figure. Hence the constrained contract curve must pass through the segment *eh* but not through *e* or *h*. It follows that the equilibrium which is at point *e* is not a constrained Pareto optimum.

Before turning to the analysis of coupons equilibria, I want to comment on a theorem which is the counterpart of the so-called second basic theorem of welfare economics in fix-price economies. This states that every constrained Pareto optimum point could be sustained as an equilibrium if we were free to alter the initial endowment within the PE domain.

To see this, first note that if the constrained contract curve passes through w, then w is a Dreze equilibrium. It immediately follows that if in a fix-price economy a point *e* is *a* constrained Pareto optimum, then if we change the initial endowment to *e*, *e* would be an equilibrium.

7.6 COUPONS EQUILIBRIA

A very interesting and alternative method of rationing is via the use of coupons. The advantage of this method is that it can achieve constrained Pareto optimality (Dreze and Muller 1980; Hahn 1978). The best way to understand a coupons economy is to superimpose it on the kind of economy discussed thus far. So suppose now, over and above the existing endowments, each of the two individuals are handed over c coupons.[10] In addition, coupon prices are announced for each of the three goods. Let q_i be the coupon price of good i. These can be positive, negative, or zero. A coupon price represents the price of a good in coupons. Thus, if an agent buys one unit of good i, he has to pay q_i coupons. This is in addition to the money prices. Hence, from now on, to buy a good an agent has to pay in two currencies: money and coupons. Similarly, when supplying goods, he receives both currencies. If, for instance, agent i sells off all his endowment of goods 1 and 2, he receives $w^i_1 q_1 + w^i_2 q_2$ coupons and $w^i_1 p_1 + w^i_2 p_2$ money. It will also be assumed throughout that the coupon price of money is zero. That is, $q_0 = 0$.

It follows that in an economy with endowment fixed at w and price at p, if each individual is given an additional endowment of c coupons and $q \equiv (q_0, q_1, q_2)$ is the coupon price of goods, a consumer i can choose any consumption vector (x^i_0, x^i_1, x^i_2), which satisfies the following two conditions:

$$p_0 x^i_0 + p_1 x^i_1 + p_2 x^i_2 = p_0 w^i_0 + p_1 w^i_1 + p_2 w^i_2, \qquad (7.1)$$

$$q_1 x^i_1 + q_2 x^i_2 \leqq c + q_1 w^i_1 + q_2 w^i_2. \qquad (7.2)$$

In terms of geometry, let 012 in Figure 7.9 be the normal budget set given by Eq. (7.1), which we have been using all this time. What Eq. (7.2) does is draw a straight line, like AB, across this space, such that the area that satisfies both Eqs. (7.1) and (7.2) is given by $OBA2$. By varying (c, q), the line AB could be made to fall anywhere. This will be called the *coupon line*. Note that, as long as $c > 0$, the feasible side of the coupon line is always the one in which the endowment point w is located. If $c = 0$ (the Hahn special case), either side could be feasible, depending on the values of q_1 and q_2. We shall put an arrow on AB to mark the feasible side as shown in Figure 7.9. When we consider both players a and b, as in the PE-domain, we shall use a broken line marked $\hat{A}\hat{B}$ to denote b's coupon's constraint. The feasible side of b will obviously be in the opposite direction from that of a. Hence we follow the convention of using the arrow to mark a's feasible side.

Following Dreze and Muller, for an economy described by fixed endowment and price vectors, we define a *coupon equilibrium* as a point e in the PE-domain, a coupons endowment, c, and a coupons price vector, q, such that, given the additional restriction imposed by c and q, each agent chooses to consume at point e. A coupons equilibrium is described as Hahn-type if $c = 0$.

Figures 7.10 and 7.11 illustrate two alternative coupons equilibria. In Figure 7.11 we have a Hahn-type coupons equilibrium at e. If AB is the coupon line, it is clear that a's best consumption point that is feasible is given by e. This is also B's

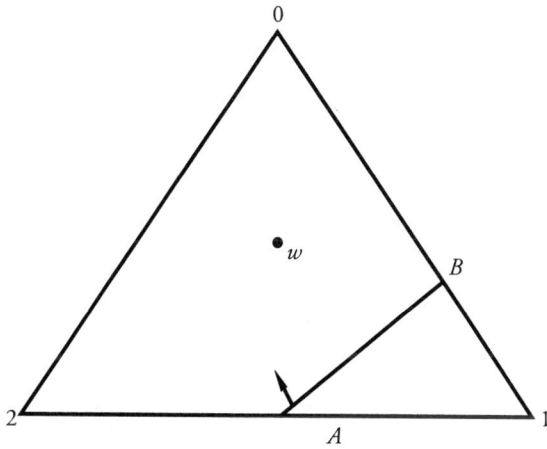

Figure 7.9: Budget Set and Coupon Line

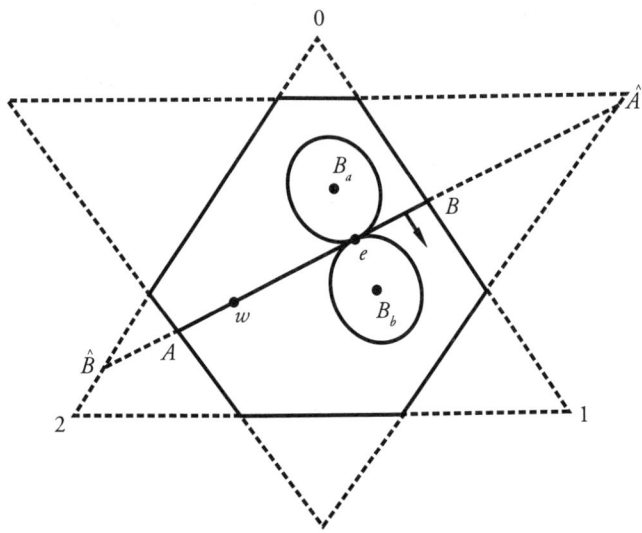

Figure 7.10: Coupons Equilibrium

chosen point. Hence this is a coupons equilibrium. Since this coupon line AB (or $\hat{A}\hat{B}$) goes through w, $c = 0$, and the equilibrium is Hahn-type.

This figure makes intuitively clear what is indeed formally true, that a Hahn-type coupons equilibrium is always a constrained Pareto optimum. This is because a Hahn-type coupons equilibrium in a three-good economy is similar to a Walrasian equilibrium in a two-good economy.

While this is an encouraging result, as Dreze and Muller point out, an economy does not always have a Hahn-type coupons equilibrium. This is very easy to show with our geometric method.

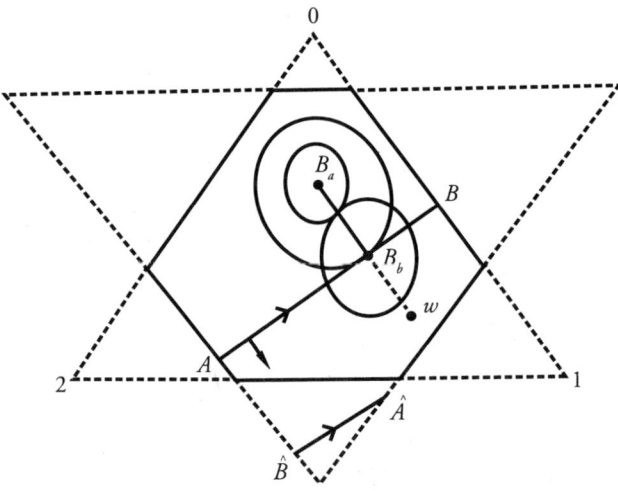

Figure 7.11: Hahn-type Coupons Equilibrium

Let us assume that the PE-domain and the agents' indifference curves are such that, in the case illustrated in Figure 7.10, the constrained contract curve is a straight line. But suppose that the initial endowment point, instead of being where it is in Figure 7.10, happens to be on the straight-line extension of the contract curve as shown in Figure 7.11. It is immediately obvious that it is not possible to draw a straight line through w such that both agents treating this as the coupon line will choose the same consumption point. In brief, a Hahn-type equilibrium does not exist.

Nevertheless, a coupon's equilibrium does exist. To see this, consider player b's constrained bliss point, B_b. Through this point draw a straight line which is tangential to a's indifference curve which goes through B_b. Let this line be marked AB. Let AB be treated as agent a's coupon line. This automatically defines player B's coupon line. This is shown as $\hat{A}\hat{B}$. Given these coupon lines, agent a would choose to consume at B_b as well. Hence we have described a coupons equilibrium. From the way this proof is constructed, it seems clear that a coupons equilibrium will always exist. This, as we know from Dreze and Muller's formal derivation, is indeed true.

The geometry also suggests that a coupon's equilibrium is always a constrained Pareto optimum. This is also formally true, though it does hinge on the assumption of strict convexity of preferences (which has been assumed to be true throughout this paper).

It is interesting to note that the above analysis also suggests that if the constrained bliss points of the two players do not coincide (i.e., $B_a \neq B_b$) and (e, c, q) is a coupon's equilibrium with c > 0, then, at equilibrium, one of the two players must be at his constrained bliss point.

7.7 Conclusion

The purpose of this paper was to develop a geometry for interpreting non-Walrasian general equilibrium theory. This could be especially valuable because of the critical role that non-Walrasian economics has played in providing microfoundations for macroeconomics. It is hoped that the geometry developed here will be of more than just heuristic value and will provoke new research in the same way that the Edgeworth box has stimulated ideas within the domain of Walrasian economics.

There are several directions that one may pursue from here. Note, for instance, that there must exist at least one point on the constrained contract curve which is not just *constrained* Pareto optimal, but Pareto optimal. A question deserving future investigation is whether we can devise reasonable rationing schemes such that the equilibrium will be driven to such an optimal point. *A priori* it seems that two-part tariffs or nonlinear pricing can play a role in devising such schemes. Other possible directions of research include the introduction of production and comparative statics to evaluate the effects of changes in exogenous variables and policy parameters.

76   ■■   WELFARE, LAW, AND GLOBALIZATION

NOTES

1. For an alternative geometric method, see Stoneman (1979).
2. Though I have developed the geometry for these papers, it should also be useful in studying related ideas which appear in, for instance, Barro and Grossman (1971), Balasko (1979), Benassy (1975, 1982), Grandmont and Laroque (1976), Malinvaud (1977), Muelbauer and Fortes (1978), and Younes (1975).
3. I am using the term 'budget set' to refer to the northeastern *surface* of the feasible set defined by the budget constraint and non-negativity constraints.
4. This follows from my assumption of strict convexity of preference. Relaxing the 'strict' part of the assumption can complicate the analysis considerably (see Grandmont 1977b).
5. It may be noted that there are other ways of defining 'no-forced trading'. For instance, 'no-forced trading' may be defined as any quantity restriction on an agent, i, which does not take away his right to consume his endowment, w^i, if he so wishes.
6. It could also transform an agent from a net supplier of good 2 to a net demander of good 2. That is, it is possible for the agent's 1-rationed locus to go from B_a to some point on the line segment $A'w$. I will not consider such a case here explicitly because it does not change the results, and the little additional geometric complication is easy to handle.
7. It is also arguable that Kornai's 'forced substitution' is a more complicated concept. For one, Kornai allows for the fact that in the event of being rationed in a market, a consumer will not only turn to other goods but may take to queing or postponement of purchase. Moreover, if a shortage is chronic, consumers may adjust so well to the situation that there may be few overt behavioural manifestations of the shortage.
8. Dreze's (1975) original definition requires also that for each good, the quantity ration imposed on the individuals' net trade is of a fixed magnitude, quite irrespective of who the individual happens to be. The slight generalization that I present here has been discussed in the formal literature (see, Grandmont 1977a, 1977b).
9. While this requirement (that both supply and demand for the same good must not be rationed at the same time) seems at first sight unquestionable, the 'Hungarian school' reminds us that this may indeed be a strong assumption. This is yet another point of divergence between the Hungarian model of shortage and the literature I am discussing. As Kornai (1982, p. 38) notes, 'It frequently happens at the microlevel that neither the collectivity of sellers nor that of buyers can perfectly fulfill their intentions'. Turning to our framework, once we bring this kind of an argument in, it is possible that the 'equilibrium' will settle in the interior of the set of quasi-equilibrium points.
10. Note that each agent's coupon endowment is the same. In Dreze and Muller's terminology the equilibrium that I define below is therefore a *uniform* coupons equilibrium. I make this assumption only for simplicity. The coupons economy described by Hahn (1978) has $c = 0$.

REFERENCES

Balasko, Yves, 1979, 'Budget-constrained Pareto efficient allocations', *Journal of Economic Theory*, 21, 359–79.

Barro, Robert J. and Herschel I. Grossman, 1971, 'A general disequilibrium model of income and employment', *American Economic Review*, 56, 82–93.

Benassy, Jean-Pascal, 1975, 'Neo-Keynesian disequilibrium theory in a monetary economy', *Review of Economic Studies*, 42, 503–23.

—— 1982, *The Economics of Market Disequilibrium*, Academic Press, New York.

Bohm, Volker and Heinz Muller, 1977, 'Two examples of equilibria under price rigidities and quantity rationing', *Zeitschrift fur Nationalokonomie*, 37, 165–73.

Dreze, Jacques H., 1975, 'Existence of an exchange equilibrium under price rigidities', *International Economic Review*, 16, 301–20.

Dreze, Jacques and Heinz Muller, 1980, 'Optimality properties of rationing schemes', *Journal of Economic Theory*, 23, 131–49.

Grandmont, Jean-Michel, 1977a, 'The logic of the fix-price method', *Scandinavian Journal of Economics*, 79, 169–86.

—— 1977b, 'Temporary general equilibrium theory', *Econometrica*, 45, 535–72.

Grandmont, Jean-Michel and Guy Laroque, 1976, 'On temporary Keynesian equilibria', *Review of Economic Studies*, 43, 53–67.

Hahn, Frank H., 1978, 'On non-Walrasian equilibria', *Review of Economic Studies*, 45, 1–17.

Kornai, Janos, 1982, *Growth, Shortage and Efficiency*, Basil Blackwell, Oxford.

Malinvaud, Edmond, 1977, *The Theory of Unemployment Reconsidered*, Basil Blackwell, Oxford.

Muelbauer, John and Richard Fortes, 1978, 'Macroeconomic models with quantity rationing', *Economic Journal*, 88, 788–821.

Stoneman, Paul, 1979, 'A simple diagrammatic apparatus for the investigation of a macro-economic model of temporary equilibria', *Economica*, 46, 61–66.

Younes, Yves, 1975, 'On the role of money in the process of exchange and the existence of a non-Walrasian equilibrium', *Review of Economic Studies*, 42, 489–501.

PART II
Law, Rights, and Well-Being

8 The Role of Norms and Law in Economics
An Essay on Political Economy

8.1 INTRODUCTION

The three-hour stretch of road between Hazaribagh and Dhanbad in eastern India is as desolate as it is beautiful. One winter evening, some half a dozen years ago, as I was travelling this route by taxi to catch a train from Dhanbad to Calcutta, I was lucky—or, I suppose, unlucky, depending on one's point of view—to be stopped by a road block created by a gang of youngsters wielding *lathis*[1] and swords. In front of us, also stopped by the ramshackle road block, was a truck, and some of the youngsters were talking to the truck driver. From the sight of some distant lanterns I figured that we were close to a village. My taxi driver looked very nervous as he waited for the youngsters to come to our car. He told me that they were hoodlums, collecting illegal money by threatening to beat up passengers and drivers. He asked me not to speak and to leave it all to him. Eventually, a bearded young man walked up to our car regally and asked me to lower my window glass. He spoke courteously and explained that he was collecting *rangdari* tax. He had a wad of papers in one hand (the other held a lathi) and he explained that after we paid the money, which, he added firmly, we would have to, he would even give us a receipt.

I had read about the institution of rangdari tax occurring in some parts of India. The 'tax' is an illegal collection made by gangs in remote rural areas where the hand of the law is lax. The reason I felt lucky about the incident was that this

This essay draws on my book *Prelude to Political Economy* (OUP, 2000), which was still a manuscript when this was published. In writing this I have benefitted from the comments of Patrick Emerson, Michael McPherson, Andy Rutlen, and Ednaudo Zambrano.

First published in Keates, D. and J. Scott (ed.), *Schools of Thought: Twenty-five Years of Interpretive Social Science*, Princeton University Press (2001).

experience is very rare for an urban Indian, and it subsequently made me think hard about the meaning of law and norms, and I owe a part of this chapter to the incident.

Though it is unimportant for my present chapter, I must nevertheless finish the story. My taxi driver, despite the cold sweat, was not the one to give up. He got into an argument and was soon asked to get down from the car and talk to the boss, who stood with others a little farther away. Several minutes passed before he returned, the bamboo road block was removed, and as we sped away towards Dhanbad, he explained how we got away without paying. His arguments in the beginning fell on deaf ears, he said. Then he suddenly changed tack and explained that I was a visitor from Delhi who had come to see rural Bihar, and it would create a very bad impression on me if I were forced to make a payment. This appeal to regional pride clicked and, like some visiting ambassadors, we were allowed to go without paying local dues.

There are several features of this little incident that shed light on the functioning of an economy and also cast shadows on our textbook models. First what the youngsters were offering us was, at a certain level of abstraction, like any exchange. If we wanted our arms intact and heads not bruised, we would have to pay them some money. In other words, they were *selling* nonviolence. And for most people, like the trucker ahead of us (and I, for that matter), thought, it was a good bargain. A small sum of money in exchange for no bodily harm seemed well worth it. But note that what they were selling was what in most societies is treated as belonging to the buyer's endowment. If I wanted *my* arm unharmed, I would have to pay him. In textbook economics we usually treat individual endowments as beyond the reach of others. But in reality, individuals often encroach on each others endowments, selling to i what in most societies would be considered as belonging to *i*. This happens between powerful landlords and poor serfs; between big countries and small countries; between big corporations and small companies. Evidently, the theorem that individuals, left to themselves, lead to an efficient society, is predicated on the *assumption* that agents respect each other's endowments. But to the extent that they do not do so in reality, this claim that individual rationality is enough to create an efficient society is false or, at best, remains to be established.

The other matter on which the incident sheds light is the meaning of the law. Virtually all accounts in the Indian press have described the rangdari tax as illegal extortion. Yet it is impossible not to notice how analogous it is to a regular tax. It is not paid by people voluntarily but needs the threat of punitive action. The fact that the extortion was taking place so close to a village makes it plausible that it has some legitimacy in the eyes of the villagers.[2] In all likelihood a part of the money is spent on local village welfare, with the remainder being used by the tax collectors on themselves. This is analogous to the uses to which government tax revenue is put. In brief, the institution that I had chanced upon that evening was pretty much like a local government. It is considered illegal only because it commits acts that, in the eyes of what we consider the real government, are illegal, though in essence the actions are similar to the real government's own actions. Hence, as per common usage, the institution of the rangdari tax cannot be thought of as supported by the law and the threat of state penalty; it is supported by norms and informal threats. But what, really, is the difference between a law and a norm? There are several differences to be sure, but at some level they are indistinguishable from each other. The latter is a nontrivial claim and is one of the central theorems of this chapter. It will be called the *core theorem*. It expresses a viewpoint that can have important influence on the way we conceptualize the role of law in economics, as will be argued in a later section. It is not a theorem in the sense of geometry or even axiomatic economics, which can be mathematically proved, but a point of view to which I aspire to convert the reader through examples, arguments, and persuasion. It is formalizable, but only potentially so.

The core theorem and the discussion around it are related to the research in economics often called the new or positive political economy,[3] and is part of the older 'institutional economics'. My method of analysis, relying on game-theoretic constructions, is similar to the method used in this new literature. But at the same time my central claims, embodied in and stemming from the core theorem, diverge from the view that is taken in the literature on political economy. Moreover, I do not share the confidence of this new literature, with which economists—ready with their median voter theorems and techniques of optimization—have rushed to explain the rise and fall of nations: why some dictators ruin nations and others bring prosperity; why one government loses the election and another one does not; and why democracies appear when they do and why they do not when they do not (hindsight being never too far away from these analyses). I do not think we are in a position to answer such large questions. But I know that a group of people all praising one another for their understanding of these questions and at the same time trying to outdo one another can create a 'cult effect', where knowledge is replaced by illusion. This essay has a much more limited objective: to expose some flaws in our thinking that lie at the base of conventional economics and even the new political economy, and to provide the preliminaries for a large programme that lies mostly ahead.

If I owe a part of my interest in this area of research to rural Bihar, I also owe a part to the Institute for Advanced Study, Princeton. I came to the institute with a fledgling interest in power and the politics of oppression. I had argued in my paper *'One Kind of Power'* that we needed to move away from traditional dyadic economics to the economics of triads if we were to incorporate the role of power and influence in our models.[4] During the year that I spent at the institute, 1985–6, I was fortunate to have been able to collaborate with two remarkable economists, Eric Jones and Ekkehart Schlicht. We brought our respective skills to write a critique—in the original sense of the word 'critique', that is, as evaluation—of the new institutional economics. In writing the paper that came out of this collaboration,[5] I learned a lot about historical methods and institutional economics. We were touching on several issues of political economy, which was then far from being a discipline in vogue.

In the next two sections I shall comment on social norms, and law and economics. Finally, I shall try to present what I believe is a new approach to the study of political economy.

8.2 SOCIAL NORMS

Assumptions in economics have been at the receiving end of a lot of attention. They have been reviled for their unrealism, admired for their elegance, the mainspring of jokes, appreciated for their explanatory powers, and dismissed as untenable. All this attention, however, has been directed at the *explicit* assumptions, such as the transitivity of preference or the convexity of technology. What has gone virtually unnoticed and therefore eluded criticism are the *implicit* assumptions. Yet the most untenable assumptions often belong to this category. One such assumption is the existence of social norms. Much of economics has been written up as if social norms do not matter. This is empirically false, as virtually all economists and certainly other social scientists will agree. What is more interesting is that it is, in all likelihood, *analytically* flawed as well. That is, a norms-free economics may not be possible. Hence, when we write up a model with no reference to norms and institutions, we are nevertheless using norms and institutions, but doing so unwittingly.

This is best illustrated by the act of exchange. According to the first principles of economics, two agents will exchange or trade goods if the following assumptions are true: (a) each individual prefers having more goods to less; (b) each person satisfies the law of diminishing marginal utility;[6] and (c) the initial endowment of goods is lopsided, for example, one person has all the butter and the other all the bread. To many economists, (a)–(c) are indeed sufficient conditions for trade to occur. What they do not realize is that these are sufficient only when the agents are already embedded in a certain institutional environment and characterized by adequate social norms.[7] For one, exchange is greatly facilitated by the ability to communicate or, even better, to speak a common language. And given that language is after all an evolved social convention,[8] trade and exchange are predicated on social conventions.

The importance of these implicit requirements for trade can be inferred from some experiments in economics, which were conducted for a different purpose. Experiments have shown that rats do prefer more to less—a fact that I suspected well before I read experimental economics. Furthermore, experiments have established that rats also satisfy the law of diminishing marginal utility or, more precisely, have convex preference. This was established by some innovative experiments conducted by Kagel *et al.* on white albino rats, belonging to the—this for the connoisseur—Wistar and Sprague-Dawley stock.[9] So, rats do satisfy our assumptions (a) and (b). All that remained to be done to check the exchange hypothesis was to give different kinds of food to different rats, which would fulfil assumption (c), and see what the rats did. It seems some relentless researchers did just that.[10] They presumably placed two rats at some distance, with each possessing a different food item. The researchers discovered that, though these rats satisfied assumptions (*a*)–(*c*), they did not, alas, indulge in trade and exchange. I feel I could have predicted this from my occasional encounter with rats, but it is anyway good to have these things experimentally confirmed.

Facetious though it may sound, the above account does amount to a very substantial critique of traditional economics. It shows that even in models which seem transparently free of any requirements of norms and institutions, that is not the case. Market-related activity, trade, and other economic functionings have to be embedded in institutions and social norms.[11] If we refuse to embed our models consciously, we will still be doing so, only unwittingly. And given that the latter is not such a wise approach, it is important that we recognize the role that social norms play and try to build these in consciously and in keeping with reality.

Before venturing to discuss different kinds of norms and critiquing mainstream economics, I want to put in a word of caution. That mainstream economics has ignored social norms is quite evident; one has simply to browse through a few random books of economics to verify the claim. That social norms are an important part of reality is also obvious enough. But these two facts are not reason enough for criticizing mainstream economics. Something can be an important part of life but not important for the research one is involved in. Indeed, for some of the core concerns of mainstream economics the social norm was not germane. Moreover, economists were wary of using a concept that was so vaguely defined that it could be used to explain almost everything, thereby falling into a tautological trap, a danger that has been pointed to by Solow.[12]

This justification for keeping norms out of our analysis, however, has grown weaker over the years. As economists have reached out to addressing larger questions, concerning political economy and law and economics, the silence on social norms has become less defensible. Moreover, with the rise in game theory, we have within our ambit methods for formalizing and giving more rigorous definition to different concepts of norms.

For the purposes of economic analysis, norms are best divided into three kinds: rationality-limiting norms, preference-changing norms, and equilibrium-selection norms.[13] By a 'rationality-limiting norm' I mean a norm that stops us from doing certain things or choosing certain options irrespective of how much utility that thing or option gives us. Thus most individuals would not consider filching another person's wallet even if it were lying unattended, not by speculating about the amount the wallet is likely to contain, the chances of getting caught, the severity of the law, and so on, but because they consider stealing wallets as something that is *simply not done*.

In traditional economics, the 'feasible set' of alternatives facing an individual (from which the person can pick one) is defined in terms of technological or budgetary feasibility. Thus a consumer's feasible set is the collection of all the combinations of goods and services that the consumer can purchase given his or her income. From the above discussion it should be evident that a rationality-limiting norm further limits the feasible set, because now certain alternatives may be infeasible to an individual not just because it is technologically infeasible (like walking on water) or budgetarily infeasible (like buying a Jaguar car) but because it is ruled out by the person's norms. Indeed, a person with norms may let go on options that could have enhanced his utility,[14] and thus such a person would be considered irrational in terms of traditional economics. Basically, such norms limit the domain over which the rationality calculus is applied.

Elsewhere I have taken the line that we can, at least partially, understand why some norms exist and some do not, in terms of *evolutionary* stability.[15] According to this argument, we do not see any society with the norm that one must not eat proteins simply because such a society would perish along with its norm. Similarly, we do not find any society where stealing anything from anyone is considered legitimate because such a society would soon be in complete chaos, become impoverished, and wither away.

On Forest Home Drive in Ithaca there is a bridge on which two cars cannot cross at the same time. When we were small we were told how in the Andes there are pathways along steep mountains that are so narrow that two persons cannot cross; and so when two persons found themselves face to face on one of these paths, the one with the quicker draw survived by shooting the other person and continued on his journey. In Ithaca a different norm is used. Cars pass in little convoys, three or four at a time, and the convoys from the two directions alternate. That is, after the third or fourth car ahead goes, one just stops and waits for an oncoming convoy and then starts once again. This stopping and waiting is against one's self-interest, so it is indeed a rationality-limiting norm, but the reason we find some norm of this kind and not the Andean custom of a shoot-out is that it is evolutionarily more stable. This is also the reason that the 'Andes custom' probably exists nowhere. A society practicing this norm would not survive and so neither would the norm.

Some may argue that instead of thinking that such norms limit individual rationality, we can simply redefine our utility function so that what I described above as normatively infeasible is described as an option that gives a very low utility, perhaps negative infinity. But that would reduce utility theory to a sterile tautology. In reality, moreover, there are certain things we would love to do but our norms get in the way. We would not have to ask the lord to deliver us from evil if evil gave us such disutility that it was no temptation to start with.

This does not mean that norms never change our preference or utility function. Certain norms do get internalized. There are many individuals whose religion requires them to be vegetarian, and they tell you that they find nonvegetarian food revolting anyway. More often than not, this is no coincidence; a religious norm adhered to over a stretch of time often gets internalized so that one begins to actually prefer what the norm requires one to do. This can explain why one finds systematic variations in taste across regions and nations. What starts out as a norm or a custom can, over time, become part of one's preference. Such a norm may be referred to as a 'preference-changing norm'.[16] Since such a norm works through an individual's preference, it can be ignored by traditional economics, which treats preferences as primitives. The only reason for being aware of this kind of a norm is that it can give us an understanding of how some of our preferences are formed.

This chapter, however, is concerned with neither rationality-limiting norms nor preference-changing norms but rather with norms that have no effect on individual preference nor the feasible set from which a person chooses, but those that help coordinate actions across human beings. Consider the norm of driving on the right in the United States. It is true that this norm is additionally fortified by the law, but it is arguable that even if this were just a norm or custom and not the law, people would still drive on the right.[17] This is because this norm, once it is in place, happens to be entirely compatible with self-interested behaviour. In the absence of such a norm, there are at least two possible equilibriums—everyone drives on the left or everyone drives on the right. The norm simply helps people to *select* an equili- brium. It is for this reason that I call such a norm an 'equilibrium-selection norm'.

According to this terminology, Akerlof's conception of caste is that of an equilibrium-selection norm.[18] In my model of totalitarian states,[19] people mimic loyalty to the totalitarian regime not because that is their preference but because the expression of loyalty is an equilibrium-selection norm. If others show loyalty to the regime it is in your self-interest to also show loyalty to the regime. Since this can be true for all individuals, the entire display of loyalty in some totalitarian states can be superficial, an exercise in mimicry from which no *individual* would want to deviate.

David Lewis's idea of a 'convention' is also close to this kind of social norm. More recently, Cooter,[20] in discussing the connection between norms and law, has identified norms entirely with equilibrium-selection norms. He describes a 'social norm' as an 'effective consensus obligation', and he goes on to identify a consensus obligation with an equilibrium of a game.[21]

Since most of this chapter focus on norms of this kind, from here on the term *norm* should be taken to mean an equilibrium-selection norm, unless explicitly stated otherwise.

8.3 LAW'S ECONOMY

The standard view of law in economics and related social sciences is of something that changes the set of strategies open to an individual, or the 'payoff function' of the individual. If the law does not permit emitting pollutants into the atmosphere, then the payoff that I expect when I build a factory that freely emits pollutants into the atmosphere will be different from the payoff I would expect from the factory if the law of the land had nothing against pollutants. In the former case, in addition to the profits from sales, I would have to calculate the probability of being caught and fined and adjust that against my expected profit in order to get to the expected payoff.

This view is predicated on a conception of the economy as a game. In other words, each individual in the economy is supposed to have a (feasible) set of 'strategies', or actions, open to him or her. The payoff that each individual, or 'player', receives depends on the strategies chosen by all the players—often referred to as a 'tuple of strategies'. The payoff is a number that expresses the net utility that a player receives from the state of the world that emerges when every player has picked a strategy from his or her set of strategies. The rule that summarizes the payoff received by each player for every possible tuple of strategies is called a 'payoff function'. This is the view taken, implicitly or explicitly, in virtually all works of law and economics. It is an idea often associated with Pigou and referred to as the 'Pigovian view'[22] and is quite explicit in, for instance, Baird, Gertner, and Picker[23] and Benoit and Kornhauser.[24] It is possible to contest this view of the economy, but that is not my purpose here; indeed, it seems to me to be an adequate model for most purposes.[25] What I want to focus on is the role of law in such an economy.

As just explained, according to the traditional view, a law is something that changes the 'economy game' by altering the payoff functions of players (or by limiting the set of strategies open to a player[26]). In other words, according to this view, a new law typically alters the payoff that a person expects from certain actions. Thus Baird, Gertner, and Picker observe, 'We can capture the change in the legal rules by changing the payoffs'.[27] And given that the payoffs are an integral part of a game, a law is treated as something that changes the game.

This has an immediate appeal. Consider a new law that raises the income tax rate. The payoff that one now expects to earn from eight hours of work will be less than what one would have earned from the same action or strategy earlier. Likewise with the example of the pollution law above. I shall, however, argue that, while this ubiquitous view of law serves well for some limited purposes, it is fundamentally flawed. The law needs to be understood very differently if we are to get a better grip on reality while building models of economics.

8.4 LAW AND ECONOMICS: CRITIQUE AND A NEW APPROACH

The standard view of the role of law in an economy would be right if it were the case that the economy game is one that is played only by the 'nongovernmental' individuals in society. That is, if the police, the tax collectors, and the judges were agents exogenous to the game, who mechanically went about doing what the law required them to do, then indeed for the other people in the society (that is, for the players of the game) a law would be something that determined the game by fixing the payoff function; and so a change in the legal regime would amount to a change of the game.

But in reality those who work for government—the police, the district judge, the tax collector, the bureaucrat, the individuals in the pollution control department, and so on—are also individuals with their own motivations, dreams, striving, and cunning. Hence they are also players and should not be treated as exogenous to the economy game. This fact, in itself, is now recognized in the new literature on economics and government.[28]

What is not always recognized is that this throws a wrench in the traditional models of law and economics. Moreover, even those economists who recognize the significance of endogenizing the 'law enforcer' balk at taking this idea all the way to its natural conclusion; and they tend to err on the side of the traditional approach in their instinctive moments.

Note that whether a particular law is there or not, the policeman's, the tax collector's, and the judge's sets of strategies remain the same. And if everybody behaves the same way, whether or not the law is there, everybody must get the same payoff. Hence, the law cannot change the payoff function.

Consider, for example, the case of antipollution law. *Whether or not the law is there,* the strategies open to the policeman include (a) arrest a person who emits pollutants and (b) not arrest a person who emits pollutants; the strategies open to the judge include (1) punish the policeman who arrests a person who pollutes the atmosphere and (2) punish the policemen who does not arrest a person who pollutes the atmosphere.[29] Now if—whether or not the law is there—the person, the policeman, and the judge behave the same way, then the person, the police-man, and the judge will get the same payoff. Hence, the game played by *all the individuals* in the economy is unaffected by the law.

If the enforcers of the law or the agents of the state automatically enforce the law, then a new law does affect the payoff function and therefore the game *played by the rest of the citizens.* But once everybody, including the enforcers of the law, are included in the game (as they should be), a law is nothing but some ink on paper. There being or not being such ink on paper cannot alter the game. This rather unusual conclusion, which is elaborated upon later in this section, is baffling at first sight. But this merely reflects the fact that the standard approach, though flawed, is deeply ingrained in modern social science thinking.

To digress for a moment, consider the new literature on rights and liberty, which expresses rights as game forms.[30] There has been much controversy about whether this is the correct way to describe rights. I would argue that, according to this conception of rights, a change in the structure of rights changes the sets of actions open to individuals and therefore changes the game. But it is not clear why a new rights assignment will change what I *can* do, even though it may well change what I *will* do. I argue that granting a person, i, a right to do something, call it x, must mean that if i does x, then another person, j, will not have the right to do something (for instance, punch i's nose). Of course, s not having the right to do something, in turn, must mean that if j does that thing, then others will acquire the rights to certain actions (typically punitive actions against j) to which they otherwise would not have had a right.[31] This is discussed in greater detail in Appendix B.

The above discussion may give the impression that law does not have any effect on society, that it is a chimera, but such an impression would be wrong. The law does not affect the payoff functions of the individuals or of the game, but it can influence the *outcome* of the game. It does so by creating focal points and by giving rise to beliefs and expectations in the minds of the individuals.[32] Thus, in the above example, the policeman can choose between (a) and (b) and the judge between from among (1) and (2), but the policeman may *believe* that the judge will choose (1) if there is no antipollution law in the state and (2) if there is an antipollution law in effect. Hence, this may prompt the policeman to choose (a) if, and only if, the law is there. This in turn may mean that no one will pollute the atmosphere if, and only if, there is an antipollution law in effect. Hence, the outcome of the game may well get influenced by the law. But note that the law works here entirely through its influence on people's beliefs and opinions. A central thesis of this chapter is that it cannot be otherwise. Law's empire, tangible and all-encompassing as it may seem, is founded on nothing but beliefs.

Of course, we will need to check that a particular outcome is self-enforcing (i.e., an equilibrium solution) before we can say that the outcome will occur given the law. But the important point is that a law *can* affect the *outcome,* and that in the final analysis, the law and the state are simply a self-supporting structure of beliefs and opinions. Hence, the order that one finds in very different kinds of collectivities—ranging from the totalitarian state to what anthropologists, in their zoological moments, call the acephalous society—are *self*-enforcing outcomes.

What is a self-enforcing outcome or a reasonable equilibrium solution for a game is itself a controversial question. Over the last two decades solution concepts have proliferated rapidly.[33] But it would be foolish to get drawn into that debate here. Hence, without further justification, we shall treat the set of *Nash equilibrium* outcomes[34] as the self-enforcing set.[35] So from here on, a reference to an 'equilibrium' outcome is always to a Nash equilibrium.

Many games have the problem of there being too many Nash equilibria. Consider a game in which you and another player will each have to choose one number (without letting the other player see what you are choosing) from among 3, 7, 9, and 100. If both of you choose the same number, each of you gets $ 1,000; if you choose different numbers, neither gets anything. In this game the following pairs of choices are the only Nash equilibria: (3,3), (7,7), (9,9), and (100,100). If you were playing this game, your essential problem is to try to guess what the other player will do. What complicates the guess is that what the other player does will depend on what (s)he guesses you will do. One way of guessing is to try to see if a particular strategy is salient or 'focal', and to employ it in the expectation that the other player will do the same. If such a salient outcome exists, it is called *a focal point*,[36] and predicting a focal outcome often turns out to be a good prediction. This method has no rigorous explanation but works through human psychology. In the above game, for instance, most human beings would choose 100. It is a large number, it is well-rounded, and somehow it stands out.

Nebulous though this method is, it works fairly well and has been used to great convenience. At Heathrow Airport there is an arbitrary place with a large sign above it saying Meeting Point. If you plan to meet a friend at Heathrow Airport and fail to decide in advance where to wait for the friend, then in this game there are millions of Nash equilibria. As long as both of you choose the same place you have a Nash equilibrium. It does not matter where that place is. The value of the sign is that it creates a focal point among all the possible Nash equilibria. You would typically choose to wait under the sign and so would your friend. There is no hard reason for doing that, but you would expect the other player to do so, and that becomes a reason enough. Putting up the sign Meeting Point does not change the game that you and your friend are forced to play by virtue of having forgotten to decide where you will meet, but it nevertheless influences the outcome. The writing on the paper that constitutes law is like the signboard in Heathrow. In itself it is quite a vacuous thing, but it creates expectations in the minds of individuals as to what the others will do; it creates focal points and thereby influences the outcome.

Suppose now the airport authority at Heathrow, in trying to be helpful and not have people walk too far, puts up twenty signboards saying Meeting Point at different locations in the airport. You may then decide that it is futile to wait under one of these (because it is not clear which one you should wait under), and remembering that your friend is a bookworm, and he knows that you know that he is a bookworm, and you know that he knows that you know that he is a bookworm, and so on, you may go to the store, Books Etc., and wait for him there. In anticipation of this, he may also choose to go the bookstall. Whether he does so or not, in this case the well-meaning signboards fail to influence behaviour and the outcome of the game. This can happen with the law as well. Poorly drafted legislation or legislation that takes inadequate cognizance of individual incentives can fail to have effect on people's behaviour or can have unintended effects by actually causing confusion. To avoid such poor-quality legislation, we have to first understand how and when the law works in the first place. For that we have to cast aside the widespread view that law changes the payoff functions and, hence, the game.

Recall that social norms (of the equilibrium-selection variety) also are simply a mechanism for players to coordinate onto an equilibrium (or some outcome .within a certain set of equilibria). It follows that actions and behaviour (and therefore outcomes) that are enforceable by law are also enforceable by social norms. Since an outcome that is enforceable by the law is an equilibrium, we can always imagine norms (which lead to beliefs) that sustain the same actions, behaviour, and outcome.[37]

To take an example, consider a society in which the law allows you to drive on any side of the road, but the norm is to drive on the left. Since we have seen that if such a law were there then it would be enforceable, it follows that this norm is also enforceable. This is an easy example because it is empirically transparent. In some remote parts of India, the hand of the law is so weak that it is indeed the case that, in effect, there is no law about which side of the road to drive on. Yet people do drive on the left because once the norm is in place, there is no reason for one to violate it.[38]

Obedience to a tyrant also is best explained along these lines, since no one really fears the hurt the tyrant can *himself* bring upon one. I have previously made use of a triadic model explaining this.[39] The subordinate's fear of the tyrant, based on what the subordinate expects other subordinates to do to him should he disobey the tyrant, is what Hume was talking about when he wrote: 'No man would have any reason to *fear* the fury of a tyrant, if he had no authority over any but from fear; since, as a single man, his bodily force can reach but a small way, and all the farther power he posses *must be found on our own opinion, or on the presumed opinion of others*'.[40] (The second set of italics is mine.)

This brings us to the central proposition of this chapter, which I've called the 'core theorem'.

CORE THEOREM: *Whatever behaviour and outcomes in society are legally enforceable are also enforceable through social norms.*

This theorem has two immediate implications or corollaries.

FIRST COROLLARY: *What can be achieved through the law can, in principle, also be achieved without the law.*

SECOND COROLLARY: *If a certain outcome is not an equilibrium of the economy, then it cannot be implemented through any law.*

Let us begin with the first corollary and, in particular, with some examples. In India till fairly recent times, and in some parts even now, a widow was expected to lead a life of general abstinence: do not eat nonvegetarian food, wear black and white clothes, avoid close relationships with men, and so on. This social norm used to be adhered to very strictly in many parts of India. To an outside observer, unfamiliar with India, this would appear to be a practice enforced by law, just like in some Islamic states where the women are required by law to wear the *chador*. But this appearance would be deceptive because there is no law that achieved this remarkable conformity in India. The conformity was achieved entirely through a system of sanctions and threats of ostracism, the threats themselves being given by individuals who feared that if they did not give such threats, they themselves would be ostracized. So this is an example of behaviour that we would expect to be caused by the law but is actually the result of social norms. The rules of caste are another example. The core theorem, in particular, the first corollary, challenges the myth that norms are somehow spontaneous and natural, while laws are intrusive and unnatural.

Turning to a different setting, consider a researcher who is given the task of finding out the extent to which the press is free in different countries. The typical thing this person will do is find out what kinds of legal restrictions each country places on its scribes. (S)He may also check on more general laws and statutes, such as the First Amendment in the United States, which guarantees freedom of speech to individuals and, therefore, also to the press. It has been found that in some countries the state persecutes its critics even when the law does not disallow criticism; and so this researcher may go a step further and check the record of state persecution of journalists and television commentators. (S)He would then somehow combine all this information to decide in which nations the press is the most free and in which nations the least. To most of us, at least at first sight, this seems like a reasonable procedure.

In light of the core theorem, however, it turns out that this method of research can yield seriously flawed results, because the method presumes that the only curb on press freedom can come from the nation's laws and the state. But the theorem tells us that what the state can do, individual citizens, going about their daily chores, can also achieve. So it is not enough to observe the law and state or governmental action.

One may try to rebut this criticism by arguing that there are practical limits to what we can study; so when we look for whether certain freedoms are guaranteed in a certain nation, it is only natural to study the nation's law and governmental behaviour. Suppose we agree to this rebuttal. Then, of course, we have to use this criterion for all studies of a similar nature. Now suppose the researcher were asked to study the amount of freedom that the widow has in different nations. (S)He would then have to say that the Indian widow is no less free than widows elsewhere in the world because she faced no legal or governmental restrictions on her behaviour. This would then also be true of India several decades ago, when in some parts of the country the widow was expected to commit *sati*—burn herself on the dead husband's pyre. Most of us would agree that the woman climbing on to the pyre was not, typically, committing a voluntary act. The voluntary act conclusion would be a folly stemming from the erroneous presumption that it is only the state that can curb individual voluntariness.

Newspapers and magazines come under all kinds of social—and in particular nongovernmental—pressures. If a newspaper criticizes a wealthy business lobby, it can face debilitating cuts in advertisements, and so it may feel compelled not to criticize the lobby. If it criticizes its government during an international crisis and the people of that nation are sufficiently nationalistic, it may face a boycott by general readers, and fearing this, it may decide not to criticize the government. Once these extralegal constraints are taken into account, certain rankings become ambiguous. Between, for instance, China and the United States, it may be relatively easy to conclude that the latter has a more free press, even without studying social control, because the state is *so much* more repressive in China; but between the United States and India, the answer is less obvious. In terms of the *law,* the US media are probably more free than the Indian ones, but the social and business sanctions seem to be greater on US newspapers and television channels. This is not just because of the pressures of political correctness, but there seems to be a wide recognition among corporations, lobbies, and power brokers in the United States that the control of opinion and information is an important ingredient for profit and survival. Even if my empirical conjecture about China, the United States, and India is false, it still remains true that merely studying legal controls may be inadequate, not just for determining press freedom but the freedom of the widow or the low caste.[41]

Freedom of speech is similarly problematic. When you say that you believe that individuals should have the freedom to say what they want or what they believe in, the main problem, to my mind, is not the moral status or appeal of that statement but to understand what it means. If by the above declaration you mean that the set of feasible actions available to an individual should include his ability to make different speeches, then your commitment to free speech is pretty meaningless. It is based on the same flawed view of an economy that underlies some of the literature on rights that I have discussed above and examined in greater depth in Appendix B. Having a freedom or a right must be interpreted as other people not having certain freedoms or rights after you have exercised that freedom.[42] And unless it is made at least partly clear what restriction one is willing to put on other people's freedom when guaranteeing a certain freedom to one person, the declaration that you believe in that freedom remains ambiguous.

An individual's freedom of speech can be curbed by the state; but it can also be curbed by the voluntary, atomistic actions of ordinary citizens.[43] Some societies are temperamentally more prone to sanctioning one another's speech and behaviour, of being less tolerant of what one considers to be deviancy. If we are committed to maximizing the freedom of speech and recognize that such freedom is not just a matter of law but also social norms, we may have to contend with the even more difficult problem that arises from the possibility that one person's exercising of his or her freedom of speech can result in the curtailment of another's freedom.[44]

Another example of how the meaning of 'freedom' can quickly become complicated occurs in the context of labour markets. Most people believe that slavery is coercion but that modern labour markets are voluntary. Those who study developing societies agree that bonded labour is unfree but that wage labour is voluntary. But once one goes beyond contemporary, industrialized society to consider examples of labour markets from primitive societies or by-gone eras, the dividing line between what is free and what is not is not so clear, as one encounters institutions that appear strange to the modern observer.[45] Moreover, on returning to contemporary markets, after such a journey, the dividing lines that had earlier seemed obvious also appear less sharp. It is true that in the light of the core theorem, individual freedoms become vastly more difficult to compute. But that cannot be reason enough for confining our attention to the law and the behaviour of government when studying individual freedom.

Let me now turn to the second corollary. According to this, if we have a law, the adherence to which entails out-of-equilibrium behaviour on the part of some individual, then such a law is doomed to fail; it can never be enforced. This is because according to the core theorem, the law can achieve only what a social norm can achieve. And since a social norm simply selects an equilibrium, no law can induce a nonequilibrium outcome.[46] Attempts to induce such an outcome would either result in the law being inconsequential or have unintended effects on the economy. Ellickson's claim that there can be 'order without law' is now easy to understand, as is the converse of that claim: disorder despite law.[47]

Considering the core theorem, the following question must arise: In what way is the law different from norms, since up to now we have shown how, in certain important respects, they look very similar? To answer this we have to recognize that the economy, described as a game, ignores a lot of information concerning prior beliefs and histories, which is a part of the real economy. In reality, even before a specific law is enacted there exists a predefined set of roles for various players concerning the way they should relate to the law, *whatever the law is*. The players are, of course, free to violate these rules, but they are nonetheless there. Thus the traffic policeman is supposed to follow the rule that he should stop drivers who violate the traffic laws. This instruction to the traffic police remains in force *no matter what the traffic laws are*. The ordinary citizen is supposed to follow the rule that he or she respects the orders of the traffic police.[48] The judge is supposed to follow the rule that he should punish the person who violates the law, and this remains valid no matter what the law is. Even if the speed limit is changed, the judge's rule remains the same. These prior rules and institutions may be referred to as 'quasi-laws', 'quasi-norms', or as 'standing orders'. The qualifier *quasi* reminds us that on their own they may not have any bite. The rule that the policeman should stop a car that breaks the speed limit is not an operational law till the speed limit is specified. But once the speed limit law is specified, the quasi-laws come to life. The speed limit law thrown in with the preexisting quasi-laws is much more than a law that simply says that a driver must not cross 65 mph. It is a law (or a set of laws, if we want to emphasize its reach) that specifies behaviour rules for various people—the driver, of course, but also the policeman, the magistrate, and frequently, the ordinary citizen (who, for instance, may not obstruct a policeman carrying out his duty). The role of quasi-laws is illustrated with an example in Appendix A.

Given that all modern societies have predefined rules or standing orders for people with respect to the law, which are independent of what the actual law is, this means that when new laws are enacted, the set of supporting activities and behaviour by the various citizens do not have to be specified separately each time. It is this preexisting structure of rules and instructions, along with the expectations in people's mind, that these will be adhered to as long as they are not against the adherer's self-interest, which makes it possible for the laws to be implemented. For any law, the full ramification of what it implies for individual behaviour is enormous. Suppose Montana enacts a new speed limit legislation. This does not ask just drivers to behave in a certain way; it also asks traffic wardens to behave in a certain way, judges to behave in a certain way, and so on.[49] The existence of preexisting rules (and, therefore, expectations) for laws is what makes a law different from a norm. If the Montana speed limit were to be introduced as a *norm,* all the supporting behaviours by the various agents would have to be specified, since norms do not have the advantage of preexisting rules and expectations. So, though for each implementable law there is also a norm that would yield the same outcome, the full statement of that norm would be enormously complicated.

The histories of norms and laws are also different.[50] Usually (though not always) social norms appear through long processes of evolution. Similar acts repeated over time can become a norm. To quote Ullmann-Margalit: 'Norms as a rule do not come into existence at a definite point in time, nor are they the result of a manageable number of identifiable acts. They are, rather, the resultant of complex patterns of behavior of a large number of people over a protracted period of time'.[51] Even some very sharply defined social norms and customs, such as the caste system or eating habits of different peoples, have such distant and diffused origins that there may be no agreement among historians as to where they came from. The law, on the other hand, is normally a product of deliberate choices, with dates of their enactment frequently known. Of course there are exceptions. The laws that certain tribes follow often merge into what we think of as norms; even in modern societies there are laws that emerged from common customs. This is true, for instance, of English common law, and the US practice of relying on interpretive principles and judicial rulings.[52] Conversely, there are some norms which are deliberate decisions.

But norms are difficult to change, because norms do not have the paraphernalia of preexisting rules, which can be used to usher in a new norm. On the other hand, norms may well be more robust than the law, because just as most norms were not deliberately instituted, it is difficult to deliberately discard them.

8.5 CONCLUDING REMARKS

In the previous sections I have discussed how to correctly model the role of law in an economy. There are, however, situations where we may willfully choose to reject the correct method, just as economists often do a partial equilibrium analysis where, strictly speaking, they should be doing a general equilibrium analysis. Indeed, it is possible to view the standard literature on law and economics as something akin to partial equilibrium analysis. It *presumes* law-abiding behaviour on the part of the law enforcement officers. Even in more sophisticated models that allow for bribery and other kinds of lapses, ultimately (and often implicitly) there is a layer of enforcement that is assumed to be automatic. This can work within limits, and if we are lucky, those who are assumed to do the job automatically actually find it in their interest to do so. But surely, instead of working with models that rely on our keeping our fingers crossed, it would be better to approach modelling law and economics as suggested here.

If one does adopt this approach to law and economics, it will have implications for several related areas of research, notably, the study of government and the state. In general, economists have been quite cavalier in modelling government. It has usually been treated as an exogenous agent or a puppet organization, carrying out the advice of economists (thereby providing a raison d'etre for policy economists). Even when economists have gone beyond this, they have generally taken a simplistic or mechanical view of government.[53] This essay draws our attention to the fact that both the enforcers of the law and, for want of a better word, the enforcees need to be modelled together, as strategic agents, having volition and choice. Such a construction will not be easy and will not happen all at once, but it is a target worth keeping in mind.

APPENDIX A: LAW AND ENFORCEMENT

This appendix illustrates formally some of the principles discussed in the main text. I proceed here entirely through an example.

For most games that economists talk about, it is possible to define a larger game by adding on to it the possibility of punitive actions after the end of the main game. Thus chess is a game, but at the end of a game of chess, I can sock my opponent in the nose, he can sock me back, and so on. So for every game G, we can define an 'expanded game' G_E, which appends to G a string of punitive actions.

I shall consider a very simple game G. This game consists of one player, called player 1, who has to choose any action from the set [0,1]. His payoff function is as follows. If he chooses $x \in [0,1]$, then he gets a payoff of x. We could think of an action as the amount of pollution generated by him. This takes a value between 0 and 1. The more he pollutes, the more profit he earns. Call this game G. If this was all there was, player 1 would pollute up to level 1.

Now consider the expanded game G_E. In period 0 of G_E, player 1 plays the above game G. In period 1, player 2 can choose between P (punish the other player) and N (not punish the other player). In period 2, player 1 chooses between P and N; in period 3, player 2 chooses between P and N; and so on *ad infinitum*. Suppose in period t (≥ 1), player i has to move. Then if i plays P, player j ($\neq i$) earns $-B$ (where $B > 0$) in that period and i earns 0; and if i plays N, both earn 0. In other words, punishment hurts and inflicting a punishment is costless and joyless (it will be interesting to modify this assumption). Both players have a discount factor of $\delta \in (0,1)$.

What we are interested in checking is how much pollution can be controlled through legislation. To keep the analysis simple, we shall assume that there is the following 'preexisting quasi-law'. This is simply a contingent definition.

At any time period $t \geq 1$, agent i's chosen action will be called *illegal* if he chooses P (i.e., punishes the other player), though player j's move at time $t - 1$ was legal (i.e., not illegal) or he chooses N, though j's move at $t - 1$ was illegal.

In words, what we are saying is this: (a) it is illegal to punish someone who has done nothing illegal and (b) it is illegal to not punish someone who has violated the law. We can think of other kinds of quasi-laws—for instance, we may think of dropping (b).

This preexisting quasi-law has no bite till we specify a law regarding what constitutes an illegal move in game G, and that is the reason I refer to it as quasi-law. Consider a possible law, which I will call 'the pollution law'. In period 0, if player 1 chooses any action greater than α, where α is a given number in $[0,1)$, then 1's action is *illegal*.

The pollution law, coupled with the preexisting partial law, is a well-defined law—let us call it a 'legal system'—which allows us to classify every action in every play of the game as either legal or illegal. Given a play of the game, a person is described as law abiding if he or she makes no illegal moves.

Note that the legal system that we are considering is parameterized by α. We want to investigate the values of α for which the legal system enforceable in the sense that there exists a Nash equilibrium outcome where everybody is law abiding.

Observe that a new pollution law or, for that matter, a new legal system, leaves the strategy sets and payoff functions (and therefore the game) unchanged. As in the main text, the legal system *can* nevertheless influence behaviour by affecting everybody's expectation about everybody else's behaviour, as long as it is enforceable.

To check this, suppose both players are law abiding. In particular, let us suppose that player 1 decides to play α in period 0 (i.e., the highest possible legal move) and be law abiding throughout. To check if this is an equilibrium, we have to verify that no one stands to benefit by deviating unilaterally.

If both are law abiding, 1 gets a payoff of α and 2 gets a payoff of 0. Clearly, 2 cannot do better through any deviation since 0 is the highest she can earn in this game. Consider 1's strategy if 1 decides to deviate from being law abiding. It is easy to see that the best deviation is to play 1 in period 0 and from then on to make only legal moves. That will of course invite punishment from player 2 [since s(he) is law abiding] in period 1. After then, it is not worthwhile for player 1 to play P because that, and only that, will prompt player 2 to play P in the following period. Hence, if 1 deviates from being law abiding, 1's highest possible payoff is $1 - \delta\beta$. Thus, 1 will not deviate if, and only if, $\alpha \geq 1 - \delta\beta$.

It follows that the only pollution laws that are enforceable are ones that permit people to pollute up to some level at least as high as $1 - \delta\beta$. If $1 - \delta\beta > 0$ and the pollution law sets $\alpha \in [0, 1 - \delta\beta)$, then a behaviour in conformity with the law cannot be enforced in any way. Since the law cannot change the game, a pollution level below $1 - \delta\beta$ is impossible in this society.

APPENDIX B: LIBERTY, RIGHTS, AND GAMES

The literature on liberty and rights that emerged from social choice theory[54] has gradually moved to a representation of rights as game forms.[55] This representation of rights has been the source of some controversy. The discussion of norms and law undertaken here suggests a new line of criticism of this approach. According to the rights-as-game-form approach, every assignment of individual rights translates into a game; thus a change in rights is represented by a change in the game being played by the players, since rights determine the strategies available to a player. Thus if i does not have the right to steal j's wallet, that option is typically omitted from i's available strategies or actions.

It is arguable, however, that i's not having a right to do something does not mean i cannot do that thing, so a change in rights should not be thought of as causing a change in the feasible set of actions. Stepping back from these academic debates, let us ask ourselves what it means to say: 'Person i does not have the right to steal j's wallet'. It means that if i does steal j's wallet, then someone else (j or a policeman) has the right to take some punitive action (y) against i, which otherwise that person would not have the right to do. Thus, i having a right to action z means that if i chooses z, then someone else, say j will not have a right to some action y that punishes i. This interdependent character of rights is very similar to the law described in Appendix A. Hence, a rights structure can influence what happens in the game, but it does not do so by influencing the game itself.

The above idea of interdependence of rights has been stressed by several writers. Hart writes that 'to have a right entails having a moral justification for limiting the freedom of another person and for determining how he should act'.[56] More pertinently, Lyons argues: 'When others are under an obligation to me and threaten to default, there are actions I might appropriately take which I would not otherwise be justified in taking'.[57]

To illustrate a rights structure of this kind, consider the game G_E in Appendix A. Think of action P as punching the other person in the nose or some such action to which no one has a right normally, but could acquire a right by virtue of the person in question doing something wrong or hurting one's right in the first place. Think of the two players in the game as neighbours and the actions open to player 1 in period 0 (given by the set [0,1]) as different levels of pollution from some activity in his backyard, which has a negative externality for player 2 but gives happiness to 1.

A *rights structure* can be defined by a basic right parameterized by α and other (contingent) rights. Let us suppose that by this society's values, 1 has the *right* to choose pollution levels up to α, but no more. If he chooses a pollution level $s > \alpha$, then in period 1, player 2 acquires the right to choose P.

If, however, in period 0, 1 chooses $s \leq \alpha$ and in period 1, 2 chooses P, then in period 3, 1 acquires the right to choose P, and so on. More formally, this may be stated as follows.

At time $t \geq 1$, if it is player i's move, we shall say that i chooses an action to which he or she *does not have a right* if that action happens to be P and if in period $t-1$ agent j ($\neq i$) had chosen an action to which he or she had a right (i.e., did not choose an action to which he or she did not have a right). This, coupled with the initial assumption that 1 in period 0 does not have a right to choose $s > \alpha$, where $\alpha \in [0,1]$, defines a rights structure.

Now when the game is played we can evaluate the *outcome* as one that does or does not respect the rights structure. So in this formulation, it is not a game that can satisfy or violate rights, but it is the outcome that can be put to this test. And, as with equilibrium-selection norms and law, a rights structure can potentially be enforced only if it is such that the intersection between the set of outcomes that satisfy the rights structure and the set of equilibrium outcomes is nonempty. No amount of policing or state intervention can change this fact, since the enforcers are already a part of the game and cannot do anything that was not already a part of their strategy sets.

NOTES

1. Rods, iron or wooden—I did not manage to find out which.
2. This fits in well with the 'self-help' view of law so perceptively described by D. Black in 'Crime as Social Control', *American Sociological Review*, 48 (1983): 34–45. Thus while at one level collecting money by issuing threats is criminal, we must remember that 'there is a sense in which conduct regarded as criminal is often quite the opposite' (p. 34). Black goes on to provide a variety of examples from different societies of 'moral' crimes. A similar view from a historical perspective emerges from E. P. Thompson's classic essay on the moral economy of the crowd: 'The Moral Economy of the English Crowd in the Eighteenth Century', *Past and Present*, 50 (1971).
3. This is not to be confused with the 'political economy' of the nineteenth century, which was the older name of economics till the advent of neoclassical economics.
4. This was later published in *Oxford Economic Papers*, 38 (1986): 259–82.
5. K. Basu, E. Jones, and E. Schlicht, 'The Growth and Decay of Custom: The Role of the New Institutional Economics in Economic History', *Explorations in Economic History*, 24 (1987): 1–21.
6. Strictly speaking, what we need is the convexity of preference. But since that turns out to be equivalent to the law of diminishing marginal utility, if the utility function happens to be additively separable, I shall here use the more familiar condition.
7. M. Granovetter, 'Economic Action and Social Structure: The Problem of Embeddedness', *American Journal of Sociology*, 91 (1985): 481–510.
8. This view of language, which I adhere to, is not undisputed, but it has the respectability of age, dating at least as far back as to the writings of David Hume. Warneryd has formalized this point of view in terms of evolutionary game theory; see K. Warneryd, 'Language, Evolution and the Theory of Games', in *Cooperation and Conflict in General Evolutionary Processes*, eds. J. L. Casti and A. Karlquist (New York: John Wiley, 1995).
9. J. H. Kagel et al., 'Experimental Studies of Consumer Demand Behavior using Laboratory Animals', *Economic Inquiry*, 8 (1975): 22–38.
10. Warneryd, 'Language, Evolution and the Theory of Games'.
11. I argue in my article 'On Misunderstanding Government: An Analysis of the Art of Policy Advising', *Economics and Politics*, 9 (1997): 231–50, that this is an implicit assumption in standard general equilibrium theory; and that once this is recognized it becomes possible to interpret the first fundamental theorem of welfare economics quite differently from what is usual.
12. R. M. Solow, 'Mass Unemployment as a Social Problem', in *Choice, Welfare and Development*, eds. K. Basu, P. K. Pattanaik, and K. Suzumura (Oxford: Clarendon Press, 1995), p. 318.
13. See my 'Social Norms and the Law', in *The New Palgrave Dictionary of Economics and the Law*, ed. Peter Newman (London: Macmillan, 1998). A more elaborate, though similar classification, occurs in R. A. Posner, 'Social Norms and the Law: An Economic Approach', *American Economic Review*, 87 (1997): 365–69.
14. The norm of reciprocity, as described by R. Sugden, which involves self-interested behaviour subject to some moral constraints, is a rationality-limiting norm; see his 'Reciprocity: The Supply of Public Goods through Voluntary Contributions', *Economic Journal*, 94 (1984): 772–87.

15. See K. Basu, 'Civil Institutions and Evolution: Concepts, Critiques and Mode', *Journal of Development Economics*, 46 (1995):19–33; and K. Basu, 'Notes on Evolution, Rationality and Norms', *Journal of Institutional and Theoretical Economics*, 152 (1996): 739–50.

16. An example of a preference-changing norm, along with its implications for economic policy, occurs in A. Lindbeck, S. Nyberg, and J. W. Weibull, 'Social Norms and Economic Incentives in the Welfare State', *Quarterly Journal of Economics* 1/4 (1999): 1–35.

17. This explains why the police have to be vigilant in enforcing the stop-sign rule or the speeding rule but not about drive-on-the-right rule. The first two are laws that are not in peoples' self-interest (they may, of course, be in *their group* interest).

18. G. Akerlof, 'The Economics of Caste and of the Rat Race and Other Woeful Tales', *Quarterly Journal of Economics*, 90 (1976): 599–617.

19. K. Basu, 'One Kind of Power', *Oxford Economic Papers*, 8 (1986): 259–82.

20. R. D. Cooter, 'Law from Order' (University of California, Berkeley, 1997, mimeographed).

21. Greif constructs an equilibrium-selection explanation of certain kinds of cultures and cultural beliefs that can explain different trading institutions: A. Greif, 'Cultural Beliefs and the Organization of Society: A Historical and Theoretical Reflection on Collectivist and Individualist Societies', *Journal of Political Economy*, 102 (1994).

22. J. M. Buchanan, 'The Coase Theorem and the Theory of the State', *Natural Resources Journal,* 13 (1973): 580–94.

23. D. G. Baird, R. H. Gertner, and R. C. Picker, *Game Theory and the Law* (Cambridge, MA: Harvard University Press, 1995).

24. J. P. Benoit and L. A. Kornhauser, 'Game-Theoretic Analysis of Legal Rules and Institutions', Economic Research Reports, #96–30 (1996), C. V. Starr Center for Applied Economics, New York University.

25. Dixit in his recent work on the political process in an economy also views the interaction between agents as a game. While not going into such a formal contraction, Bhagwati and O'Flaherty nevertheless adopt a game-theoretic approach in which individuals who constitute government are also players with their own objective functions: A. Dixit, *The Making of Economic Policy: A Transaction-Cost Politics Perspective* (Cambridge, MA: MIT Press, 1996); and B. O'Flaherty and J. Bhagwati, 'Will Free Trade with Political Science Put Normative Economists Out of Work?' *Economics and Politics*, 9 (1997): 207–20.

26. One way of limiting the strategies open to a player is to assume that all the strategies are still available, but some of them give a payoff of negative infinity and thus would never be adopted. Hence, once we allow for the payoff function to be altered, there may be no need for separately assuming that the strategy set can be shrunk.

27. D. G. Baird, R. H. Gertner, and R. C. Picker, *Game Theory and the Law,* p. 15.

28. See J. Bhagwati, R. Brecher, and T. N. Srinivasan, 'DUP Activities and Economic Theory', in *Neoclassical Political Economy*, ed. D. Collander (Cambridge: Ballinger, 1984); D. Friedman, 'A Positive Account of Property Rights', in *Property Rights*, eds. E. F. Paul, F. D. Miller, and J. Paul (Cambridge: Cambridge University Press, 1994); R. L. Calvert, 'Rational Actors, Equilibrium and Social Institutions', in *Explaining Social Institutions*, eds. J. Knight and I. Sened (Ann Arbor: University of Michigan Press, 1995); A. Dixit, *The Making of Economic Policy;* K. Basu, 'On Misunderstanding

Government: An Analysis of the Art of Policy Advising', *Economics and Politics*, 9 (1997): 231–50; and R. Gibbons and A. Rutten, 'Hierarchical Dilemmas: Social Order with Self-Interested Rulers' (Cornell University, 1997, mimeographed). An interesting attempt to apply some of these theoretical ideas to political events and phenomena occurs in R. H. Bates and B. R. Weingast, 'Rationality and Interpretation: The Politics of Transition' (Harvard University, 1996, mimeographed). For an endogenous but evolutionary view of institutions, see A. Schotter, *The Economic Theory of Social Institutions* (Cambridge: Cambridge University Press, 1981).

29. The game, properly defined, would require specifying what the remaining strategies are, the sequence of moves, and other such details, but since I do not aim to analyse the game in any detail, it is all right to leave the description at this level of generality.

30. W. Gaertner, P. K. Pattanaik, and K. Suzumura, 'Individual Rights Revisited', *Economica*, 59 (1992): 161–78; A. Sen, 'Minimal Liberty', *Economica*, 59 (1992): and R. Deb, 'Waiver, Effectivity and Rights as Game Forms', *Economica*, 61 (1994): 167–78.

31. More formally, a 'rights structure' is simply a specification of a subset of actions from among all the actions open to the relevant player at that information set. The interpretation is that the player has a '*right*' to choose an action only from the specified subset.

32. This point of view builds on the legacy of David Hume. In his essay 'Of the First Principles of Government', Hume had puzzled about the sources of state influence on social and economic outcomes. Thus he wrote: 'Nothing appears more surprising to those who consider human affairs with a philosophical eye, than the easiness with which the many are governed by the few'. And he reaches the remarkable conclusion that those who rule do so only by the force of opinion: 'It is therefore on opinion only that government is founded; and this maxim extends to the most despotic . . . governments, as well as to the most free and most popular', David Hume, *Essays: Moral, Political and Literary* (reprint, Indianapolis: Liberty Fund, 1987), p. 32.

33. See, for instance, M. J. Osborne and A. Rubinstein, *A Course in Game Theory* (Cambridge, MA: MIT Press, 1994).

34. A *Nash equilibrium* outcome of a game is a choice of strategy by each player (i.e., a tuple of strategies) such that no individual can do better by *unilaterally* deviating to some other strategy. Thus, once a Nash equilibrium outcome is expected by all players, the outcome becomes self-enforcing.

35. Left to myself, I would prefer to use the coarser solution concept of 'rationalizability', but for the present purpose it is simpler to rely on the much more widely used solution of Nash equilibrium. It is worth noting that in some situations, despite there being a rationalizable equilibrium and even a Nash equilibrium, there may be no reasonable way of predicting the outcome. See my 'On the Nonexistence of a Rationality Definition for Extensive Games', *International Journal of Game Theory*, 19 (1990): 33–44; and 'The Traveler's Dilemma: Paradoxes of Rationality in Game Theory', *American Economic Review*, 84 (1994): 391–95.

36. See T. C. Schelling, *The Strategy of Conflict* (Oxford: Oxford University Press, 1960).

37. For a real-life illustration of how norms and beliefs translate into action and policy, see P. Katzenstein, 'Coping with Terrorism: Norms and Internal Security in Germany and Japan', in *Ideas and Foreign Policy*, eds. J. Goldstein and R. O. Keohane (Ithaca, NY: Cornell University Press, 1993).

38. In this case, the norm is imported from the cities where it is the law and is enforced.

39. K. Basu, 'One Kind of Power'.

40. Hume, *Essays*, p. 34.

41. Bernstein's engaging study of the diamond industry makes it amply clear that the advice carries over even to the more microeconomic domain of specific industries and markets; see L. Bernstein, 'Opting Out of the Legal System: Extralegal Contractual Relations in the Diamond Industry', *Journal of Legal Studies*, 21 (1992): 115–57.

42. I am fully aware that this is a circular definition. But it is not meaningless for that reason, as is demonstrated in Appendix B.

43. G. Loury, 'Self-Censorship in Public Discourse: A Theory of "Political Correctness" and Related Phenomena', *Rationality and Society*, 6 (1994): 428–61.

44. O. M. Fiss, *The Irony of Free Speech* (Cambridge, MA: Harvard University Press, 1996).

45. S. L. Engerman, 'Coerced and Free Labor: Property Rights and the Development of the Labor Force', *Explorations in Economic History*, 29 (1992): 1–29.

46. Following Leibnitz, Steiner has argued that for a set of rights to be implementable, it is necessary that the rights be 'compossible', that is, the set of social outcomes or states where each of these rights are satisfied be nonempty. By the same argument, he would no doubt argue that for a set of laws to be implementable, a necessary condition is that they be compossible. Viewed in this light, the second corollary can be thought of as simply taking this argument further and claiming that for a set of laws to be implementable, the intersection of the set of outcomes that satisfy these laws and the solution set of the economy game must be nonempty. A similar extension is possible for rights, as illustrated in Appendix B. See H. Steiner, *An Essay on Rights* (Oxford: Blackwell, 1994), pp. 2–3.

47. R. C. Ellickson, *Order without Law: How Neighbors Settle Disputes* (Cambridge, MA: Harvard University Press, 1991).

48. Ask yourself why you would not stop driving if an ordinary civilian (perhaps mad), pretending to be a traffic warden, asked you to stop. The reason is not what we expect this person to do to us; neither the traffic warden nor the civilian would do anything directly to us. Moreover, both may report the license plate number to the police department. The difference is not in that. The difference is in how we expect others to react to this person. When the licence plate number is reported to the police department, we expect very different kinds of action on the part of the police, depending on who the report comes from, a traffic warden or a mad person.

49. Some of these preexisting rules may not even have the status of quasi-laws. They may be more in the nature of norms. Thus, for the successful implementation of a law, it may be important for the law to be embedded in a certain structure of norms. Cooter's claim that 'state law builds upon preexisting social norms', though based on a different kind of argument, has some parallels to the position being taken here; see Cooter, 'Law from Order', p. 2.

50. See Ellickson, *Order without Law*.

51. E. Ullmann-Margalit, *The Emergence of Norms* (Oxford: Clarendon Press, 1977), p. 8.

52. See J. Ferejohn, 'Law, Legislation, and Positive Political Theory', in *Modern Political Economy*, eds. J. S. Banks and E. A. Hanushek (Cambridge: Cambridge University Press, 1995).

53. R. Hardin, 'Economic Theories of the State', in *Perspectives on Public Choice,* ed. D. C. Mueller (Cambridge: Cambridge University Press, 1997).

54. Among others, see A. Sen, 'The Impossibility of a Paretian Liberal', *Journal of Political Economy,* 78:1 (January–February 1970): 152–7.

55. Gaertner, W., P. K. Pattanaik and K. Suzumura, 'Individual Rights Revisited', Deb, 'Waiver, Effectivity and Rights as Game Forms', *Economica,* Vol. 59, (1992), 161–77.

56. H. L. A. Hart, 'Are There any Natural Rights?', in *Rights,* ed. D. Lyons (Belmont, CA: Wadsworth Publishing, 1979), p. 19.

57. D. Lyons, introduction to *Rights,* p. 5.

9 The Right to Give up Rights

9.1 INTRODUCTION

If we grant individuals rights over certain spheres of decision-making, it seems natural that we ought also to grant them the right to give up these rights. It has been argued that, if this principle is respected in social decision-making, then many of the fundamental paradoxes of liberty would disappear.

Beginning with Sen's (1970) work on the inconsistency between Paretianism and libertarianism, there have been many investigations in this area, and some have tried to show that our notion of liberty is, in itself, inconsistent (Farrell 1976; Gibbard 1974; Suzumura 1978). In his widely discussed paper, Gibbard argued that many of these paradoxes would vanish if we granted individuals the right to give up rights and assumed that (α): an individual will waive his right if he expects to be 'better-off' by doing so.[1] Let us refer to this escape route from the liberty paradox as the *meta-rights approach*.

By formalizing (α) in alternative ways, different special cases of the meta-rights approach can be explored. The best known of these special cases is Gibbard's own formalization. Gibbard assumed that an individual will waive his right over (x, y) if he prefers x to y and y to z and from x there exists a chain up to z, with the connections being made of the Pareto principle and other people's rights. This is stated rigorously in the next section and is referred to as the *Gibbardian waiver profile* (GWP).

This particular formalization of the meta-rights approach soon came under attack from Kelly (1976), who argued that the GWP is incentive-incompatible[2] in

This paper was written while the author was a visiting fellow at CORE, Louvain-la-Neuve, and the Centre d'economie mathematique, Brussels. For some very valuable comments and criticisms, the author is grateful to two anonymous referees.

First published in *Economica*, Vol. 51 (1984).

the sense that there exist situations where, by waiving rights in the way Gibbard suggests, individuals could actually harm themselves. So Kelly proposed an alternative formalization. Suzumura (1980), in turn, has shown that Kelly's approach runs into difficulties as well. Then, by examining other assumptions about situations where individuals would waive their rights, Suzumura has established a number of impossibility results.

One begins to suspect that there is some basic problem with the meta-rights approach itself. And indeed there is. This chapter tries to show that all reasonable interpretations of the meta-rights approach are incentive-incompatible and, hopefully, it thereby obviates the need for analysing particular formalizations separately. More precisely, it is shown that, if individuals are allowed to waive their rights *voluntarily*, then there is no guarantee that they will do so in a way that resolves the liberty paradox. Thus Gibbard's theorems do not really bear out his own motivation.

The present work has another objective: to circumvent two rather pervasive shortcomings in the literature. The first one, noted by Sen (1976), is concerned with the interdependence of meta-rights. Clearly, there can arise situations where, if i waives his right, the reason for *j* to waive his right disappears.[3] This interdependence is not permitted in Gibbard and in much of the literature that followed. I quote Sen (1976, p. 224):

The difficulty arises from each person deciding what right is "useless" for him on the basis of some presumption as to what rights the others would exercise, but, one person's decision not to exercise his right (on the supposed grounds of its being "useless" when others exercise their rights) renders erroneous another person's conviction that his right is "useless" (based on that person's assumption that others will exercise their rights). This problem of interdependence and of "correctable miscalculation" proves to be a deep one for the "pragmatic interpretation" of the Gibbard system... .

A second criticism stems from the fact that implicit in the GWP and other waiver profiles assumed by Kelly (1976) and Suzumura (1980) is the assumption that the SWF is Paretian and libertarian. This follows from the fact that all these approaches look for chains using the Pareto criterion and liberty rights. Thus, if we are discussing an SWF that is libertarian and not necessarily Paretian, it is not clear why an individual would wish to waive his rights because there exists a chain from *x* to *z* linked by other people's rights *and the Pareto principle*. It is therefore an anomaly—not nonexistent in the literature—to consider simultaneously a libertarian (and not necessarily Paretian) SWF and a Gibbard-type rights-waiving system. Similarly, if more properties are imposed on the SWF, then corresponding changes should be permitted in the waiver profiles. For example, if an SWF is libertarian, Paretian, and also 'equitable' (in some sense), then we should look for chains that may have some connections based on the equity criterion as well in deciding whether to waive a right or not.

Both these criticisms can be overcome by assuming that an individual's decision to waive his rights depends on how others waive their rights and on the nature of the SWF being used.

9.2 RIGHTS AND META-RIGHTS

Following Suzumura's (1980) notation and definitions, suppose that $N(\infty > \# N \geq 2)$ is the set of individuals. Given that X_i ($\infty > \# X_i \geq 2$) is the set of characteristics of 'personal' concern to $i \in N$, the set of social states is defined as[4]

$$X = \prod_{i \in N} X_i.$$

Let $\mathscr{R}$ be the set of all n-tuples of orderings (i.e., transitive, reflexive, and complete binary relations) on X. $R \in \mathscr{R}$ implies $R = (R_1, \cdots, R_n)$ where R_i is the weak preference ordering of individual i. The asymmetric and symmetric parts of R_i are denoted by $P(R_i)$ and $I(R_i)$. Let D_i be the set of all pairs (x, y) where x and y are social states which differ only in their ith component.[5] Hence if (x, y) belongs to D_i, then the issue of choosing between x and y is personal to individual i. We therefore say that, if $(x, y) \in D_i$, then i has the *natural right* to decide between x and y;[6] and the n-tuple $(D_1, \cdots, D_n)$ is referred to as the *natural rights system*. When we say that an individual i has natural rights over {x, y}, we mean $(x, y) \in D_i$ and $(y, x) \in D_i$.

In keeping with Gibbard (1974), we allow individuals meta-rights, that is the right to waive their natural rights if they so wish. Let $S \subset X$. Then $D_i \cap S \times S$ are the pairs in S over which i has natural rights. Let W_i^S be the pairs over which he waives his natural rights in S. Hence $W_i^S \subset D_i \cap (S \times S)$. Let $\mathscr{L}$ be the collection of all nonempty subsets of X. A *waiver profile*, W, is defined as follows:

$$W \equiv \{W^S\}_{S \in \mathscr{L}} \equiv \{W_i^S\}_{S \in \mathscr{L}, i \in N}$$

A *social choice rule* (SCR) *or* a 'voting scheme' is a function C, which, *for* all nonempty sets of alternatives S, for all n-tuples of preference orderings R and for all waiver profiles W, specifies a subset of S containing no more than one element. Thus, using $\mathscr{W}$ to denote the set of all waiver profiles, we have

$$C: \mathscr{L} \times \mathscr{R} \times \mathscr{W} \rightarrow \{T \subset X \mid \# T \leq 1\}.$$

If, given a waiver profile W, it is true that for all S, for all R, C is nonempty, then the SCR is said to be *decisive* (given W).

The two central ethical axioms may now be stated. The Pareto principle says that if everybody prefers x to y, y must not be chosen in the presence of x.

AXIOM P: *The SCR, $C(\cdot)$, satisfies axiom P iff*

$$[(x, y) \in \bigcap_{i \in N} P(R_i)] \rightarrow [x \in S \rightarrow y \notin C(S, R, W)].$$

The liberty axiom is stated directly introducing the idea of meta-rights. It says that, if an individual has natural rights over (x, y) which he has not waived, then, if he prefers x to y, y must not be chosen in the presence of x.

AXIOM L: *The SCR, $C(\cdot)$, satisfies axiom L iff*

$[\textit{for some } i \in N, (x, y) \in P(R_i) \cap D_i \backslash W_i^S] \rightarrow [x \in S \rightarrow y \notin C(S, R, W)].$

In most of the existing literature an individual's decision to waive his rights depends neither on the SCR nor on how others waive their rights. It depends only on the preference profile, R, in question. Both the pervasive shortcomings in the literature mentioned in the introduction section can be shown to originate from this restrictive assumption. Consequently, in this chapter we shall permit individuals to base their decision to waive rights on the nature of the SCR, the preference n-tuple in question, and also on how others exercise their meta-rights. Hence, given $C(\cdot)$ and R,

$$W_i^S = W_i^S (W_1^S, \cdots, W_{i-1}^S, W_{i+1}^S, \cdots, W_{\#N}^W). \qquad (9.1)$$

If individuals have no meta-rights, that is if $W_i^S = \phi$ for all S and for all i, then Gibbard has shown that, let alone P, there is no decisive SCR satisfying L. He identified the absence of meta-rights as the principal culprit in the original liberty paradox (Sen, 1970). He went on to argue that individuals ought to have the right to give up rights and that that would remove the conflict between the principles of Pareto and liberty. Clearly, the resolution of the paradox depends on how individuals exercise their meta-rights, that is on the specification of Eq. (9.1). Let us first consider Gibbard's own assumption.

The Gibbardian Waiver Profile (GWP), $\hat{W}$, is defined as follows:

$$(y_r, y) \in \hat{W}_i^S \text{ iff there exists } y_1, \cdots, y_{r-1} \text{ in } S \text{ such that } (y, y_1) \in R_i, y \neq y_1$$

and

$$\forall t \in \{1, \cdots, r-1\}: (y_t, y_{t+1}) \in \left[\bigcap_{j \in N} P(R_j) \right] \cup \left[\bigcup_{j = N \backslash \{i\}} \{D_j \cap P(R_j)\} \right] \qquad (9.2)$$

Gibbard proved that, given a GWP, there exists a decisive SCR satisfying Axioms P and L. The attractiveness of this theorem depends on how the GWP is interpreted. Consider the following description by Suzumura (1980, p. 411) of the intuitive appeal of Gibbard's specification:

An ingenious proposal crystallized in Gibbard's third libertarian claim [i.e. axiom L based on the waiver profile, $\hat{W}$] is to make individual's libertarian rights *alienable* in the cases where *the exercise of one's libertarian rights brings him into a situation he likes no better than the situation that would otherwise have been brought about.* [my italics]

If this is where the intuitive appeal of the meta-rights approach lies—and this is indeed where it should lie—then, unfortunately, Gibbard's theorem lacks intuitive appeal. However, instead of establishing this specifically for the GWP, I shall prove a more general theorem of which this is a corollary.

To motivate this, note that the new *approach* opened up by Gibbard, what I call here the meta-rights approach, is broader than Gibbard's *theorem*. After all, the theorem uses one particular formalization of Eq. (9.1), namely the GWP. So perhaps there are other specifications of Eq. (9.1) that are incentive-compatible. And anyway, the GWP does have important weaknesses.

First of all, as noted by Kelly (1976) and also by Suzumura (1980, p. 413), there may exist a 'sequence $\{z_1, z_2, \cdots, z_\lambda^*\}$' which seems to repair in the eyes of the individual i the damage caused upon him by a sequence $\{y_1, y_2, \cdots, y_\lambda\}$'. In that case the individual would not waive his right, unless there is yet another sequence that nullifies the effect of $\{z_1, z_2, \cdots, z_\lambda^*\}$.

Even with this correction, one difficulty remains (and this criticism applies to Kelly [1976] and Suzumura [1980] as well). Note that the GWP is characterized by a presumption on the part of each individual that others never waive their rights, because otherwise Eq. (9.2) would have to read

$$\forall t \in \{1, \cdots, r-1\}: (y_t, y_{t+1}) \in \left[\underset{j \in N}{\cap} P(R_j) \right] \cup \left[\underset{j = N \setminus \{i\}}{\cup} \{(D_j \setminus W_i^S) \cap P(R_j)\} \right].$$

In other words, the functions [Eq. (9.1)] are constant mappings in the GWP and also in the waiver profiles defined and discussed by Kelly and Suzumura. This difficulty, as mentioned above, has been noted by Sen (1976).

To rectify this, we have to look for a specification of Eq. (9.1) that allows for interdependence in waiver sets. But the next result is a general one, which shows that all such effort is bound to be futile because no matter how Eq. (9.1) is specified, a resolution between Axioms P and L necessarily entails involuntary rights-waiving, in some sense. The definition and theorem that follow establish this.

A matter of notation first. Given a waiver profile W, define W_{-iS} as a waiver profile that differs from W (if at all) only in the (i, S)th element. In other words, if $W_{-iS} = \bar{W}$, then $\bar{W}_j^K = W_j^K$, for all $(j, K) \neq (i, S)$.

DEFINITION 1: Given an SCR $C(\cdot)$ and a waiver profile W, individual i is subject to *involuntary rights-waiving* if it is *not* the case that for all S and R,

(i) for all W_{-iS}, $C(S, R, W_{-iS})$ is nonempty and
(ii) $(x, y) \in R_i$ where $\{y\} = C(S, R, W_{-iS})$ and $\{x\} = C(S, R, W)$.

THEOREM: *For all decisive SCRs satisfying P and L, given any waiver profile, W, there must exist an individual who is subject to involuntary rights-waiving.*

PROOF: Given that $x_i \in X_i$ and $\hat{x} = x_3, \cdots, x_n$ and $\hat{x}' = x'_3, \cdots, x'_n$, let a, b, c, and d be distinct social states defined as follows:

$a = (x_1, x_2, \hat{x})$,
$b = (\bar{x}_1, x_2, \hat{x})$,
$c = (x_1, x_2, \hat{x}')$,
$d = (x_1, \bar{x}_2, \hat{x}')$.

Note that $(a, b) \in D_1$ and $(c, d) \in D_2$. Let $K = \{a, b, c, d\}$ and $R \in \mathcal{R}$ be such that its restriction on K is as follows (in descending order of preference from left to right):

$$R_1|K: c\ a\ b\ d$$
$$\forall j \in N \setminus \{i\}, R_j|K: b\ d\ c\ a$$

Any waiver profile in K must be one of the following four kinds:

Case I: $W_1^K = \{a, b\},^7$ $W_2^K = \{c, d\}.$
Case II: $W_1^K = \phi,$ $W_2^K = \phi.$
Case III: $W_1^K = \{a, b\},$ $W_2^K = \phi.$
Case IV: $W_1^K = \phi,$ $W_2^K = \{c, d\}.$

Case I, it is easy to check, is the GWP. Let us consider this case first. Note that by Axiom P,

$$d, a \notin C(K, R, W).$$

Since both 1 and 2 have waived their respective rights in K, Axiom L is ineffective. hence, if the SCR satisfying Axioms P and L is decisive, then

$$\text{either } \{b\} = C(K, R, W) \text{ or } \{c\} = C(K, R, W).$$

Without loss of generality, assume the former. Define W_{-1K} by replacing W_1^K with $\bar{W}_1^K$, where $\bar{W}_1^K = \phi$. Then either

$$\{c\} = C(K, R, W_{-1K}) \text{ or } \phi = C(K, R, W_{-1K})$$

since b is now ruled out by the exercising of 1's rights. Hence individual 1 is subject to involuntary rights-waiving.

Case II is one where no rights are waived. This is the same situation as in the original Sen (1970) paradox: by Axiom P, a and d cannot be chosen. By Axiom L, b and c cannot be chosen. Hence $C(K, R, W) = \phi$. Thus the SCR is not decisive.

Now consider case III. Here a and d are ruled out by Axiom P, and c is ruled out by Axiom L. Hence, given a decisive SCR, it must be that $C(K, R, W) = \{b\}$. But $C(K, R, W_{-1K}) = \phi$, where W_{-1K} is got by replacing W_1^K with $\bar{W}_1^K = \phi$, for the same reason as in case II. Hence individual 2's rights-waiving described in case III does not dominate all other rights-waiving schemes open to him. So he is subject to involuntary rights-waiving.

Case IV is symmetric to case III.   *Q.E.D.*

It is interesting to note that in case I, which is the Gibbardian case, we showed that each individual's specified waiver set is not dominant by comparing it with a case where he does not waive any of his rights. Hence, not only have we shown that Gibbardian rights-waiving is nonoptimal, but we have shown, in particular, that it entails forcing people to forgo their rights.

9.3 Involuntariness and Equilibrium

It is important to appreciate that definition (9.1) is *one* formalization of the intuitive notion of involuntariness. There can be others. To see this it is useful to visualize the above problem in game-theoretic terms. Suppose society is about to choose from a set S, which is a subset of X. The SCR and the preference profile R are known. Individual i's problem is to choose a waiver set W_i^s in order to do as well as possible in terms of his preference relation. This is, however, not a well-defined game, because the choice set corresponding to some strategy n-tuples,

$\{W_i^S\}_{i \in N}$, may be empty. And since it is reasonable to suppose that a person's preference over $\{x\}$ and ϕ—in brief, over $\{x, \phi\}$—is incomplete, it is possible to interpret the remark that 'given $\hat{W}_i^S$ for all $i \neq j$, $\hat{W}_j^S$ is j's best strategy' in different ways. Two reasonable interpretations are: (i) 'there does not exist $\bar{W}_j^s$ such that the chosen social state corresponding to $(\hat{W}_1^S, \ldots, \hat{W}_{j-1}^S, \bar{W}_j^S, \hat{W}_{j+1}^S, \ldots, \hat{W}_n^S)$, where n denotes $\# N$, is preferred by j to the social state corresponding to $\{\hat{W}_i^S\}_{i \in N}$'; (ii) 'the social state chosen, given $\{\hat{W}_i^S\}_{i \in N}$, is considered by j to be at least as good as any of the social states that would be chosen if he changed his $\hat{W}_j^S$ to some other waiver set'. Since his preference over $\{x, \phi\}$, for any $x \in X$, is incomplete, this interpretation requires that the social choice set be nonempty for all strategy n-tuples $(\hat{W}_1^S, \ldots, \hat{W}_{j-1}^S, W_j^S, \hat{W}_{j+1}^S, \ldots, \hat{W}_n^S)$ that we can get by varying W_j^S.

If the pay-off matrix was well-defined, meaning thereby that the social choice set was never empty, the two interpretations, (i) and (ii), would coincide. It is not difficult to see that in this paper the definition of involuntary rights-waiving is based on interpretation (ii). That is, a person is involved in involuntary rights-waiving if his waiver set does not satisfy optimality in the sense of (ii). [It should be clear at this point why we assumed that $\# C(S, R, W) \leq 1$. Otherwise, to define a person's 'best' waiver set, we would need to extend his preference ordering on X to a preference ordering on the power set of X; and this—as is becoming increasingly clear—can be a formidable task.]

Using this game-theoretic approach, the theorem of Section 9.2 may be stated as an impossibility result. Given an SCR and a preference profile, let $\{\hat{W}_i\}_{i \in N}$ be described as *Nash consistent* over S if there does not exist any individual who is subject to involuntary rights-waiving over S [i.e., if (ii) is true for all $j \in N$]. The notion of Nash consistency is similar to the well-known concept of Nash equilibrium and would be exactly the same if the pay-off matrix were well-defined. Now we are in a position to restate our theorem: there does not exist any decisive SCR or waiver profile that satisfies Axioms P and L and Nash consistency on every $S \subset X$.

While I have used a Nash-type equilibrium concept, and have, therefore, assumed that each person supposes that others do not respond to his change of strategies, it is possible to use other solution concepts where individuals have 'conjectures' about the response of others. I do not, however, pursue this line here.

Once our theorem is restated in terms of Nash consistency, it acquires a certain similarity to Sen's (1983, p. 26) claim that

in a situation exemplifying the conflict between the Pareto principle and individual liberty, there might exist no equilibrium at all — with some states being rejected by the Pareto-improving contract and the others being rejected by individual decisions over their personal spheres ... The impossibility of the Paretian liberal—interpreted in terms of descriptive choice—leads to a game with an empty 'core'.

While this result and the theorem in this chapter are both about the nonexistence of equilibrium, it is important to distinguish between the two. Sen's equilibrium is a situation in which, among other things, a Pareto-improving contract is not possible. Thus a Pareto suboptimal state cannot qualify as an equilibrium. In our framework, individuals act atomistically and the search is for a Nash-type equilibrium, which of course may be Pareto suboptimal (as is the case with the Nash–Cournot equilibrium in the standard duopoly model). What we establish is that even such a Pareto suboptimal equilibrium does not exist.

Finally, a comment on this chapter's exclusive focus on the pragmatic interpretation of Gibbard (see note 1). What is interesting is the interconnection between the pragmatic and ethical interpretations. As Sen (1976) defines it, the ethical interpretation is concerned with how the conflict—between Pareto and liberty—*ought* to be resolved via rights-waiving. The pragmatic approach, on the other hand, examines whether, once people are free to give up their rights, the conflict continues to exist or not. Thus, an attempt to solve the pragmatic problem is equivalent to trying to determine whether or not the ethical problem exists at all. To that extent, this chapter does have some bearing on the ethical question. What it shows is that the ethical problem remains given the Gibbardian system of rights-waiving; more importantly, it will remain no matter what system of voluntary rights-waiving is considered.

9.4 PERSISTENCE OF THE ORIGINAL PARADOX

This concluding section draws attention to an apparently perplexing feature of the meta-rights approach, concerning the 'strength' of axioms.

In most areas of social choice, as we impose additional axioms on the social welfare function (e.g., SWF the Pareto axiom, the nondictatorship axiom, etc.) we, for quite obvious reasons, move closer to an impossibility result. In the approach adopted in this chapter this need not be so. Consider an SCR satisfying Axiom L. It has been argued above that it is reasonable to allow each individual's waiver profile to depend on, among the other things, the nature of the SCR. For example, a Nash-consistent waiver profile is quite clearly characterized by this kind of interdependency. In this framework it is not clear that additional axioms move us closer to an impossibility result. This is for the interesting reason that, with the imposition of each new axiom on the SCR, the waiver profile gets altered and this in turn changes the strength of the liberty axiom. Thus in this area it is misleading to make a remark like 'Axiom L itself gives us an impossibility result, without having to invoke any other axiom (e.g. P).'

In this section I explore a related issue which also clarifies further the relation between Sen's (1970) original framework and that of Gibbard (1974).

Consider 'weakening' the liberty axiom to 'minimal libertarianism' in the sense of Sen (1970). That is, suppose that only two individuals have natural rights and that too over one pair each. In particular, assume that 1 has rights over $\{x, y\}$ and 2 has rights over $\{z, w\}$, and assume that x, y, z, and w are distinct elements.[8] The *axiom of minimal liberty* is the same as Axiom L but is based on this restricted domain of natural rights. That is, if individual 1 does not waive his right over $\{x, y\}$, then he is decisive over these elements; and similarly for individual 2.

What is interesting is that this 'weakening' of the liberty axiom strengthens the impossibility result implied by our theorem. Note that ours is an 'existential' result: it asserts that there exists a problem in the form of involuntary rights-waiving *somewhere* in the domain of the SCR. If we now replace Axiom L by the axiom of minimal liberty, then it turns out that in *every* situation where the Sen paradox is rescued by Gibbardian rights-waiving there must exist involuntary waiving of rights.

Let $\hat{D}_1 = \{(x, y), (y, x)\}$ and $\hat{D}_2 = \{(w, z), (z, w)\}$; that is, $\hat{D}_1$ and $\hat{D}_2$ constitute the natural rights system. Recall that Sen's original paradox arises if there exists a sequence $(z_1, \cdots, z_r)$ in X such that

$$(z_t, z_{t+1}) \in \left[\bigcap_{i \in N} P(R_i)\right] \cup [\hat{D}_1 \cap P(R_1)] \cup [\hat{D}_2 \cap P(R_2)], \tag{9.3}$$

where t belongs to a modular number system (mod r).[9] In the ensuing discussion all the subscripts t, t', t'' should be treated as belonging to such a number system.

When R is such that a cycle of the above kind occurs, we shall refer to it as a *potential paradox situation*. Gibbard showed that, once individuals are allowed to waive their rights and given that they do so in a particular way (namely, as specified by the Gibbardian waiver profile, the paradox vanishes.

Assume that given a certain $R \in \mathcal{R}$, $(z_1, \cdots, z_r)$ constitutes a potential paradox situation. This implies that there must exist t' and t'' such that

$$(z_{t'}, z_{t'+1}) \in \hat{D}_1 P(R_1) \tag{9.4}$$

and

$$(z_{t''}, z_{t''+1}) \in \hat{D}_2 P(R_2) \tag{9.5}$$

In other words, at least two steps in the cycle must be based on rights. If not then at least one individual's preference must be intransitive. This is easy to see. Suppose (9.5) is false. Then (9.3) and (9.4) imply that, for all t, it is the case that $(z_t, z_{t+1}) \in P(R_1)$, which implies intransitivity in 1's preference.

So (9.4) and (9.5) must be true. Since x, y, z, and w are distinct, neither $z_{t'+1} = z_{t''}$ nor $z_{t''+1} = z_{t'}$. This implies, given (9.3),

$$(z_{t'+1}, z_{t'+2}) \in \bigcap_{i \in N} P(R_i)$$

and

$$(z_{t''+1}, z_{t''+2}) \in \bigcap_{i \in N} P(R_i)$$

From this it follows, given Gibbardian rights-waiving, that both individuals will waive their rights.[10] Then, given a decisive SCR satisfying the Pareto criterion and the liberty criterion, the only elements that can be chosen from the set $\{z_1, \cdots, z_r\}$ are $z_{t'+1}$ and $z_{t''+1}$. This is because by the Pareto criterion all other elements are ruled out, and since all rights have been waived, the liberty axiom is inconsequential.

Without loss of generality assume that $z_{t'+1}$ is the chosen element.[11] If individual 1 had not waived his right, then $z_{t'+1}$ could not have been chosen. Hence the chosen set would have to be $z_{t''+1}$ or ϕ. Since individual 2 has rights over $(z_{t''}, z_{t''+1})$, the sequence from $z_{t''+1}$ to $z_{t'+1}$ that is $z_{t''+1}, z_{t''+2}, \cdots, z_{t'}, z_{t'+1}$ must be linked by individual 1's rights or the Pareto criterion. Hence by the transitivity of individual 1's preference $(z_{t''+1}, z_{t'+1}) \in P(R_1)$. Hence, either (i) or (ii) of definition (9.1) is violated. Since we had started from an arbitrary potential paradox situation, this establishes the disturbing proposition that, given minimal libertarianism, in *every* situation that is a potential paradox situation, the Gibbardian solution *necessarily* involves involuntary rights-waiving.

NOTES

1. In this chapter, attention is restricted to a 'pragmatic' interpretation of the 'alienable rights system' as opposed to an 'ethical' one (see Sen 1976).
2. This is not meant in the sense of strategic manipulation as in Karni (1978).
3. While this paper allows for strategic behaviour, it should not be confused with Gardner's (1980) analysis of strategic consistency. In Gardner's paper, an agent goes in for strategic behaviour regarding which preference to reveal—the real one or a false one. In the present paper, it is assumed that real preferences are known. Strategic behaviour comes in at the stage of meta-rights, i.e., in the decision as to whether to waive a right or not. There is no 'dishonesty' problem here. Thus the issues raised in this paper would persist even if everybody's mind were known. To that extent the present paper, though explicitly concerned about pragmatic issues, highlights normative difficulties as well.
4. Of course social states may differ in 'nonpersonal' characteristics as well, and in a more complete description these ought to be brought in (see Suzumura 1980), but for the present purpose it is adequate to treat all 'nonpersonal' characteristics as constant and therefore irrelevant.
5. Formally, $D_i = \{(x, y) \in X \times X \mid x_j = y_j, \text{ for all } j \neq i\}$.
6. Note that $(x, y) \in D_i$ implies $(y, x) \in D_i$.
7. This should be more correctly written as $W_1^K = \{(a, b), (b, a)\}$, since rights over pairs are always symmetric.
8. We may ignore here the fact that social states are elements of the Cartesian product of personal characteristics, and treat them as primitives instead.
9. A modular number system (mod r) consists of r distinct numbers $1, \cdots, r$. All other integers are treated as equal to one of these numbers by the following rule. Let n be any integer and let $n = a + kr$ where $a \in \{1, \cdots, r\}$ and k is an integer (note that for each n there is a unique such a). Then we say $n = a$, or more correctly $n = a$ (mod r). This number system has the advantage that $r + 1 = 1$. Hence, when we want to refer to (x_r, x_1), we may write, instead, (x_r, x_{r+1}). This obviously simplifies notation when discussing cycles.
10. This is proved as follows. Consider the sequence from to $z_{t'+2}$ to $z_{t'}$, i.e.,

$$z_{t'+2} \, z_{t'+3} \cdots z_{t'-1} z_{t'}.$$

From (9.3) it follows that each adjacent pair in this belongs to either of the three:

(i) $\underset{i \in N}{\cap} P(R_i)$; (ii) $\hat{D}_1 \cap P(R_1)$; (iii) $\hat{D}_2 \cap P(R_2)$

(iii) is ruled out by (9.4) and by the fact that each individual has rights over only one pair. Hence if 1 claims his right, then by the Pareto axiom and individual 2's rights we can go all the way up to $z_{i'+2}$. But $(z_{i'+1}, z_{i'+2}) \in P(R_1)$. Hence, in accordance with the definition of the GWP, individual 1 waives his rights over $(z_{i'}, z_{i'+1})$. A similar proof is possible for 2.

11. A symmetric proof of involuntary rights-waiving on the part of individual 2 is possible if the chosen element is $z_{i''+1}$.

REFERENCES

Farrell, M. J., 1976, 'Liberalism in the theory of social choice', *Review of Economic Studies*, 43, 3–10.

Gardner, R., 1980, 'The strategic inconsistency of Paretian liberalism', *Public Choice*, 35, 241–52.

Gibbard, A., 1974, 'A Pareto-consistent libertarian claim', *Journal of Economic Theory*, 7, 388–410.

Karni, E., 1978, 'Collective rationality, unanimity and liberal ethics', *Review of Economic Studies*, 45, 571–4.

Kelly, J. S., 1976, 'Rights-exercising and a Pareto-consistent libertarian claim', *Journal of Economic Theory*, 13, 138–53.

Sen, A. K., 1970, *Collective Choice and Social Welfare*, Oliver and Boyd, Edinburgh.

—— 1976, 'Liberty, unanimity and rights', *Economica*, 43, 217–45.

—— 1983, 'Liberty and social choice', *Journal of Philosophy*, 80, 5–28.

Suzumura, K., 1978, 'On the consistency of libertarian claims', *Review of Economic Studies*, 45, 329–42.

—— 1980, 'Liberal paradox and the voluntary exchange of rights-exercising', *Journal of Economic Theory*, 22, 407–22.

10 The Economics and Law of Sexual Harassment in the Workplace

Some years ago, a marshland (subsequently named Salt Lake) adjoining the city of Calcutta was developed by the local government and sold as small plots, at a subsidized price, to people who were not rich and who might not otherwise have been able to afford their own property. Politicians masterminding the plan worried that, unless some special precautions were taken, the rich would soon buy the land from the original owners, thereby 'depriving' them of their plots. Thus, a law was announced which prohibited the sale of land by the original owners in Salt Lake. When I tell economists about this law, they usually laugh at the folly of politicians. If an owner wants to sell land, it must be that the owner expects to be better off by doing so; it is thus hardly a favour not to allow the sale.

However, economists do not laugh when it is pointed out that under current US law, a firm cannot offer a job contract in which the pay is high and the benefits good—but the employer reserves the right to sexually harass the worker. The fact that the worker who accepts this job must find the cost of sexual harassment to be less than the benefits associated with the job does not seem a reason enough to allow such a contract. Though in some ways the two examples may appear similar—both involve two adults choosing to make an exchange that seems to

In writing this paper, I have benefited from conversations with and comments from Heather Antecol, Abhijit Banerjee, Alaka Basu, Stephen Coate, Stanley Engerman, Francisco Ferreira, Avik Ghosh, Gayatri Koolwal, Glenn Loury, Ghazala Mansuri, Mandar Oak, John Roemer, Debra Satz, Amartya Sen, Timothy Taylor and Michael Waldman and from seminar presentations at Cornell University's Department of Economics, Boston University's Institute on Race and Social Divisions, the MIT-Harvard Economic Theory Workshop and the World Bank. I would like to record my indebtedness to Ernest Haffner, Jennifer Kaplan, John Schmelzer and Patrick Ronald Edwards of the office of the Equal Employment Opportunity Commission in Washington, D.C., for their advice and help.

First published in Journal of Economic Perspectives, Vol. 17 (2003).

have no obvious negative externality on others—most people perceive some crucial difference. However, when it comes to enunciating just what the difference is, we often make hand-waving references to how some exchanges are 'obnoxious' or some contracts 'unconscionable'.

This chapter seeks to spell out an economic principle of why certain kinds of contracts, such as one where a worker is subjected to sexual harassment, may need to be legally banned. I begin with some empirical context, discussing the magnitude of sexual harassment in the workplace and the evolution of the law. The focus will be on US experience, since the United States has played a pioneering role in curbing workplace harassment and the American law has been a model to many nations that have recently drafted sexual harassments laws (like Bangladesh) or are in the process of drafting such a law (like India). I then offer a theoretical model to explain why society may desire legislative intervention to control sexual harassment in the workplace. It is important to unearth the underlying *principle* for such rules, since it can influence a host of labour market policies, such as stopping workers from being exposed to excessive health hazards and having statutory limits on the hours of work. This economic approach is then used to critique the current law and government policy, both with regard to sexual harassment and in other matters of labour rights and standards.

10.1 THE CONTEXT

In the United States, charges of sexual harassment are usually handled under Title VII of the Civil Rights Act of 1964, which prohibits employment discrimination on the basis of race, colour, religion, sex, or national origin. The Equal Employment Opportunity Commission (EEOC), which went into effect in 1965, was established to enforce and administer this statute.

10.1.1 The Incidence of Sexual Harassment

In 2001, 15,475 cases of sexual harassment were filed in the United States, shown in Table 10.1. This total combines charges filed with the EEOC and various state and local level Fair Employment Practices Agencies. Since all cases that end up in the courts are filed with the EEOC or one of the state and local agencies, this total offers a fair measure of the number of charges that are serious enough to enter the legal process. The total number of sexual harassment charges rose sharply from 1992 to 1995, and has remained at roughly the same level since then. This was probably a lagged effect of the confirmation hearing of Clarence Thomas for the US Supreme Court, which brought high visibility to issues of sexual harassment in the workplace and also the passage of the Civil Rights Act of 1991. This Act, which became effective in November 1991, allowed victims to claim compensatory damages for suffering caused by discrimination. Prior to the 1991 act, the only forms of relief available were 'injunctive relief'—a court injunction that the discriminatory act be stopped—or, in the most egregious cases, that the harassee, who may have been denied promotion or dismissed, get back pay.

Table 10.1 also presents the number of cases 'resolved' each year. (These figures

Table 10.1: Sexual Harassment Charges Filed with Equal Employment Opportunity Commission and Fair Employment Practice Agencies, 1992–2001

	1992	1993	1994	1995	1996	1997	1998	1999	2000	2001
Total charges	10,532	11.908	14,420	15,549	15,342	15,889	15,618	15,222	15,836	15,475
Charges filed by men (%)	9.1	9.1	9.9	9.9	10.0	11.6	12.9	12.1	13.6	13.7
Resolutions	7,484	9,971	11,478	13,802	15,861	17,333	17,115	16,524	16,726	16,383
Merit resolutions	2,019	2,524	2,713	2,709	2,882	3,253	3,576	3,840	4,724	4,768
Total charges where the basis was the person's sex (% of total charges)	96.7	96.6	97.6	96.9	97.4	97.1	97.1	97.0	97.1	97.5

Source: Equal Employment Opportunity Commission, National Data Base.

can exceed the number of charges filed in any particular year because of overhang of charges from previous years.) A lack of resolution does not mean that the case did not have merit since many cases are closed for administrative reasons, such as failure of the charging party to respond to EEOC communications. Of the cases that are resolved, of particular interest are the 'merit resolutions', because in these cases the allegations seemed to have enough merit that the charges led to outcomes favourable to the charging parties.

The final row of the column shows cases where the harassment may be sexual in nature but the *basis* of the harassment is not the plaintiff's sex, but some other attribute, such as the person's race or religion. Hence, in 2001, 2.5 per cent of sexual harassment charges involved cases where the basis of harassment was not the plaintiff's sex.

Of course, Table 10.1 should not be treated as a measure of the actual incidence of sexual harassment in the workplace. The table does not include charges that do not reach the EEOC and the state and local level Fair Employment Practices Agencies, because they are filed and resolved within firms and corporations. Moreover, there is surely a large number of cases where the victim does not file charges at all. On the other hand, there must be some bloating of numbers caused by false charges.

A few sector-specific studies offer some sense of the percentage of people who face sexual harassment in the workplace. The largest study was based on a survey of active-duty women in the US Armed Forces in 1995. A questionnaire was sent to 49,003 individuals and there were 28,296 returns, of which 22,372 were from women. According to this study, 70.9 per cent of active-duty women had faced some form of sexually harassing behaviour over the previous one year (Antecol and Cobb-Clark 2002). This figure seems very high, which may reflect the special circumstances of the armed forces or be caused by the survey design.

One of the largest studies from the civilian sector in the United States was conducted by the US Merit Systems Protection Board in 1980, 1987, and 1994 among federal employees all over the country (USMSPB 1995). In 1994, a questionnaire was sent out to nearly 13,200 federal employees and about 8,000 were returned. This study found that 44 per cent of the women employees and 19 per cent of the male employees had faced some form of sexual harassment in the previous two years. These numbers were only slightly higher than in the previous surveys. In a similar survey in 1987, 42 per cent of women employees and 14 per cent of male employees claimed to have faced some form of sexual harassment; still further back in a 1980 survey, the figures were 42 per cent of women employees and 15 per cent of male employees. The study also found that only 6 per cent of those who faced harassment actually lodged a complaint (and, no doubt, many fewer filed a legal charge).

Since sexual harassment is notoriously difficult to define, self-reported surveys like these have to be treated with caution and more work remains to be done in measuring the extent of sexual harassment (Welsh 1999). For example, it is a plausible but unproven hypothesis that the quantity of sexual harassment in the workplace has declined in recent years and decades, but that an increasing awareness and empowerment of women workers has led to a greater percentage of cases being reported in surveys. The statistics we observe are a mixture of these countervailing forces. But despite the uncertainties that surround estimates of the extent of sexual harassment, the figures are sufficient to indicate that actual charges reported in Table 10.1 are the tip of the iceberg of instances of sexual harassment in the workplace.

10.1.2 The Concept of Sexual Harassment in Law

One reason for the different definitions is that 'sexual harassment' as a legal concept is only about 25 years old. According to Farley (1978), the concept was 'discovered' in 1974 in the course of discussions in a class on women and work in Cornell University (see also Crouch 2001). We have come a long way since then. In the years after the passage of the Civil Rights Act of 1964, federal courts typically refused to view sexual harassment as a form of employment discrimination under the meaning of the statute. However, in the case of *Barnes v. Costle* (561 F.2d 983 [D.C. Cir. 1977]), a federal circuit court held that sexual remarks and solicitations, when linked to threats about being fired, constituted sexual harassment. MacKinnon (1979), who was one of the attorneys for Barnes, then published a pioneering book which argued that sexual harassment itself, with or without a threat of being fired, constituted employment discrimination on the basis of sex in the meaning of the Civil Rights Act of 1964.

In 1980, the EEOC issued 'Guidelines on Discrimination Because of Sex', which declared sexual harassment a violation of Title VII of the Civil Rights Act of 1964. Along with offering some guidelines for establishing criteria for determining when sexual harassment has occurred, these guidelines split sexual harassment into two categories: 'quid pro quo' harassment, whereby a refusal to grant sexual favours were met with blocked promotion or frozen wages or outright dismissal from work; and 'hostile environment' harassment, which took the form of sexually abusive language or gestures which made some workers feel humiliated and discriminated against.[1]

The courts soon followed the lead of the EEOC. In the case of *Bundy v. Jackson* in 1981 (641 F.2d 934, 942 n. 7 [D.C. Cir. 1981]), a hostile environment alone was for the first time recognized by a federal appeals court as a form of harassment. In *Bundy*, a female employee of the Department of Corrections in Washington, DC, was repeatedly invited by her supervisor to describe her sexual experience. When she complained about these comments to a senior manager, he took it lightly, saying that the feelings of the supervisor were understandable. The court upheld Bundy's charge that the innuendo and implicit threats created an intimidating and hostile atmosphere, and were unlawful, *even though she had not suffered any tangible loss*, such as the withholding of salary increments or promotion.[2]

In 1986, the US Supreme Court affirmed this distinction in *Vinson v. Meritor Bank* case (477 US 57 [1986]), in which Michelle Vinson, a trainee-teller, was repeatedly propositioned by Sidney Taylor, a vice president of the bank. After resisting for some time, she relented for fear of losing her job and was subjected to repeated unwanted sexual relations for over four years. In this case the court did not find that the worker had suffered in terms of pay or promotions; in fact, the court did not even find it necessary to decide whether a sexual relationship between worker and manager had happened at all. The court held that a hostile work environment alone was a violation of employment discrimination under the Civil Rights Act of 1964.

10.2 A MODEL OF SEXUAL HARASSMENT IN THE WORKPLACE

In building a model of labour markets for analysing sexual harassment, I want to motivate the exercise with a conceptual puzzle. One of the basic principles used by economists to guide policy decisions is the 'principle of free contract', which asserts that when two or more consenting adults agree to a contract or an exchange that has no negative externalities on uninvolved individuals, then government has no reason to intervene and prohibit such a transaction.[3] This principle is the product of two more fundamental ideas: the Pareto principle and consumer sovereignty. The Pareto principle asserts that if a change is such that at least one party is better off and others are no worse off, then that change is desirable. Consumer sovereignty asserts that each (adult) individual is the arbiter of that individual's own welfare.

Now consider a case where a firm, either by virtue of its reputation for harassment or by writing down an explicit contract, ensures that a potential employee knows that she will be sexually harassed on the job. If she nevertheless accepts the job, then, by the principle of free contract, there seems to be no economic case for stopping such a contract. From this it seems a short step to argue that government should not use the law to stop sexual harassment in the workplace. It should be left to the individuals involved to be worked into the terms of employment contract appropriately. Given the heterogeneity among human beings, some will agree to work for lower wages but want a guarantee of no harassment; others may prefer higher wages, while relinquishing the right not to be harassed.[4]

However, I will argue that what is described as a 'short step' in the above paragraph is deductively invalid, because there is a difference between a single contract or a small number of contracts of a particular kind and a large number of contracts of that same kind. The former may be morally permissible, but the latter not. Even if single contracts are Pareto-improving, we cannot automatically conclude that such contracts in general should be legally permitted. I shall call this the 'large numbers argument' and begin by constructing a model that validates this insight.[5] The model consists of a straightforward adaptation of the standard demand–supply model of labour.

10.2.1 A Model of Wages With and Without a Ban on Sexual Harassment

Consider a market in which every firm and every worker is a price taker. I shall assume that employers are the potential harassers and the workers the potential harassees. I am therefore ruling out the problem of one worker (for instance, a supervisor) harassing another worker (for instance, a trainee). In reality such harassments occur, and under the current US law the employer has 'vicarious liability' and can be held responsible for not having taken adequate measures to prevent such harassment.

Assume that we are in a legal regime where firms are allowed to harass workers as long as that possibility is made clear at the time of employment. Treating harassment as a zero-one concept, it is clear that at most two kinds of contracts will come to prevail—one where the owner retains the right to harass his workers (contract H) or one where the owner guarantees no harassment (contract N). Hence, at most two wages will come to prevail: w_H for jobs with harassment and w_N for jobs with no harassment Let θ be the amount of benefit or 'perverse gratification', measured in output units, that the employer gets from being able to harass an employee. Since firms are free to offer N or H contracts, the following condition must hold:

$$w_H - \theta = w_N.$$

Note that in equilibrium, firms will be indifferent between the two kinds of contracts, because the gratification of harassment is just offset by the higher wage that has to be paid. I shall use $D(w_N)$ to denote the aggregate demand for labour by all firms when the wage associated with a no-harassment job is w_N.

Let us now turn to the workers. They find harassment painful; however, the extent of pain will differ across individuals. Measuring pain in money terms (i.e., in the same units as wages), let us think of c_i as the pain of harassment as perceived by worker i, and use c_{max} and c_{min} to denote the pain levels felt by, respectively, the worker who finds harassment the most painful and the worker who finds it the least painful. The interesting case occurs when

$$c_{max} > \theta > c_{min},$$

and, in what follows, I will assume this condition holds true. It means that there exists at least one worker whose pain exceeds the gratification the employer gets from harassing and at least one worker whose pain is less than the gratification.

This condition ensures that both H and N contracts will occur in the market if the law permits harassment contracts. That is, some workers can be persuaded to accept harassment contracts by virtue of the higher wage, but not all workers. This condition seems empirically plausible and, in the absence of more evidence, the natural assumption to use.

Before proceeding further, it is worthwhile commenting on some indirect empirical evidence which supports the assumptions of this model. Of course, there is harassment in the workplace even though it is prohibited by the law. Antecol and Cobb-Clark (2002) in their study of active-duty personnel in the US found strong evidence that sexual harassment led to diminished job satisfaction— a fact for which there is a lot of evidence from studies in psychology (for instance, Schneider, Swan, and Fitzgerald 1997). On the other hand, carefully constructed bivariate probit analysis shows that sexually harassing behaviour does not lead to a heightened desire to quit military employment. While Antecol and Cobb-Clark present this finding as a bit of a puzzle, if the women went into this job knowing that they would face harassment, then the finding of diminished job satisfaction but no accompanying desire to quit the job are perfectly compatible, and in keeping with this theoretical model.

Concerning labour supply we will make the usual assumptions. Each worker's labour supply, s, is an upward-sloping function of the net wage that she earns. If worker i chooses an N contract, the amount of labour supplied by her is given by $s(w_N)$, and if she chooses an H contract, the supply is $s(w_H - c_i)$. Recall that though the latter worker gets a wage of w_H, her net wage is less because she has to deduct the cost of harassment from it.

Equilibrium in this labour market is defined in the usual way: The wage rates, w_N and w_H, associated with the N and H contracts, respectively, constitute an *equilibrium* if, given these wages, for each type of contract the demand for labour equals supply. I shall denote the equilibrium wages by w_N^* and w_H^*.

Consider next a labour market in which there is a law prohibiting sexual harassment (and this law is fully enforced). In such a legal regime there will be only one equilibrium wage, equilibrium being defined as the wage rate for which aggregate demand equals supply. Let me denote the equilibrium wage by w^*. What is easy to prove, and of central interest here, is that the equilibrium wage, when harassment is illegal, will exceed the equilibrium wage, associated with a no-harassment job in a regime which permits harassment (Basu 2000); in the terms of the model, $w_N^* < w^*$.

The intuition behind this result is straightforward. Consider the legal regime where no harassment is permitted and the wage is w^*. Now suppose government revokes the law banning sexual harassment. Assume for a moment that, in this new legal regime, the wage that is paid for no-harassment jobs, namely, w_N, happens to be equal to w^*. In that case the total demand for labour in this new regime will be equal to the demand for labour in the old regime, that is, $D(w^*)$. Remember that, given w_N, we know that w_H will be equal to $w_N + \theta$ and employers are indifferent between the two kinds of contracts. What will labour supply be in this new regime? All those who choose no-harassment contracts will supply the same amount of labour as in the old regime, since $w^* = w_N$. Consider now a person i who chooses a harassment contract. The reason for such a choice must be that the wage they receive, after subtracting their personal costs of harassment $w_H - c_i$, exceeds w_N. But this means that a person who makes such a choice faces a higher net wage than in the old regime. Hence, all such people will supply more labour than in the old regime. This means that aggregate supply of labour exceeds aggregate demand in the new regime. Hence, the wage w_N, described above, cannot be an equilibrium. In particular, the equilibrium wage for no-harassment jobs will have to be lower, so that $w_N^* < w^*$. It follows from this result that all those people who take up jobs with no-harassment guaranteed, in a regime where there is no ban on harassment, are better off when the economy switches over to a regime in which harassment is prohibited outright by the law.

Indeed, it also turns out that many workers who, in a regime with no ban on harassment, choose jobs with harassment clauses would also prefer a regime where no-harassment is permitted. That is, there will exist workers, who, when confronted with a choice between the low-wage job with no-harassment and the high-wage job with possible harassment, prefer the latter, that is, $w_H^* - c_i > w_N^*$. However, compared to both of these choices, they would prefer a job with the intermediate wage that would come to prevail in an economy where there is an outright ban on sexual harassment, that is $w^* > w_H^* - c_i > w_N^*$. Thus, a ban on sexual harassment benefits those who would otherwise have chosen the no-harassment contract and also a number of those who would have chosen the harassment-allowed contract.[6]

Hence, even if a single pair of agents entering a harassment contract is Pareto-improving, prohibiting sexual harassment in general does not result in a state that is Pareto-inferior to the one that would occur in the absence of such a law, but rather involves a set of tradeoffs where some groups benefit and others suffer. This outcome does not mean that we already have a case *for* banning harassment, but simply that the case against a ban, on purely Paretian grounds, no longer exists.

10.2.2 Moral Judgement

To go from here to a case for a ban one needs to combine the above positive result with some normative conditions. I shall argue that, once we are in the domain where alternatives cannot be ranked purely on Paretian grounds, there is a case for breaking away from welfarism and looking instead at what lies *beneath* people's preferences.

We often hold views of approval or disapproval about other people's preferences. Thus, we may consider a person's preference for racism unacceptable, whereas a person's preference for alcohol (though not good for his health) fine. The argument here hinges on distinguishing between two kinds of preferences that we consider acceptable or legitimate: maintainable preferences and inviolable preferences. A particular preference is *maintainable* if a person has the right to the preference, even though the person may have to pay a price for having such a preference. On the other hand, an *inviolable* preference is something that a person has a right to have and, in addition, no one should have to pay a price for it (Basu 2000 elaborates further). For instance, a person's preference for working only two days a week is a maintainable preference. No one can object to it. Of course, the person with that preference will be poorer and cannot expect society to make amends for it, but he has a right to that preference. On the other hand, the preference not to be sexually harassed, we can argue, is inviolable. Not only do people have a right to such a preference but no one should have to pay a price for having such a preference. What is being claimed here is that this normative evaluation of preferences should come into play when there is no ranking based on Paretian grounds.

The categorization of preferences completes the argument. If a person is forced to choose a low-paying job because she has a strong aversion to being sexually harassed (as we showed above), we may have reason to take legal action so that this does not happen. What is being argued for is a blending of consequentialism and rights or, more correctly, the moral worth of preferences.[7]

It is important to note that the case for banning sexual harassment is being justified here only partly by the suffering of those who get harassed. The model draws attention to all those who do worse in the labour market when a ban on sexual harassment does not exist, a matter of some policy significance and one to which I will return later.

Before moving on to policy matters, let me briefly address a theoretical question that arises from the above model: How is it possible for each single contract to be Pareto-improving but a large set of such contracts to be not so? The way this was established above was by a standard competitive equilibrium argument, in which each agent is assumed to be a price taker. Hence *one* more worker signing a harassment contract is assumed not to affect wages, and, therefore, the welfare of other workers. On the other hand, when a large number of agents sign such contracts, this affects wages and, through that, can have an adverse effect on the welfare of other workers. Though this point is not always made explicit in standard economics texts, this kind of competitive equilibrium analysis turns out to be *formally* valid only in economies with an infinite number of agents. Is there any way of justifying this within the more realistic paradigm of a finite number of agents? This question, which is also one of the core concerns of Parfit's (1984) much-discussed work in philosophy, is pursued further in Basu (2002). It is shown, for instance, that one way to make this 'large numbers argument' formally consistent is to assume that human beings do not have endlessly fine perception. That is, they cannot distinguish between very fine differences. Such an argument is based on recognizing that individual preferences are typically not transitive. This seems strange at first sight because the assumption of transitivity is so familiar to economics. Yet it is realistic to assert that an individual will be indifferent between n and n+1 grains of sugar, for every integer n, but may well prefer one spoon of sugar to three spoons. This clearly implies a violation of transitivity.[8]

10.3 SEPARATING SEXUAL HARASSMENT FROM SEX DISCRIMINATION

The economic perspective on sexual harassment legislation, developed above, comes with its own suggestions for the kind of law that ought to be used to control harassment in the workplace. This perspective, which is generally ignored in the large literature and many legislative debates on workplace harassment, has a lot to offer.

The key to the US legislation regarding sexual harassment, as discussed earlier, has been to view it as a form of sex discrimination in employment (Crouch 2001; LeMoncheck and Sterba 2001; Mackinnon 1979; Schultz 1998). This American model has influenced legislation worldwide; Husbands (1992) presents a comprehensive account of sexual harassment law in different nations. For example, Britain's Sex Discrimination Act of 1975, which closely parallels Title VII in the United States, recognizes sex discrimination to be unlawful and also that discrimination occurs if a person treats a woman less favourably than he treats or would treat a man. In India, sexual harassment is for the first time being recognized as unlawful. India does not have a special law against sexual harassment and has had to be content with the use of tort and criminal law to deal with the problem (Nussbaum 2004). However, a recent writ of mandamus (a Court order that serves as law till formal legislation is adopted) issued by India's Supreme Court in the context of a judgement in a rape case, *Vishakha and Others v the State of Rajasthan and Others, 1997* (6 Indian Supreme Court 241 [1997]), provides guidelines for controlling and punishing harassment in the workplace. This well-crafted writ draws on both the Indian constitution and American legislative practices, as it stresses the discriminatory aspect of harassment and the principle of vicarious liability (Haspels et al. 2001), mentioned above.

However, I shall argue that this tying up of sexual harassment with sex discrimination, though it has played an important role historically, is now becoming a hindrance. There should be strong laws to prevent discrimination and strong laws to prevent harassment. But it would be unfortunate if the only way to establish sexual harassment is to categorize it as a form of discrimination, because this approach raises a number of problems (Abrams 1994; Hajdin 2002). Employment discrimination by sex has traditionally meant men discriminating against women; for example, consider Farley's (1978) definition: 'Sexual harassment is best described as unsolicited nonreciprocal male behavior that asserts a woman's sex role over her function as worker'. But sexual harassment is a more complex topic.

First, men's claims of sexual harassment are increasing; from the EEOC data reported in Table 10.1, men's claims now account for 13.7 per cent of all sexual harassment charges, up from 9.1 per cent in 1992. The coupling of sexual harassment with sex discrimination does not work neatly to protect people in such cases.

Second, in the United States there is a significant amount of same-sex harassment, another situation in which law based on discrimination according to sex often does not provide adequate protection to sufferers—and when it does, it is only because judges and lawyers interpret the law according to its likely intent rather than what it actually says (Talbot 2002). According to the study by the US Merit Systems Protection Board (USMSPB 1995), 21 per cent of men who report being harassed were harassed by other men. However, in 1998 the US Supreme Court cleared up some confusion in lower courts by ruling in *Oncale v. Sundowner*

Offshore Services, Inc., (523 US 75 [1998]), in a case involving a man who experienced a hostile working environment as a result of harassment by other men working on an oil rig in the Gulf of Mexico, that Title VII should be viewed as including same-sex harassment.

A third category which is difficult within the sex discrimination framework involves the problem of the boss who harasses both men and women with equal vigour and thus does not harass anybody because of his or her sex (Paul 1990; Epstein 1985).

A final category is those who are harassed not because of their sex, but because of their sexual orientation. This problem was exemplified by a case concerning Medina Rene, an openly gay person who worked as a butler at the MGM Grand Hotel, Las Vegas, and for years was harassed by co-workers. His complaint to the hotel resulted in no action and so he took the matter to the court. A US district court ruled that Rene's harassment was not based on his sex but on his sexual orientation and so was not covered under Title VII (Abelson 2001). However, the ruling was overturned on September 24, 2002 by the US Court of Appeals of the Ninth Circuit, which ruled in *Rene v. MGM Grand Hotel, Inc.* (No. 98-16924) that Rene could bring suit (Talbot 2002, p. 57; decision online at <http://www.ca9.uscourts.gov>).

The linkage from sexual harassment to employment discrimination law occurred through a set of understandable historical connections. The original sexual harassment cases in the late 1960s and early 1970s focussed on 'quid pro quo' harassment in which men threatened women with loss of job or reduction in pay unless the women participated in sexual activity. But these situations left out the issues raised by 'hostile environment' sexual harassment. As Lipper (1992, p. 301) pointed out, in the past, 'while a woman who had been physically assaulted, i.e. grabbed, touched or kissed, might have prevailed under tort theories, one who had been the object of sexual jokes would be unlikely to be compensated for her resulting anguish'. Such situations often had no clear injuries from the viewpoint of traditional employment discrimination law, and so they had historically been ignored and victims left without legal recourse.

The advantage of treating a hostile work environment as workplace discrimination was that it provided a preexisting legal avenue—to wit, laws against discrimination—for addressing the problem. But once a hostile work environment has come to be recognized as wrong, it is not clear why it should not be considered wrong *per se*, that is, whether or not it can be classified as workplace discrimination. Society may wish to have separate legal provisions for harassment that is motivated by discrimination, because this phenomenon is pervasive, and also because we may want to punish both the harassment and the discrimination. But it seems odd to confine the scope of bringing harassment charges only to cases where the harassment is prompted by discrimination.

In an influential paper, Superson (1993) defended the connection between harassment and discrimination by arguing that sexual harassment should be viewed as 'an attack on the group of *all* women, not just the immediate victim' (p. 49). One problem with this argument is that it is not clear why an attack on one, say, black, gay, woman is to be viewed as an attack on all *women*, and not all blacks, all gays, or, for that matter, all moral beings? A second problem is that sexual harassment may not always arise from a feeling that the victim is inferior, as this argument suggests. It seems more sensible to categorize harassment by the effect on the harassee, no matter what motivates the harasser. However, in Superson's emphasis on discussion of social mechanisms whereby one person's victimization affects others, her argument is very much in the spirit of this chapter, which emphasizes how such effects can spread through the workings of the market.

Indeed, the economic model of sexual harassment presented here suggests that even if a harassment-allowing contract was freely accepted by both sides, society would have reason to ban harassment because of the costs that it imposes on those who choose the no-harassment contract. This economic approach suggests that the existing interpretation of sexual harassment may not be going far enough. The existing legal provisions are not cognizant of the losses of those who are not harassed because they may have taken otherwise-inferior jobs, where there is the assurance of no-harassment, or have remained unemployed. Such workers cannot seek compensation under existing sexual harassment laws, because they are not harassed physically or even environmentally. But they nevertheless pay a price. Hence, perhaps surprisingly, the economic approach takes us to a more wide-spread interpretation of what constitutes the harm of sexual harassment.

Finally, it is worth noting that in the above model it was assumed that sexual harassment is well-defined and can be the subject of enforceable contracts. It is possible, however, to argue that it is difficult to write contracts about harassment, and try to construct an argument for a legal ban on the basis of this difficulty. While this is a line that needs to be explored, one must not prejudge its conclusions. Even if complex contracts are not possible, it is often possible to write simple contracts, such as, 'I relinquish all rights that I have under Title VII'. If the rights under Title VII are meaningful, then relinquishing those rights must be meaningful. Hence, in principle, it is possible to grant individuals the right to give up those rights. But the fact that people may want to use more nuanced contracts about what workplace conduct is acceptable, which in turn may be harder to monitor and enforce, does lead to open questions deserving future investigation.

10.4 REFLECTIONS ON LABOUR STANDARDS AND RIGHTS

The approach to analysing sexual harassment in this chapter provides an instrument for analysing other kinds of labour market problems, such as occupational safety, child labour, and labour rights and standards in general. In all of these cases, it is possible to compare an across-the-board government regulation with a market in which workers sort themselves into different jobs according to their preferred combinations of pay and work environment.

Consider the problem of hazardous jobs. Should workers be allowed to opt for such jobs, on the argument that some workers may find poverty more grueling than the pain of such a job? In answering or examining this question, the focus is invariably on the workers who take up such jobs; for instance, Cohen's (1987) thought-provoking paper on this subject has this focus. I am arguing that in considering the general question of whether such jobs should be allowed or whether firms should be compelled by law to take safety measures, a crucial constituency is not the workers who currently have such jobs, but those, especially poor workers, who do not. Moreover, one needs to consider the constituency of those who will choose a high-wage hazardous job over a low-wage safe job, but who would prefer a medium-wage safe job.[9]

Of course, the arguments presented here should not be taken as a defense of all labour market interventions. One hugely contentious debate on labour market policy in recent times concerns international labour standards (for instance, Basu 2001; Bhagwati 1995; Engerman 2003; Krueger 1997; Satz 2003). Is there reason for international authorities to try to enforce certain minimal standards in labour markets, such as prohibiting child labour, or stopping workers from doing hazardous work, or imposing a minimum wage, or allowing workers to unionize? On some of these matters we may reach a straightforward conclusion on the basis of the standard externality or multiple equilibria arguments. For example, in the case of child labour, Basu (1999) argues that there are two possible equilibria, a low-wage equilibrium with child labour and a preferable high-wage equilibrium where children attend school, and so, if an economy happens to be caught in the less-desirable equilibrium, a natural case arises for prohibiting the behaviour that occurs in that equilibrium so that the economy is deflected to the other equilibrium. But there are cases where there may not be more than one equilibrium and the externality principle may not work, at least not without stretching the principle beyond recognition. In such cases, it will be worthwhile trying to apply the large-numbers argument along with the normative criterion discussed above.

As a particular example, consider what may be called the 'maquiladora dilemma'. A maquiladora, or an export-processing zone is an area often on the border of a country where goods are often imported, processed in some way, and then shipped back across the border, free of export duties. In some countries, if a worker wants to work in a maquiladora, that worker is required to give up certain labour rights that are guaranteed in all other sectors of the economy. Many labour rights advocates argue against such exemptions (ILO 2000; and the proceedings of an ILO meeting at <http://www.ilo.org/ public/english/region/asro/bangkok/ download/epz.pdf>). However, if a worker chooses to give up certain rights in order to work in an export-processing zone, it must be because the worker expects to be better off by doing so. On standard Paretian grounds, there seems to be no reason to stop a worker from making such a choice.

The only way to justify such an intervention is to check where the large-numbers argument takes us. That is, find out how the existence of such export-processing zones may affect the wages and work conditions of those who do not work in the zones, and if some of those who work in the zones prefer that the zones be outlawed. It seems clear that the spillovers of sexual harassment reduce the welfare of other workers who may not be direct victims, but it is an open question how the spillovers from export-processing zones affect workers who are not in those zones. Finally, the argument that it is unfair to force workers who prefer not to be sexually harassed into low-wage jobs has considerable moral force; the corresponding argument that it is unfair to force workers who prefer that, say, their job be guaranteed by the government to last in perpetuity into lower-wage jobs has somewhat less moral force. Of course, listing these arguments does not settle the maquiladora question, but it suggests how the analyst might proceed in analysing labour standards.

NOTES

1. Juliano (1992) discusses the history of the 'hostile environment' claim.
2. As a digression, note that the 'hostile environment' clause suggests that what consti-tutes sexual harassment may well have a cultural element to it and so may, reasonably, differ across time or across nations. In the United States, defendants in harassment cases have sometimes tried to use the First Amendment (Schauer 2004), arguing that, for instance, *Playboy* posters in the workplace should be allowed as a form of freedom of expression. Even if we were to contest this argument, most people would recognize that there is a line where not all art or expression that is somehow related to sex or gender and that offends a single person should be grounds for a sexual harassment suit. The point here is that different societies may wish to draw the line in different places.
3. This principle has a long intellectual history and has been subjected to repeat scrutiny. Mill (1848) favoured this principle, though was concerned about some special cases, such as a labourer's right to enter into voluntary slavery. For contemporary discussions in the context of labour markets, see Kanbur (2002); Sunstein (2001); Trebilcock (1993); and Zimmerman (1981).
4. Observe that in this scheme each worker has the right not to be harassed, and in addition they have the right to give up that right. As Sunstein (2001) lucidly demonstrates, giving a person the right to give up a certain right, call it R, is not the same as not giving the right R (see also Basu 1984).
5. An important precursor of the large-numbers argument is Parfit's important work (1984), where he tries to argue that we cannot morally evaluate a *set* of acts on the basis of our moral judgement about each act contained in the set; to think otherwise is to make a mistake in 'moral mathematics' (see also Genicot 2002; Neeman 1999).
6. One effect of harassment that is important in reality but has been ignored here is that it reduces the harassed worker's productivity. In addition, harassment can have substantial spillovers, resulting in the decline of productivity of the co-workers of those who are harassed. The USMSPB (1995) study estimated that, during 1992–4, such 'group productivity losses' because of sexual harassment in the Federal Govern-ment alone was equal to $193.8 million. In terms of the model, this spillover amounts to another reason why wages associated with contracts that guarantee no-harassment (in a regime in which there is no outright ban on harassment) may be lower and the intermediate wage that would prevail in a regime with an outright ban on harassment. A formal modelling of this could be worthwhile in the future.
7. This line of argument is close to Sen (1982), but the particular kind of blending that is being recommended here is different from his.

8. It is arguable that the notion of 'pecuniary externality', which is germane to competitive equilibrium analysis, hinges on such intransitivities.

9. In certain situations, it is possible to have an equilibrium in which firms offer only one kind of contract, where individuals have to relinquish the right not to be harassed, and the worker may take it or leave it. This gives rise to some interesting philosophical questions concerning the moral status of 'standard form' contracts that many large and powerful enterprises use. Wertheimer (1996, chapter 2, pp. 45–6) discusses this problem in the context of *Henningsen v. Bloomfield*, (32 N.J. 358, 161 A.2d 69 (1960), in which Mrs. Henningsen, who was injured while driving her newly purchased Plymouth car, sued for damages, whereas Bloomfield Motors and Chrysler argued that by signing the standard purchase order, Mr. Henningsen, her husband, had absolved Bloomfield Motors and Chrysler of any responsibility. The question was whether by signing beneath the fine print of the *standard form* he could be described as having got into the contract *voluntarily*. It seems arguable and the New Jersey Supreme Court did indeed take the line, that 'freedom of contract' applies most clearly to contracts that are the result of 'free bargaining of parties' rather than the standardized mass contracts used by enterprises with strong bargaining power. What Wertheimer does not discuss is another dimension to this problem which arises from the fact (to which we have been alerted by the new behavioural economics) that human choice is highly dependant on what is presented as the default option. Workers can be made to contradict their own choice by altering what is presented as the standard form.

REFERENCES

Abelson, R., 2001, 'Men, increasingly are the ones claiming sexual harassment by men', *New York Times*, June 10, 1.

Abrams, K., 1994, 'Title VII and the complex female subject', *Michigan Law Review*, August, 92, 2479–540.

Antecol, H. and D. Cobb-Clark, 2002, 'The sexual harassment of female active-duty personnel: Effects on job satisfaction and intentions to remain in the military', Claremont McKenna College (Mimeo).

Basu, K., 1984, 'The right to give up rights', *Economica*, 51, 413–22.

—— 1999, 'Child labour: Cause, consequence and cure with remarks on international labour standards', *Journal of Economic Literature*, September, 37, 1083–119.

—— 2000, *Prelude to Political Economy: A Study of the Social and Political Foundations of Economics*, Oxford University Press, New York and Oxford.

—— 2001, 'Compacts, conventions and codes: Initiatives for higher international labour standards', *Cornell International Law Journal*, 34(3), 487–500.

—— 2002, 'Sexual harassment in the workplace: An economic analysis with implications for worker rights and labour standards policy', Massachusetts Institute of Technology, Department of Economics, Working Paper No. 02–11; available at < http://papers.ssrn.com/abstract_id=303184 >.

Bhagwati, J., 1995, 'Trade liberalization and "fair trade" demands: Addressing the environment and labour standards issues', *World Economy*, 18(6), 745–59.

Cohen, G. A., 1987, 'Are disadvantaged workers who take hazardous jobs forced to take hazardous jobs?', in Gertrude Ezorsky (ed.), *Moral Rights in the Workplace* (chapter 12), SUNY Press, Albany.

Crouch, M., 2001, *Thinking about Sexual Harassment: A Guide for the Perplexed*, Oxford University Press, New York and Oxford.

Engerman, S., 2003, 'The History and Political Economy of International Labour Standards', in Basu et al. (eds.), *International Labor Standards: History, Theories and Policy Options*, Blackwell Publishing, Cambridge, MA.

Epstein, R., 1985, *Takings: Private Property and the Power of Eminent Domain*, Harvard University Press, Cambridge, MA.

Farley, L., 1978, *Sexual Shakedown: The Sexual Harassment of Women on the Job*, McGraw-Hill, New York.

Genicot, G., 2002, 'Bonded labour and serfdom: A paradox of voluntary choice', *Journal of Development Economics*, 67(1), 101–28.

Hajdin, M., 2002, *The Law of Sexual Harassment*, Associated University Presses, London.

Haspels, N., Z.M. Kasim, C. Thomas, and D. Mc Cann, 2001, *Action against Sexual Harassment at Work in Asia and the Pacific*, International Labour Office, Bangkok.

Husbands, R., 1992, 'Sexual harassment law in employment: An international perspective', *International Labour Review*, 131(6), 535–59.

ILO, 2000, *Proceedings of the Tripartite Meeting on Labour Practices in the Footwear, Leather, Textile and Clothing Industries*, International Labour Organization, Geneva.

Juliano, A., 1992, 'Did she ask for it?: The "unwelcome" requirement in sexual harassment cases', *Cornell Law Review*, 77, 1558–92.

Kanbur, R., 2002, 'On obnoxious markets', Cornell University (Mimeo).

Krueger, A., 1997, 'International labor standards and trade', in Michael Bruno and Boris Pleskovic (eds.), *Annual World Bank Conference on Development Economics 1996*, (pp. 281–302), World Bank, Washington.

LeMoncheck, L. and J. Sterba (eds.), 2001, *Sexual Harassment : Issues and Answers*, Oxford University Press, New York and Oxford.

Lipper, N., 1992, 'Sexual harassment in the workplace: A comparative study of Great Britain and the United States', *Comparative Labor Law Journal*, 13, 293–342.

Mackinnon, C., 1979, *Sexual Harassment of Working Women: A Case of Sex Discrimination*, Yale University Press, New Haven.

Mackinnon, C. and R. Siegel (eds.), 2004, *Directions in Sexual Harassment Law*, Yale University Press, New Haven.

Mill, J. S., 1848, *Principles of Political Economy* (All references are to the 1970 edition), Penguin, Harmondsworth.

Neeman, Z., 1999, 'The freedom to contract and the free-rider problem', *Journal of Law, Economics and Organization*, 15(3), 685–703.

Nussbaum, M., 2004, 'The modesty of Mrs. Bajaj: India's problematic route to sexual harassment law', in C. MacKinnon and R. Siegel (eds.), *Directions in Sexual Harassment Law*, Yale University Press, New Haven.

Parfit, D., 1984, *Reasons and Persons*, Clarendon Press, Oxford.

Paul, E. F., 1990, 'Sexual harassment as sex discrimination: A defective paradigm', *Yale Law and Policy Review*, 8, 333–65.

Satz, D., 2003, 'Children at Work', *World Bank Economic Review*, Vol. 17.

Schauer, F., 2004, 'The "speech-ing" of sexual harassment', in C. MacKinnon and R. Siegel (eds.), *Directions in Sexual Harassment Law*, Yale University Press, New Haven.

Schneider, K., S. Swan, and L. Fitzgerald, 1997, 'Job-related and psychological effects of sexual harassment in the workplace: Empirical evidence from two organizations', *Journal of Applied Psychology*, 82(3), 401–15.

Schultz, V., 1998, 'Reconceptualizing sexual harassment', *Yale Law Journal*, 107(6), 1683–806.

Sen, A., 1982, 'Rights and agency', *Philosophy and Public Affairs*, 11(1), 113–32.

Sunstein, C., 2001, 'Human behavior and the law of work', *Virginia Law Review*, 87, 205–76.

Superson, A., 1993, 'A feminist definition of sexual harassment', *Journal of Social Philosophy*, 24(1), 46–64.

Talbot, M., 2002, 'Men behaving badly', *New York Times Magazine*, October 13, p. 52.

Trebilcock, M., 1993, *The Limits of Freedom of Contract*, Harvard University Press, Cambridge, MA.

USMSPB, 1995, 'Sexual harassment in the federal workplace: Trends, progress and continuing challenges', *A Report to the President and the Congress of the United States*, Washington, D.C., US Merit Systems Protection Board.

Welsh, S., 1999, 'Gender and sexual harassment', *Annual Review of Sociology*, 25, 169–90.

Wertheimer, A., 1996, *Exploitation*, Princeton University Press, Princeton.

Zimmerman, D., 1981, 'Coercive wage offers', *Philosophy and Public Affairs*, 10(2), 121–45.

11 Labour Laws and Labour Welfare in the Context of the Indian Experience

11.1 INTRODUCTION

Markets often work in peculiar ways. A policy that seems obviously good for some groups of people may turn out, when the dust settles and an equilibrium is established, to be detrimental to their welfare. Economics would not have been an interesting subject if this were never the case. One market where such pathologies often occur is the labour market. So it is not surprising that labour market legislation is one area where well-meaning but erroneous policies abound.

In many parts of the world, workers do not seem to have done particularly well and have in fact often lost out in relative terms. This can be because the policy-makers did not care but it can also be because they cared but misunderstood the way the labour market works and so their interventions did not work the way they expected they would.

Observers often argue that trade unions use their muscle power to get benefits for organized labour to the detriment of other workers. But in reality it is not evident that organized labour has done that well either. One reason for this is that the relation between the legal and contractual environment of a nation and the well-being of workers is sufficiently complex that trade unions do not always understand what is good for them and so do not demand what is in their interest.

The aim of this chapter is to construct plausible theoretical models, using India as the backdrop of stylized facts, to show that this may indeed be the case. It will

This is a topic on which I have, over the years, had discussions one with Gary Fields, Mrinal Datta Chaudhuri, Shub Debgupta, and Martin Rama. I would like to thank them and also Jan Svejnar for helpful comments.

First published in Janvry, Alan de and Ravi Kanbur (eds), *Poverty, Inequality and Development: Essays in Honor of Erik Thorbuke*, Kluwer, Norwell, Massachusetts (2005).

be argued that India's myriad labour laws, meant to protect labourers, may have actually hurt them. The argument will be presented in terms of a theoretical model and, as such, should be of interest to other developing and transition economies as well. I take the view in this chapter, in keeping with the normative position that Erik Thorbecke has espoused time and again (Thorbecke 2003a, 2003b) that a society or government has a special responsibility towards the disadvantaged sections of a nation. Hence, in general, labourers should be the target of welfare-enhancing government intervention.[1]

Several pieces of labour legislation in India were drafted expressly to make the laying off of labourers difficult. If an employer found that a worker (1) was shirking from putting in enough effort or (2) did not have adequate skill for the job in question, in many situations he would not be allowed to dismiss the worker (or be allowed to do so only at a considerable cost), no matter what the initial contract with the worker. I shall show that the eventual labour market equilibrium that emerges in an economy with such legislation may actually cause workers to have a lower welfare than in an economy with less protective legislation; and that *between* legislating to prevent layoffs and legislating to maintain minimum wages, the latter may be the more desirable policy from the point of view of worker welfare.

Section 11.2 presents some institutional details of the Indian labour market. In the model in Section 11.3 it will be assumed that labour effort is fixed and so only (2), above, is the relevant issue. It will be shown that in a labour market model, which *prima facie* captures the broad realities of the Indian economy, an employer's inability to dismiss workers who turn out not to possess the required skill could, in equilibrium, hurt all workers, including the unskilled. The essential argument goes as follows. If worker dismissal is disallowed or very costly, firms which need specialized skills and talents may operate on a smaller scale or, worse, close down. This would of course hurt the skilled workers and, by turning them out to the unskilled labour market, could also lower wages in the latter, thereby hurting all workers. There are other possible routes to a similar conclusion. If a firm faces a fluctuating-demand environment and is prevented by law from laying off workers, it may once again close down or function only on a small scale, thereby causing a contraction in the demand for labour and depressing wages (Basu, Fields, and Debgupta 2004). Likewise, consider the case where workers can shirk effort, that is, (1), above, is relevant. In such a situation it is possible that workers may want to be forced to work hard. That would make labour a more coveted input and could increase the demand for labour and wages so much that it would more than compensate the workers for the higher effort.[2]

Before proceeding further I want to emphasize that the modelling here is based on realistic assumptions, but assumptions all the same. Hence, by altering these we can get different results. Nevertheless, it is interesting to see that legislation which is seemingly pro-worker *may* end up hurting the same workers it is supposed to help. Even if, in reality, this is not always so, the fact that this can happen under realistic assumptions should alert us to the fact that labour laws need more careful scrutiny, theoretical and empirical, to sort out which ones *actually* help workers and which ones hurt.

11.2 THE INSTITUTIONAL BACKGROUND

Beginning with policy adjustments to stave off a foreign exchange and fiscal crisis in mid-1991, the Indian government has gone on to attempt major economic reforms during the last decade. While significant changes have been effected in several sectors (see Basu 2004), notably those dealing with international trade and investment, one area which has resisted reform is that of labour markets and labour legislation. This is a matter of some concern since it is arguable that, in the long run, reforms in this area will matter more than those in many other sectors.

Up to now, and in sharp contrast to many other Asian countries, India has failed to deploy her large labour resources to compete better in the domestic and international markets. As a consequence Indian workers remain poor, underemployed, and often unemployed.[3] On the other hand, in India a large number of labour laws have been enacted with the express purpose of protecting labour.[4] It is the aim of this chapter to suggest that these two facts may not be unrelated. To the extent that these laws are well-meaning and their intent is widely supported, it will be argued here that the condition of the Indian worker represents a major *intellectual* failure—the failure to appreciate that overt protection can ultimately do harm when the market has fully responded to the policies and laws and settled down to an equilibrium. In particular, it will be shown that enabling retrenchment and layoffs may result in larger employment and higher wages in the resulting equilibrium. Casual empiricism certainly does not contradict such a hypothesis. The East Asian and South East Asian countries where employment has grown and wages risen are also the countries which, in contrast to India, have fewer protective laws. As Edgren (1989, p. 1) notes at the start of a major International Labour Organization (ILO) study: 'legislation governing hiring and firing, minimum wages and the scope for collective bargaining differs between different parts of Asia, with the countries of the Indian subcontinent having stricter regulations of employers' rights to hire and fire, and granting unions wider scope for bargaining'.

The kind of legislation that is of central interest to me here is India's *Industrial Disputes Act*, 1947. This act, along with the amendments of 1976 and 1982, places restrictions on layoffs and dismissal by large firms. What is important is that these restrictions are exogenous in the sense that they override any contract between the employer and workers. For firms employing 50 or more workers there are predetermined compensations which the employer has to give to workers when they are laid off. For firms employing more than 100 workers, the Act requires the employer to take prior permission from government for layoffs and retrenchment of labour and for closing down the firm. And, as Datta Chaudhuri (1994) points out, 'government permission is seldom given', and in most employment-related disputes, government gets involved and treats the handling of labour as a child custody problem in a divorce suit. In India, government intervention in labour markets goes much deeper than would appear through studying the labour laws. The *Industrial Disputes Act* allows the Labour Departments of the Centre and State governments to intervene not only in labour disputes but also in anticipated labour disputes. In addition there is the problem of political and ministerial intervention (Ramaswamy 1984). Finally, as mentioned above, the judiciary often takes a custodial attitude to labour. In 1992 in a case involving a bankrupt private firm, a judge of the Calcutta High Court, Justice Hazari, argued that, if another private firm took over the firm, there would be no guaranteeing that that firm would not, in turn, go bankrupt and cause workers to be laid off. He, therefore, directed the Government of West Bengal to take over the firm and 'run it with the existing workers' (Datta Chaudhuri 1994).

It will be shown that it is likely that such laws and practices hurt not only workers who are not protected by the law (because, for instance, they work in small firms), but also the workers who are allegedly protected by law. The fact that protective labour legislation may have hurt India's overall growth and efficiency has been pointed out by many observers (see, e.g., Ahluwalia 1991; Lucas 1988; Papola 1994). They are probably right but my argument here is distinct because I am claiming that such legislation may have hurt the very constituency that it was meant to protect, to wit, labour.[5] Hence, Kannan's (1994) observation that wages in the 1980s have not in general kept pace with labour productivity, put forward as a critique of the view that increasing protective legislation has hurt growth and efficiency, and Ghose's (1994) finding that employment per unit of gross value added in manufacturing fell, monotonically, throughout the 1980s sit very comfortably with the theoretical findings of this chapter. I must stress that what I am arguing is not for firms to be given the freedom to hire and fire as they wish, but for firms and workers to have greater freedom to sign contracts concerning layoffs, retrenchment, and closure, without these being overruled by exogenously determined conditions as wantonly as they currently are.

Suppose some workers in a large firm ask their employer to pay them a higher wage and, in turn, they promise to go away without compensation whenever the employer wishes to sack them. Even if both the workers and the employer benefit by such a contract, it is unlikely in contemporary India that an employer will agree to such a contract. This is because if, after paying a higher wage for some time, the employer actually gives notice to his workers (perhaps because demand for the product has fallen off), the workers can appeal to government and government is very likely to cite the *Industrial Disputes Act* and declare such a dismissal illegal *no matter what the prior agreement between workers and employer*. Indeed there are not too many credible ways for workers to give up their *right* not to be dismissed. Over here we must distinguish between 'not resisting dismissal' and 'giving up the right not to be dismissed'. A worker can of course choose not to resist dismissal; but what is interesting is that he may not (and, in the case of India, he *is* not) able to waive the right not to be dismissed.

One kind of reform that my model in Section 11.3 prompts is to allow employers and workers to sign any contract concerning dismissal conditions and have the state or judiciary uphold such a contract. A less radical but, nevertheless desirable reform would be to leave much of the law, for example, the *Industrial Disputes Act*, as it is but to add on a clause which gives workers the right to waive the right not to be dismissed as conferred on them currently by the Act.[6] Such provisos are not unheard of. In the United States a student has the right to see the recommendation a professor writes for him or her, but (s)he also has the right to waive this right.

Before proceeding to construct a formal model, it is worth asking ourselves what is the correct market structure to assume in describing the interaction between firms and workers. Since most of the layoff and dismissal laws apply to large firms, as discussed above, a model with only atomistic firms will not be the right one. At the same time we know that there are lots of firms which do not come under the purview of laws such as the *Industrial Disputes Act* by virtue of being too small. Hence, the right model seems to be one with some large or dominant firms that are capable of affecting the market wage, along with a wage-taking fringe of smaller firms. In addition, it will be assumed that it is the large firms which need specialized or skilled labour.

11.3 MODEL: TURNOVER AND QUALITY

11.3.1 Basic Concepts

There are certain kinds of skills which are not captured by usual indices like university degrees or IQ test scores, but which nevertheless matter to the employer. The skills needed for dealing with people, for remembering little tasks to be performed, and for punctuality often fall in this category. People better endowed with these skills may not necessarily find it easier to acquire education, so even in equilibrium such skills may not be strongly correlated with the degrees and diplomas held by a worker, thus making standard job-market signaling models irrelevant for our present purpose. The only way for an employer to know exactly how much skill of the above kinds a worker possesses is to employ the worker. For reasons of modelling simplicity I shall assume that skill can be of only two levels. Hence workers are either skilled or unskilled. Let us assume that a fraction t of the labour force is skilled. Let the aggregate supply curve of labour be given by

$$s = s(w), \qquad s'(w) > 0, \qquad (11.1)$$

where w is wage.

By assuming that both skilled and unskilled workers are otherwise homogenous or that their labour-leisure choices are identical, we know that if a wage of w is fixed, the supply of skilled labour is $ts(w)$ and the supply of unskilled labour is $(1 - t) s(w)$.

Next, suppose that there are two kinds of firms—ones where the worker's skill matters (these are also the large or 'dominant' firms) and ones where they do not (these are the small or 'fringe' firms). From each skilled worker the *dominant* firms can get an output of $r > 0$. They have no use for unskilled labour, who produce 0 in such firms. On the other hand, a *fringe firm* has no special use for skilled labour. Each worker, skilled or unskilled, produces an output less than r.

Let me sketch the intuitive argument first. If the dominant firms are not allowed to lay off workers, they will be forced to make do with a labour force in which a fraction t will be skilled and fraction $1 - t$ unskilled (since the only way to ensure that you have all skilled workers is to employ people, check them out, lay off the unskilled, employ new people in their place, check them out, and so on). This will typically result in a smaller demand for labour from the dominant firms. Hence the supply of labour in the fringe job market will be greater. This will tend to push down wages. Hence, a law preventing layoffs can actually push down *all* wages. It is interesting here to note that the detailed empirical study of Fallon and Lucas (1993) reveals that demand for labour in large firms fell as legislation preventing labour dismissal was made stronger. And Fallon and Lucas (p. 269) go on to conjecture: 'This decline may be understood in terms of [...] reluctance to hire in case employees prove to be poor matches with their job demands, a mismatch which cannot readily be reversed'.[7]

In the model that follows, I shall make special assumptions to keep the algebra simple but the aim is to formalize the above general argument.

As explained in the previous section, the fringe firms will be treated as wage-takers, whereas the dominant firms are like oligopsonists in the labour market. Hence, we are considering a labour market similar to the product-model of Nichol (1930) and Stigler (1950) and subsequently extended and discussed by Basu (1993), Dixit and Stern (1982), and Encaoua and Jacquemin (1980).

Let the fringe firms' aggregate demand for labour be given by

$$d = d(w), \qquad d'(w) < 0. \tag{11.2}$$

Hence, if w is the wage, the supply of labour in excess of what is needed by the fringe firms is given by

$$\psi(w) = s(w) - d(w). \tag{11.3}$$

Let us suppose that there are k dominant firms. If these firms choose workers randomly, the expected output from a worker is $rt \, [= rt + 0(1 - t)]$. If these firms are not allowed to lay off workers then the expected output from each employed worker is rt. If, on the other hand, the firms can freely dismiss workers then we shall assume the worker productivity to be r. This is because through successive laying off of unproductive workers, the dominant firms' labour force will converge to a purely skilled group.

Each dominant firm's cost of production consists of an entry fee of $K \, (> 0)$ and the wage bill.

Our model consists of two periods. In period 1, each of the k dominant firms have to decide whether to enter the industry or not. Then in period 2 the ones that enter decide how much labour to employ, keeping in mind that their decision will affect the labour demand of the fringe firms. Our aim is to characterize the *subgame perfect* equilibrium of this two-period game. Actually, we have two games. One in which layoffs are allowed and one in which the law prevents layoffs. Our aim is to compare the equilibria in these two games.

There are, in reality, many intermediate cases. For instance, a law could make layoffs costly, instead of disallowing layoffs. As discussed above, *India's Industrial Disputes Act* does exactly that for firms employing between 50 and 100 workers.

Fortunately, such cases are easy to discuss once we have worked out the two polar models.

I shall characterize the subgame perfect equilibria by first analysing the Nash equilibria in the second period game when no layoffs are permitted (Section 11.3.2) and when layoffs are permitted (Section 11.3.3) and then turn to the outcomes in the full two-period model (Section 11.3.4). If m dominant firms enter the industry and layoffs are not permitted, the second period game is described as $G(m,N)$. The second period game with m dominant firms and layoffs permitted is called $G(m,L)$.

11.3.2 THE GAME $G(M,N)$

There are m dominant firms that have entered the industry and confront the labour supply function $\psi(w)$, described in Eq. (11.3). Let the inverse of this function be given by $\varphi(\cdot)$. If the firms hire $n_1, \cdots, n_m$ units of labour, recalling that no layoffs are allowed in this model, the profit function of firm i is given by

$$\pi_i^N(n_1, \cdots n_m) = rtn_i - \phi(n_1 + \cdots + n_m)n_i. \tag{11.4}$$

This is the model used by Encaoua and Jacquemin (1980), though in their model the dominant firms and the fringe compete in the *product* market. I shall

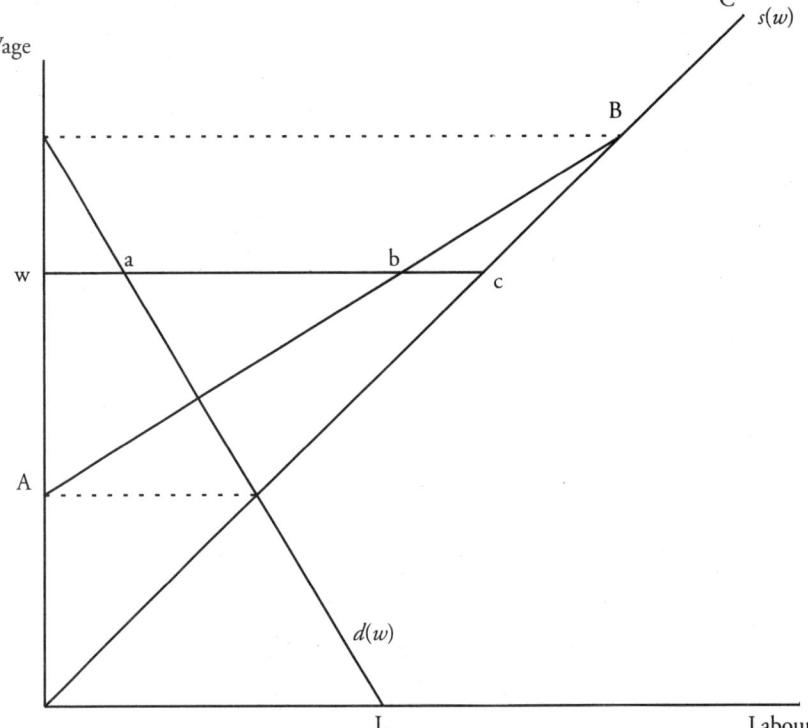

Figure 11.1: Dominant Firms and Fringe

here focus on the symmetric Nash equilibrium of this m-player game. In the examples that I consider below such an equilibrium always exists and, in fact, is the only Nash equilibrium. Hence, each firm employing n^N labour is an *equilibrium* if and only if, for all $i \in \{1, \cdots, m\}$,

$$\pi_i^N(n^N, \cdots, n^N) \geq \pi_i^N(n^N, \cdots, n^N, n_i, n^N, \cdots, n^N), \text{ for all } n_i \qquad (11.4)$$

If n^N is the equilibrium in $G(m,N)$, define[8]

$$\pi^N(m) \equiv \pi_i^N(n^N, \cdots, n^N)$$

Hence, $\pi^N(m)$ is the profit earned in the period 2 equilibrium by each firm when there are m firms and no layoffs (hence the superscript N) are allowed. In equilibrium the fringe firms hire $d(\phi(mn^N))$ labourers and pay wage equal to $\phi(mn^N)$.

While this completes the description of the outcome of $G(m,N)$, let me introduce a geometric illustration of the problem which could aid intuition later in the more complicated case.

If the aggregate supply curve, $s(w)$, and the demand curve of the fringe firms, $d(w)$, are as shown in Figure 11.1, then the supply curve faced by the dominant firms is given by ABC where AB is drawn such that for any wage w, the line wa is equal to bc. The final equilibrium is then the usual oligopsony equilibrium for m firms facing the supply curve ABC.

11.3.3 THE GAME $G(M, L)$

There are m dominant firms that have entered the industry and they are allowed to layoff workers. This means that (in the end) these firms will only employ skilled workers and get an output of r from each employed worker. Hence, the supply curve of the labour that is of relevance to these firms is given by $ts(w)$. What we have to be careful about is that if the wage that the dominant firms pay drops too low then some of the skilled labourers may prefer to go to the fringe firms. Keeping this in mind, let us now work out the aggregate demand function that these dominant firms face.

Define the wages, w and $\underline{w}$, as, respectively,

$$d(\bar{w}) = (1 - t)\, s(\bar{w})$$

and

$$d(\underline{w}) = s(\underline{w}).$$

In this model with layoffs the skilled workers can get a different wage from unskilled workers. Let w^1 be the wage earned by each skilled worker and w^0 the wage of an unskilled worker.

If $w^1 > \bar{w}$, the supply of skilled workers to the dominant firms is $ts(w^1)$ and $w^0 = \bar{w}$. If $w^1 \leq \bar{w}$, the supply of workers to the dominant firms is $s(w^1) - d(w^1)$ and $w^0 = w^1$.[9] The dominant firms will ensure that the workers they employ are skilled workers. This is feasible since

$$s(w^1) - d(w^1) < ts(w^1),$$

which is an implication of $w^1 \geq \bar{w}$.

The information in the above paragraph is summed up, by writing the supply function of skilled workers faced by the dominant firms as

$$s(w^1) = \begin{cases} ts(w^1) & \text{if } w^1 \geq \bar{w} \\ s(w^1) - d(w^1), & \text{if } w^1 < \bar{w} \end{cases} \tag{11.5}$$

If $\theta(\cdot)$ is the inverse of $S(w^1)$, the profit function of firm i is given by

$$\pi_i^L(n_1, \cdots, n_m) = rn_i - \theta(n_1 + \cdots + n_m)\, n_i, \tag{11.6}$$

the superscript L being a reminder that this is a model with layoffs.

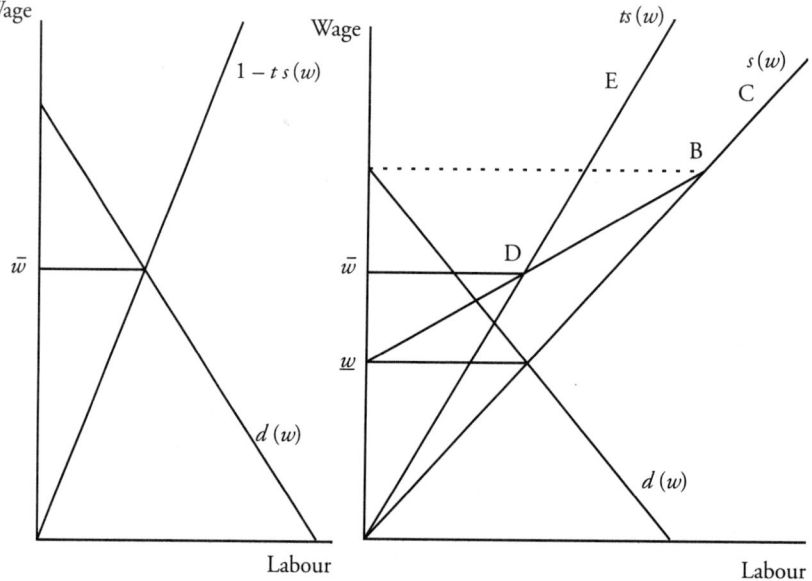

Figure 11.2: Supply Curve of Skilled Workers

Figure 11.2 illustrates Eq. (11.5). The right-hand panel of Figure 11.2, shows $d(w)$ and $s(w)$, as in Figure 11.1, and in addition illustrates $ts(w)$. The left-panel shows $d(w)$ and $(1 - t)s(w)$.

The supply curve (11.5) is given by the line ADE. Note that in the case without layoffs the supply curve was given by ABC. Hence, the supply now faced by the dominant firms is smaller. But, of course, all labourers employed by the dominant firms now are skilled labourers.

Following the exercise in Section 11.3.2, we define n^L to be *equilibrium* employment by each firm in game $G(m,L)$ if, for all i,

$$\pi_i^L(n^L, \cdots, n^L) \geq \pi_i^L(n^L, \cdots, n^L, n_i, n^L, \cdots, n^L), \text{ for all } n_i.$$

Define,

$$\pi^L(m) = \pi_i^L(n^L, \cdots, n^L)$$

Hence, $\pi^L(m)$ is the profit earned in the period 2 equilibrium by each firm when there are m firms and layoffs are permitted.

If, $\theta(mn^L) > \bar{w}$, the fringe firms pay a wage of $\bar{w}$ in equilibrium. Otherwise they pay a wage of $\theta(mn^L)$.

11.3.4 Subgame Perfect Equilibria With or Without Layoffs

It is now easy to work out the subgame perfect equilibria of the two two-period games that we have, one with the second period game $G(m, N)$ and the other with $G(m, L)$. Call the two two-period games G^N and G^L, respectively.

Consider first the case where no layoffs are allowed, that is, game G^N. Let the ordered pair, (t, n), denote the number of firms, t, that decide to enter the industry in period 1 and the number of units of labour, n, that each entrant firm employs in period 2. Recall that if a firm enters and earns π in the post-entry game, the firm's net profit is $\pi - K$, and also that k is the total number of potential entrants.

(t^*, n^*) is described as a subgame perfect equilibrium of G^N if

1. $t^* = k$ and $\pi^N(k) - K \geq 0$,

2. $t^* = 0$ and $\pi^N(1) - K \leq 0$,

3. $0 < t^* < k$ and $\pi^N(t^*) - K \geq 0$ and $\pi^N(t^* + 1) - K \leq 0$, and

4. n^* is an equilibrium in $G(t^*, N)$.

The subgame perfect equilibrium of G^L is defined in the same way with π^L replacing π^N and $G(t^*, L)$ replacing $G(t^*, N)$.

If legislation like the *Industrial Disputes Act* is repealed, the labour market game will switch from G^N to G^L. Our aim is to compare the resulting new equilibrium with the equilibrium in G^N. While all kinds of welfare changes are possible, I focus here on the relatively counter-intuitive one and demonstrate that labourers—all labourers—can be better off under G^L.

I shall demonstrate this by confining attention to a class of linear models. In particular, I shall assume

$$s(w) = bw, \tag{11.7}$$

$$d(w) = A - Bw. \tag{11.8}$$

Hence Eq. (11.4) now becomes

$$\pi_i^N(n_1, \cdots, n_m) = \left[rt - \frac{\sum n_j}{b + B} - \frac{A}{b + B} \right] n_i$$

It must be recalled that this is valid, assuming that equilibrium wage does not exceed A/B, because otherwise fringe demand is zero. That is,

$$\frac{m}{m + 1} [(b + B)rt - A] \leq \frac{bA}{B}$$

It is easy to compute the amount produced and the profit earned by each firm under symmetric Nash equilibrium or what is simply called the equilibrium in Section 11.3.2. These are given by

$$n^N = \frac{rt(b + B) - A}{m + 1},$$

(11.9)

$$\pi^N(m) = \left[\frac{rt(b + B) - A}{m + 1}\right]^2 \cdot \frac{1}{b + B}$$

(11.10)

Let us now turn to the case where layoffs are permitted. From the definition of $\bar{w}$ and using Eqs. (11.7) and (11.8) we get:

$$\bar{w} = \frac{1}{(1 - t)b + B} \cdot$$

From Eqs. (11.5) and (11.6), it follows that

$$\pi_i^L(n_1, \cdots, n_m) = \begin{cases} \left[r - \dfrac{\sum n_j}{tb}\right]n_i, & \text{if } \dfrac{\sum n_j}{tb} \geq \dfrac{A}{(1 - t)b + B} \\[3ex] \left[r - \dfrac{\sum n_j + A}{b + B}\right]n_i, & \text{if } \dfrac{\sum n_j}{tb} < \dfrac{A}{(1 - t)b + B} \end{cases}.$$

Let us suppose that the top line of the above function is relevant. Then

$$n^L = \frac{rtb}{m + 1},$$

$$\pi^L(m) = \frac{r^2 tb}{(m + 1)^2},$$

and

$$\frac{mr}{m + 1} \geq \frac{A}{(1 - t)b + B}$$

(11.11)

The last inequality merely ensures that the top line of the π_i^L function is relevant. If Eq. (11.11) is untrue, n^L and $\pi^L(m)$ will be given by Eqs. (11.9) and (11.10).

Suppose we have a situation where (11.11) holds and also $\pi^L(m) > \pi^N(m)$. That is,

$$r^2 tb > \frac{[rt(b + B) - A]^2}{b + B}.$$

In that case, we can find a $K > 0$ such that

$$\pi^L(1) > K > \pi^N(1)$$

(11.12)

or

$$\frac{r^2 tb}{4} > K > r^2 \left[\frac{rt(b + B) - A}{2}\right]^2 \cdot \frac{1}{b + B}$$

(11.13)

If, for instance, $r = 2$, $t = \frac{1}{2}$, $b = 2$, $B = 8$, $A = 8$, and $K = \frac{3}{4}$, then (11.11) and (11.12) are true [with $m - 1$ in (11.11)].

If (11.11) and (11.12) are true, there will be no dominant firms in the industry if layoffs are banned, since $\pi^N(1) - K < 0$. But if layoffs are allowed, entry is bound to occur since $\pi^L(1) - K > 0$. So if layoffs are banned all workers will be getting a wage of $\underline{w}$, where, as before, $d(\underline{w}) = s(\underline{w})$. If such a ban is revoked, some dominant firms, which need skilled labour, will enter the industry. They will of course pay their workers a higher wage. And, by virtue of their employing workers, the supply of labour to the fringe sector will fall. Hence, the fringe sector wage will rise.

By confining attention to situations where $\pi^L(1) > K > \pi^N(1)$, we here get the result that there will be no entry of firms using skilled labour if layoffs are not permitted. In that case we can assert that allowing for layoffs will cause an expansion in the sector using skilled labour thereby pulling up wages for all workers. The scenario is not at all unlike what has been observed in many developing countries, such as India. There is skilled labour and a lot of scope for using the labour, but the labour laws prevent adequate use of this labour.

In the present model we focussed on one kind of causation. Firms need to try out workers, dismiss some, employ others, and so on in order to improve the skill-level of their labour force. Laws which make layoffs impossible discourage the emergence and growth of such firms. There are other kinds of causation, for instance, that involving the workers' choice of effort which could push us towards the same conclusion. The latter is studied in Section 11.4.

11.3.5 Costly Layoff

Up to now we focussed on the polar cases of 'costless layoffs' and 'no layoff'. Now suppose there is a law that requires that firms pay a compensation of c units to every worker that is dismissed. The firm will then face a choice of either employing randomly picked workers and not laying them off in which case a fraction $(1 - t)$ of its workforce will be unproductive as in game $G(m, N)$ or incurring the compensation cost but moving towards a skilled labour force. If the firm wishes to employ n_i workers who are skilled, it could, for instance, employ n_i dismiss the $(1 - t)n_i$ who turn out to be unproductive, employ $(1 - t)n_i$ new workers, dismiss $(1 - t)^2 n_i$ of them who turn out unproductive, and so on. The total number of dismissals will be

$$\left(\frac{1-t}{t}\right) n_i = (1 - t)n_i + (1 - t)^2 n_i + (1 - t)^3 n_i + \cdots .$$

Ignoring the process by which this is achieved and treating it as instantaneous, we would write firm i's profit function as

$$\pi^i = \left[r - \theta (n_1 + \cdots + n_m) - \frac{c\,(1 - t)}{t} \right] n_i,$$

where $\theta(\cdot)$ is the same as in Eq. (11.6). Whether a firm finds it worthwhile to replace its unskilled worker with a skilled worker depends on whether

$$r - rt \geqq c\,(1 - t)/t.$$

It is quite obvious that if *c* is large enough we could have the same kind of result as demonstrated above. The dominant firms will either not enter at all or not expand, thereby causing a glut of labour on the fringe market and depressing wages.

11.3.6 Minimum Wages

The implementation of legal minimum wages is quite ubiquitous across nations. In India, the *Minimum Wages Act*, 1948, empowers the government to announce legal minimum wages and periodically revise these.[10] Failure on the part of an employer to adhere to this is 'punishable with imprisonment [up to] six months, or a fine [up to] five hundred rupees, or with both'. Despite this, in practice, the minimum wage law is frequently contravened in India.[11]

In the above model it can be shown that, unlike anti-layoff laws or custom, minimum wage laws have a desirable effect on labour welfare *if they are applied only to the dominant firms and in small measure*. There could be other, more ubiquitous reasons for having minimum wage laws, as argued in Basu, Genicot, and Stiglitz (2003), but the analysis here is confined within the structure of the above model.

This is easy to see. The usual reason why some economists argue against the use of a minimum wage is transparent by an appeal to the conventional demand and supply curve analysis of labour-market equilibrium. In such a model, a minimum

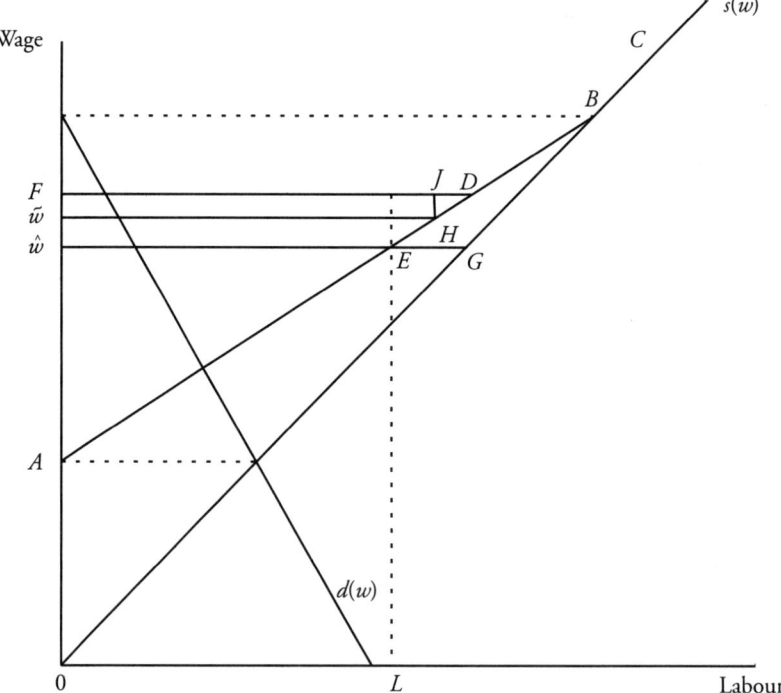

Figure 11.3: Supply and Demand

wage law can indeed raise wages but this also causes unemployment. Hence it benefits some workers but only by hurting other workers.

The above reasoning however does not carry over to the model built in this chapter, when the minimum wage law is applied to the oligopsonists. To demonstrate this with the simpler model, suppose that layoffs are not permitted. Consider the game $G(m,N)$, that is, a game with no layoffs and m large firms. Figure 11.3 reproduces $s(w)$ and $d(w)$ from Figure 11.1.

In addition, the figure marks F on the wage axis, which shows the expected amount earned by a firm from a randomly chosen labourer (i.e., $OF = rt$). Since the supply curve of labour faced by the oligopsonists is ABC, a standard analysis suggests that the oligopsonists will employ OL labourers [OL being $m/(m + 1)$ of the length of the line FD] with each oligopsonist employing OL/m labourers, and equilibrium wage being $\hat{w}$. Even the fringe sector will pay $\hat{w}$ (and employ EG labourers) in equilibrium.

Now consider a minimum wage, $\bar{w}$, a little above $\hat{w}$. That may result in some of the (large) firms to not enter the industry. If $\bar{w}$ is so high that no firm enters the industry then wage will drop to OA in Figure 11.3 and labourers will be worse off. But as long as that does not happen, that is, some large firms continue to produce, the labourers benefit because of the minimum wage legislation. If $\bar{w}$ is such that $\bar{w} > \hat{w}$ and $\bar{w}HJF = K$ (the entry cost), then any minimum wage which is below w, benefits the labourers.[12]

The above analysis suggests that if the welfare of workers is of concern to the government, it may be best to repeal anti-layoff legislation (i.e., as explained earlier, the terms and conditions of layoffs should be decided by the workers and their employer through free contracting) but to have some minimum wage requirements for large firms (not for the small fringe firms).

11.4 CONCLUSION

Modern economists are aware—or at any rate ought to be—that markets often fail, and even when they do not fail, they may result in intolerable inequities. From this truism to jump to the conclusion that government should be brought in wherever the market is expected to fail is however a fallacy. Just as markets can fail, so can governments. It is entirely possible that government will act as a handmaiden of vested interests or will simply be ineffective and falter. In the context of labour, markets do often fail. This means that we may need certain kinds of nonmarket interventions to ensure minimal labour standards and basic worker rights. Yet, the steps from this realization to its execution can be fraught with pitfalls, unless we do this on the basis of very careful theorizing and documenting of facts. The aim of this paper was to sound this warning bell.

This chapter evaluated labour laws concerning layoffs and minimum wage legislation in India and, on the basis of a theoretical model, reached the following policy conclusions. It argued that legislation or even *customary practice* which makes the laying off of labour illegal or (exogenously) costly may be harmful for the workers. The same may happen if employers are *a priori* given the freedom to fire workers at will. Instead, workers and employers should have the freedom to develop their own contract concerning the conditions for the dismissal of labour. There will be reason to set limits on the range of contracts allowable (without this, contracts run the risk of becoming so complex that workers may not fully comprehend what they are committing themselves to), but there is reason to believe that a larger freedom than what is currently permitted would benefit workers. I would expect that many firms needing specially skilled labour would, in the first place, come into existence and, second, agree to give higher wages and expand their labour force if such an amendment to existing laws were to be made.

This chapter argued that there may be a need to impose some legislative restrictions on minimum wages. In particular, workers will benefit by having some minimum wage restrictions, as long as these are not set too high and as long as these are confined to the large firms. Of course, if the layoff legislation is amended as suggested here, the market wage paid by the large firms may well outstrip the legal minimum wage naturally. This is consistent with the experience of Taiwan and Korea (Ranis 1994).

From these broad do's and don'ts, to answer specific questions, such as how high should minimum wages be, or, when allowing workers to sign contracts to give up the existing right not to be dismissed, what restrictions should be placed on the allowable range of contracts, we need to move to empirical and descriptive research. The purpose of this chapter was to provide an analytical basis for such work.

NOTES

1. This is not to deny that there are people who are even more disadvantaged (such as the unemployed, the aged, and the infirm), but as a class, workers are, in most societies, poor enough to deserve attention.

2. Though my analysis is in terms of preexisting legislation, this is closely related to the question of trade union empowerment and its effect on labour welfare (see Rama [2000] for discussion). For a general discussion of labour regulation, see Ehrenberg (1994).

3. There is, in addition, some evidence of employment growth not having kept pace with growth *per se* in the country. One reason for this is a growing tendency in firms to replace labour with capital (Nagaraj 1993; Papola 1989). For a detailed micro study of evolving labour market conditions, see Mathur (1991). The problem of layoffs in connection with the closure of firms is documented in Anant et al. (1993). For an excellent study of the whole range of labour legislation in India and its effect of worker welfare, see Singh (2000).

4. Legislation which is explicitly meant to protect labour interests include *The Trade Unions Act*, 1926; *Industrial Employment (Standing Orders) Act*, 1946; *Industrial Disputes Act*, 1976.

5 . This is also the line taken by Besley and Burgess (2003), who, taking advantage of the fact that labour legislation in India is on the 'concurrent list' (i.e., the law can, in principle, be modified by the states) do an interstate comparison. It is, however, worth keeping in mind that the variability possible across states in reality is less than what is

permitted in principle. This is the reason why the erstwhile Chief Minister of Andhra Pradesh, Chandrababu Naidu, had been arguing that labour legislation should be made entirely a state matter. I believe this is a good suggestion though with the provision that there should be laws to ensure that the competition across states does not become ruinous.

6. The distinction between not giving workers certain rights, R, and giving them the rights, R, and giving them the further right to waive R is discussed in Basu (1997).

7. If we assume some wage rigidity, these arguments can easily translate to the effect of anti-dismissal laws on unemployment. Such a law (see, e.g., Layard, Nickell, and Jackman 1994, p. 108) clearly has opposing *ex ante* and *ex post* effects, so that the net impact on employment can be ambiguous. Hence, a law making labour dismissal harder may result in a larger unemployment. Though Layard, Nickell, and Jackman reach a similar conclusion, they point out that empirical studies do not provide any clear resolution of this ambiguity.

8. We are proceeding under the assumption of there being a unique equilibrium—an assumption that holds true in the class of examples considered below.

9. If $w^1 < \underline{w}$, the supply to the dominate firms is, of course, zero.

10. The Act also specifies other details, like hours of work and over time rates.

11. Compliance with this law turns out to be far from perfect even in developed countries. Ashenfelter and Smith's (1974, pp. 333–50) study of the US labour market of the early 1970s shows that there was 65 per cent compliance. They are careful to define compliance not as percentage of labourers who get a wage above the statutory minimum, but as the percentage of workers who earn the statutory minimum wage or lose their jobs among workers who would earn less than the statutory minimum wage in the absence of a minimum wage law. Though I cannot cite evidence, it seems natural to me to expect that minimum wage regulation would be violated more widely than anti-dismissal legislation. This is because when striking a deal for a new job, it may be in the worker's interest to comply with a less than minimum wage (because otherwise he may not get the job) but when served a dismissal notice a worker has no interest in complying with the employer's demand.

12. A recent study by Card and Krueger (1994) finds that an upward revision of the minimum wage in New Jersey increased employment in the fast-food industry (see also Card 1992). As the authors point out, this is not surprising if the labour market is oligopsonistic (though in their study there are other changes which cast some doubt on the market being an oligopsony). See Drazen (1986) for an argument why minimum wage legislation may be efficient.

REFERENCES

Ahluwalia, I. J., 1991, *Productivity and Growth in Indian Manufacturing*, Oxford University Press, New Delhi.

Anant, T. C. A., S. Gangopadhyay, and O. Goswami, 1993, *Industrial Sickness in India: Characteristics, Determinants and History, 1970–1990*, Indian Statistical Institute, New Delhi (Mimeo).

Ashenfelter, O. and R. S. Smith, 1974, 'Compliance with the minimum wage law', *Journal of Political Economy*, 87.

Basu, K., 1993, *Lectures in Industrial Organization Theory*, Blackwell, Oxford and Cambridge.

—— 1997, 'Some institutional and legal prerequisites of economic reform in India', in H. E. Bakker and N. G. Schulte Nordholt (eds.), *Corruption and Legitimacy*, SISWO Publications, Amsterdam.

—— 2004, 'Indian economic reforms: Up to 1991 and since', in K. Basu (ed.), *India's Emerging Economy: Performance and Prospects in the 1990s and Beyond*, MIT Press, Cambridge, MA, and Oxford University Press, New Delhi.

Basu, K., G. Fields, and S. Debgupta, 2004, 'Alternative labour retrenchment laws and their effects on wage and employment', Cornell University (Mimeo).

Basu, K., G. Genicot, and J. Stiglitz, 2003, 'Minimum wage laws and unemployment benefits when labour supply is a household decision', in K. Basu, P. B. Nayak, and R. Ray (eds.), *Markets and Governments*, Oxford University Press, New Delhi.

Besley, T. and R. Burgess, 2003, 'Can labour regulation hinder economic performance? Evidence from India', BREAD Working Paper No. 44.

Card, D., 1992, 'Do minimum wages reduce employment? A case of California, 1987–89', *Industrial and Labour Relations Review*, 46.

Card, D. and A. B. Krueger, 1994, 'Minimum wages and employment: A case study of the fast-foot industry in New Jersey and Pennsylvania', *American Economic Review*, 84.

Datta Chaudhuri, M., 1994, 'Labour markets as social institutions in India', CDE Working Paper No. 16, Delhi School of Economics.

Dixit, A. and N. Stern, 1982, 'Oligopoly and welfare: A unified presentation with applications to trade and development', *European Economic Review*, 19.

Drazen, A., 1986, 'Optimal minimum wage legislation', *Economic Journal*, 96.

Edgren, G., 1989, 'Structural adjustment, the enterprise and the workers', in G. Edgren (ed.), *Restructuring Employment and Industrial Relations*, ILO, Geneva.

Ehrenberg, R., 1994, *Labour Markets and Integrating National Economies*, Brookings Institution, Washington, D.C.

Encaoua, D. and A. Jacquemin, 1980, 'Degree of monopoly, indices of concentration and threat of entry', *International Economic Review*, 21.

Fallon, P. R. and R. E. B. Lucas, 1993, 'Job security regulations and the dynamic demand for industrial labour in India and Zimbabwe', *Journal of Development Economics*, 40.

Ghose, A. K., 1994, 'Employment in organized manufacturing in India', *Indian Journal of Labour Economics*, 37.

Kannan, K. P., 1994, 'Levelling up or levelling down? Labour institutions and economic development in India', *Economic and Political Weekly*, 23 July.

Layard, R., S. Nickell, and R. Jackman, 1994, *The Unemployment Crisis*, Oxford University Press.

Lucas, R. E. B., 1988, 'India's industrial policy', in R. E. B. Lucas and G. F. Papanek (eds.), *The Indian Economy: Recent Developments and Future Prospects*, Oxford University Press.

Mathur, A., 1991, *Industrial Restructuring and Union Power: Micro-economic Dimensions of Economic Restructuring and Industrial Relations in India*, ILO, Geneva.

Nagaraj, R., 1993, 'Employment and wages in manufacturing industries in India', Discussion Paper 98, Indira Gandhi Institute of Development Research, Bombay.

Nichol, A. J., 1930, *Partial Monopoly and Price Leadership*, Press of Smith-Edwards Co., Philadelphia.

Papola, T. S., 1989, 'Restructuring in Indian industry: Implications for employment and industrial relations', in G. Edgren (ed.), *Restructuring Employment and Industrial Relations*, ILO, Geneva.

—— 1994, 'Structural adjustment, labour market flexibility and employment', *Indian Journal of Labour Economics*, 37.

Rama, M., 2000, 'Downsizing in the presence of monopoly rights: The road to riches', Development Research Group, World Bank (Mimeo).

Ramaswamy, E. A., 1984, *Power and Justice*, Oxford University Press, New Delhi.

Ranis, G., 1994, 'Labour markets, human capital and development performance in East Asia', Yale University (Mimeo).

Singh, J., 2000, *Some Aspects of Industrial and Labour Markets in India: Perspectives from Law and Economics*, Ph.D. Dissertation, Delhi School of Economics, Delhi University.

Shapiro, C. and J. E. Stiglitz, 1984, 'Equilibrium unemployment as a worker discipline device', *American Economic Review*, 74.

Stigler, G., 1950, 'Monopoly and oligopoly by merger', *American Economic Review*, 23.

Thorbecke, E., 2003a, 'Conceptual and measurement issues in poverty analysis', Cornell University (Mimeo).

Thorbecke, E., 2003b, 'Poverty analysis and measurement within a general equilibrium framework', in C. M. Edmonds (ed.), *Reducing Poverty in Asia*, Edward Elgar, Cheltenham, U.K.

12 Child Labour and the Law
Notes on Possible Pathologies

12.1 THE PROBLEM

Beginning a little over two hundred years ago—from the time of Robert Peel's Factories Act of 1802 in Britain—there have been repeated attempts to use legislative action to bring an end to child labour. And one of the more curious features of this phenomenon is how often it has beaten the law and persisted or even got worse (Nardinelli 1990). While child labour did, eventually, come to a virtual end in industrialized nations, it continues to be widespread in developing countries,[1] despite a plethora of legal checks. The purpose of this essay is to show that this is one area where seemingly reasonable policy interventions can backfire and there are good theoretical reasons why that may be so.

The policy with which I shall here illustrate the risk of pathological reaction is the standard one where a firm is fined a certain amount if it is found employing children. India's Child Labour (Prohibition and Regulation) Act, 1986, for instance, has precisely such a clause. Section 14 of this Act requires the government to charge a fine between Rs. 10,000 and Rs. 20,000 from a person or firm found employing children in contravention of the provisions of the Act (Government of India 1986). What will be shown here is that a small dose of an intervention of this kind can actually exacerbate the problem of child labour. If the fine for employing children is raised, child labour could increase for a while before declining. In other words, the response to the policy could be inverse-U shaped. Hence, developing countries like India, trying to legislate against child labour, has to be careful in its design of the law and in the choice of the size of the punishment. Otherwise the law could have the effect opposite to what is intended.

The paper has benefited from the comments of Abhijit Banerjee, Ron Benabou, Per Botolf Maurseth, Dilip Mookherjee, and Omar Robles.

First published in *Economics Letters*, Vol. 87 (2005), 169–74.

This is a purely theoretical chapter. The reader may thus wonder if its warning needs to be heeded, given that it is not empirically proved. My response to that is to observe that (1) there is plenty of empirical support for the main axiom on which the analysis here is founded and (2) the *negation* of the hypothesis put forward here has not been empirically demonstrated, either. In other words, the claim that an increase in the fine for child labour will cause child labour to decline has not been empirically proved. It is simply taken for granted. The chapter demonstrates that there is *no* reason for this presumption. The chapter recommends empirical research to investigate the effects of antichild labour legislation, and, till that happens, caution about the laws commonly used.

12.2 THEORY

The reason why child labour policy turns out to be intricate is because of the somewhat unusual factors that cause child labour in the first place. Child labour is intricately linked to poverty. Virtually all the worlds labouring children are located in poor countries. In the same developing country, where lots of children work, one would rarely find the child of a doctor, lawyer, or professor working. The evidence is overwhelming that poverty is a major cause of child labour and, typically, parents send children to work in order to achieve some minimal level of consumption (see Edmonds 2005; Edmonds and Pavcnik 2005; Grootaert and Patrinos 1999).[2] The counterintuitive result derived in this chapter is a consequence of this assumption.[3]

Consider a labour market in which there are several, identical households with each household consisting of one adult and m children. Each child produces a fraction γ of the labour that an adult can produce. In other words, full time work by one child is equivalent to γ units of an adult's full-time work. I shall assume that the adult always supplies labour perfectly inelastically, whereas children work only to the extent that this is necessary to achieve a critical subsistence level of consumption for the household. Let s be that critical amount of consumption.

From these assumptions it immediately follows that children will work only when adult wage is below s. Let w be the adult wage. If w exceeds s, subsistence consumption is achieved without requiring the children work. Note that, given the above assumptions, whenever adults and children are found working, it must be the case that, if adult wage is w, the wage rate for a child labourer, w^c, will be γw. Otherwise all firms would employ only children or only adults.

Let us now bring government into the picture. Suppose government announces that each time a firm is found employing a child the firm will be fined D rupees. For every child employed by a firm, let p be the probability of the firm being caught. In that case for every child employed, the firm has an expected punishment cost of pD. Hence, unless child wage is less than γw by pD, it does not make sense for a firm to employ children. It follows that the relation between child and adult wages will be

$$w^c = \gamma w - pD. \tag{12.1}$$

I refer to the variables p and D as the 'governmental variables', because these are chosen by government.[4]

Next, if w falls short of s, the household will send the children out to work. Let $e \in [0, m]$ be the number of children that the household sends to work. Since households send children to work only so as to be able to reach subsistence, it must be the case that

$$ew^c = s - w.$$

Or, equivalently,

$$e = (s - w)/(\gamma w - pD). \tag{12.2}$$

It follows that as adult wage drops, the household will send more children to work. Of course, this cannot go on endlessly because after some time the household will run out of children. Then onwards, as w drops there will be no further increase in the supply of child labour. Labour supply of each household is, therefore, $\min\{(s - w)/(\gamma w - pD), m\}$.

One more condition has to be kept in mind. As w falls, w^c will decline as well [see Eq. (12.1)]; and beyond a point w^c, will cross zero and be negative. This happens when $0 \geq \gamma w - pD$. Clearly, when w^c reaches this critical level, parents will stop sending their children to work. Working for a zero wage is not much help achieving minimal consumption targets.

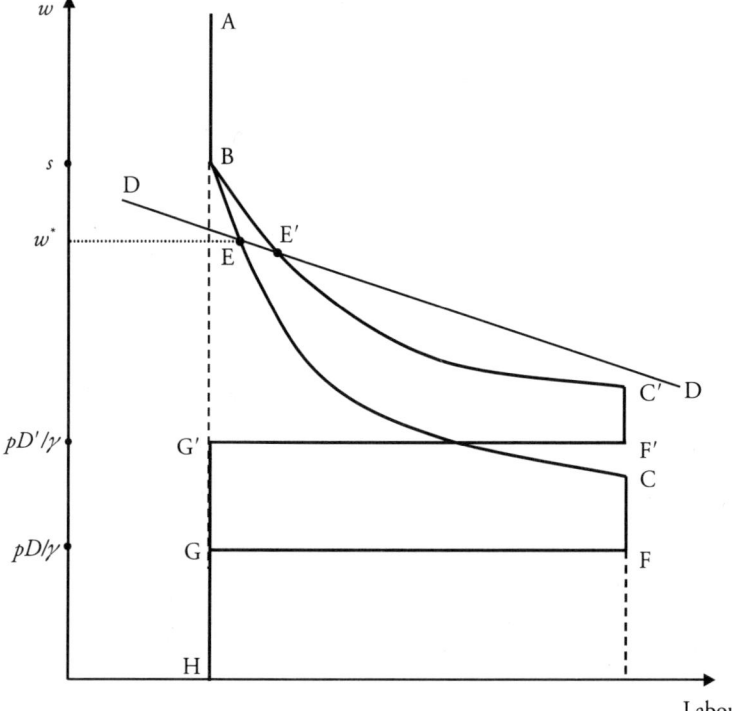

Figure 12.1: Child Labour—Legislative Penalty and Incidence

Gathering these pieces, we can now describe the household's child-labour supply as a function of adult wage and the governmental variables:

$$e = 0, \quad \text{if } w \geq s \text{ or } 0 \geq \gamma w - pD$$

$$e = min\{(s - w)/(\gamma w - pD), m\}, \qquad \text{otherwise.} \qquad (12.3)$$

These facts can be captured pictorially, as shown in Figure 12.1, where the vertical axis represents w (adult wage) and the horizontal axis represents labour, measured in adult labour equivalence units. If w is above s, then only the adult will work. Hence, the labour supply curve will be vertical as shown by the segment AB. As w drops below s, children go out to work, chasing the subsistence target. Hence, the backward bending segment BC. As w keeps falling, there will be a point beyond which there will be no more labour to supply. This explains the CF segment. Finally, as w keeps falling, $\gamma w - pD$ will eventually reach zero, and e will then fall back to zero and only the adult will be working. Hence, the supply curve of labour now reverts back to the GH segment. The full labour supply curve of labour is therefore given by ABCFGH. The sharp corners and angularity of the labour supply curve are caused by the simplifying assumptions. With more general assumptions the curve would smoothen out. But the main point is that it will have this basic feature of bulging out and then shrinking back as the adult wage rate falls.

The aggregate supply curve of labour will look the same as this curve, but for a horizontal magnification. I shall, therefore, without loss of generality, assume that this same curve is the aggregate supply curve of labour.

Many of the peculiarities of the child labour market with which the literature has been concerned, such as the possibility of multiple equilibria (Basu and Van 1998; Jafarey and Lahiri 2002; Swinnerton and Rogers 1999), can be constructed by using this kind of supply characterization. But that is not the direction I wish to pursue here. To stay away from that, let me consider the case where the demand curve is sufficiently elastic so that there is only one equilibrium. This is illustrated by the demand curve for labour DD. The market equilibrium is given by the point E, where adult wage, w^*, is below subsistence and there is a small incidence of child labour.

My concern here is with policy interventions. Consider the case where the government, starting from the case illustrated in Figure 12.1, raises the fine for employing children. (We could, also, think of a switch from no fine to some positive fine.) Let the new fine be D' and we are assuming that D' exceeds D. The effect of this on the supply curve of labour is easy to work out. It is obvious that the segment BC will move up to BC' as shown. To understand this, suppose adult wage is fixed at w^*. As the fine for child labour is raised, child wage will drop. Hence, each household will be forced to supply more children to the labour market in order to reach the subsistence target s.

Keeping in mind that e cannot exceed the total available child labour and e will be zero if w is less than pD'/γ, it is evident that the new supply curve of labour is given by ABC'F'G'H in Figure 12.1.

The important property is that for some wage levels, namely, between s and pD'/γ, the higher penalty for child labour increases the supply of child labour. And this leads to the pathological reaction that I discussed earlier, to wit, that child labour will increase as a consequence of a higher penalty for employing children.

To trace the full range of possibilities, continue to raise D. Clearly child labour will rise, and then fall, eventually reaching zero. If, for instance, D is so high that pD/γ exceeds, s, then the supply curve of labour becomes a vertical line through point H and so child labour must be zero in equilibrium.

12.3 THREE REMARKS

The household's behaviour described in the model can be deduced from the more standard formulation of an optimizing household. To see this, let X be the set of all triples, (c, K, L), such that $c \in [0, \infty)$, $1 - e \equiv K \in [0, 1]$, and $1 - E \equiv L \in [0, 1]$, where c is (as before) total household consumption, K is the leisure enjoyed by the children of the household and L is the leisure enjoyed by the adult, with E being the work done by the adult. Each household has a binary preference ordering on X and the household's aim is to maximize the preference by choosing (c, K, L) such that the triple belongs to its budget set defined by $c = (1 - K)w^c + (1 - L)w$.

I shall now define a preference ordering that will generate the behaviour described in the previous section. Let each household's preference ordering, $\succeq$, be described as follows.[5] For all (c, K, L) and (c', K', L'), if $c \geqq s$ and $s > c'$, or $c \geqq s$, $c' \geqq s$, and $K > K'$, or $s \geqq c > c'$, then $(c, K, L) \succ (c', K', L')$. If none of the above conditions is true, then $(c', K', L') \succeq (c, K, L)$.

It is easy to check that maximizing the above preference will lead to adults working as long as adult wage is non-negative and child labour supply responding to changes in w precisely as described in Figure 12.1.

Since the child labour problem is made worse by the imposition of a fine for employing children, it is natural to wonder if it would not be the case that the problem of child labour can be mitigated by subsidizing firms for employing children. The answer is no. A subsidy does not work like the reverse of a tax or a penalty.

To see this we must understand what was implicit in the previous section. Suppose that a firm decides to use C units of child labour. Clearly it can do this by employing different numbers of children. It can, for instance, employ $2C$ children with each child doing half-time work or C children with each child working full time. In most models of economics, it does not matter how the total is broken up. In the above model, with a penalty for every child that is found working in the firm, a firm will have a preference for employing as few children as possible. So if the firm decides to have C units of child labour and gets this from n children, then the cost (wage bill plus expected penalty) is given by $w^c C + npD$. Clearly, it will try to make n as small as possible; hence n will be C.[6]

The trouble with a *subsidy* for employing children is that this implicit assumption (which is valid when there is a *fine* associated with child labour) in the above model breaks down. In the presence of a subsidy for each child employed it will be in the interest of the firms to get the same volume of labour from many children and take these children to the local government office as proof of child labour and collect the subsidy.

Finally, it is worth emphasizing, as I have cautioned elsewhere, that a decline in child labour need not *always* coincide with a rise in child welfare. Hence, if one is trying to maximize welfare, one may choose not to punish child labour. That is, however, not the point that was being made in this chapter. Here we were not concerned with welfare, but simply the incidence of child labour and what was demonstrated was that, even from the point of view of this limited objective, certain obvious policy deterrents may not be worth using.

NOTES

1. According to latest ILO (2002) estimates there are 186 million child labourers in the world.
2. It must be clarified that to say that poverty causes child labour is not to deny that child labour can have other causes, such as, lack of schooling opportunity or credit, parental illiteracy (see, e.g., Baland and Robinson 2000; Bhalotra and Heady 2003; Emerson and Souza 2003).
3. Natural though this assumption is it is at the root of other unexpected results in this area (see Basu 2000; Rogers and Swinnerton 2004; Singh 2003).
4. In a political economy model, these would be the outcome of some political process (see, for instance, Doepke and Zilibotti 2005), but I shall treat these here as exogenous.
5. I use $>$ and $-$ to denote the asymmetric and symmetric counterpart of $\geq$.
6. I am assuming that a child's probability of getting caught depends simply on whether or not a child works and not on how much she works. The analysis would go through with a less extreme assumption.

REFERENCES

Baland, J.-M. and J. Robinson, 2000, 'A model of child labour', *Journal of Political Economy*, 108.

Basu, K., 2000, 'The intriguing relationship between adult minimum wage and child labour', *Economic Journal*, 110.

Basu, K. and P. H. Van, 1998, 'The economics of child labour', *American Economic Review*, 88.

Bhalotra, S. and C. Heady, 2003, 'Child farm labour: the wealth paradox', *World Bank Economic Review*, 17.

Doepke, M. and F. Zilibotti, 2005, 'The macroeconomics of child labour regulation', *American Economic Review*, 95.

Edmonds, E., 2005, 'Does child labour decline with improvements economic status', *Journal of Human Resources*, 40(1), 77–9.

Edmonds, E. and N. Pavcnik, 2005, 'Child labour in the global economy', *Journal of Economic Perspectives*, 19(1), 199–220.

Emerson, P. and A. Souza, 2003, 'Is there a child labour trap? Intergenerational persistence of child labour in Brazil', *Economic Development and Cultural Change*, 51.

Government of India, 1986, Child labour (prohibition and regulation) Act, 1986. (http://www.vakilno1.com/bareacts/childlabouract/childlabouract.htm).

Grootaert, C. and H. Patrinos, 1999, *The Policy Analysis of Child Labour*, St. Martin's Press, New York.

ILO, 2002, *Every Child Counts: New Global Estimates on Child Labour*, IPEC-ILO, Geneva.

Jafarey, S. and S. Lahiri, 2002, 'Will trade sanctions reduce child labour? The role of credit markets', *Journal of Development Economics*, 68.

Nardinelli, C., 1990, *Child Labour and the Industrial Revolution*, Indiana University Press, Bloomington, IN.

Rogers, C. A. and K. Swinnerton, 2004, 'Does child labour decrease when parental incomes rise', *Journal of Political Economy*, 112.

Singh, N., 2003, 'The impact of international labour standards: a survey of economic theory', in K. Basu, H. Horn, L. Roman, and J. Shapiro (eds.), *International Labour Standards: History, Theories and Policy*, Blackwell Publishers, Oxford.

Swinnerton, R. and C. A. Rogers, 1999, 'The economics of child labour: Comment', *American Economic Review*, 89.

13 Efficiency Pricing, Tenancy Rent Control, and Monopolistic Landlords

with Patrick Emerson

13.1 INTRODUCTION

Standard, or first generation, rent control places a ceiling on the rents that a landlord can charge. Hence, under standard rent control, excess demand for housing is common.

Arguably, more common than standard rent control is 'tenancy rent control', which allows a landlord to set the rent freely when leasing to a new tenant (subject to, of course, the tenant's right not to accept), but prevents the landlord from raising the rent or evicting the tenant.[1] It will be shown in this chapter that through the workings of the market, tenancy rent control can result in an outcome which looks *as if* there is standard rent control. That is, in equilibrium it may be in the landlord's interest to keep the rent for new tenants so low that there is excess demand for housing at that rent. The landlord voluntarily behaves *as if* there were a legal ceiling on the rental rate. This surprising result has analogies in the theory of efficiency wage (Leibenstein 1957; Shapiro and Stiglitz 1984) or the theory of efficiency interest rate (Stiglitz and Weiss 1981).

It should be clarified here that our model is one where there is tenancy rent control and landlords have monopolistic power. In fact what we are considering here is the case of a monopolist landlord—which is the other polar extreme to the case modeled in Basu and Emerson (2000). We do not claim that this case is in any way closer to the reality of housing markets than the perfect competition

The authors would like to thank David le Blanc and two anonymous referees of *Economica* for insightful comments and criticism.

First published in *Economica*, Vol. 70 (2003), 223–32.

assumption. The reality is surely somewhere in between the two models. We do believe that it is essential to understand both polar cases so that we can better comprehend the reality of modern rent control systems. The standard result of monopoly (with fixed supply) where price is raised above the market-clearing level and some of the product (housing, in this case) remains unsold (vacant), arises in our model. Of course, a monopolist with a fixed supply may not sell all of its good; this is also standard and arises in our model as well. But what is surprising is that, under certain parametric conditions, we can have the opposite case, where the landlord sets the rent so low as to give rise to excess demand for housing.

The assumption of perfectly competitive or monopolistically competitive housing suppliers is a common one in the theoretical literature on rent control (see Arnott 1995, for a thoughtful review of the literature.) But there is a fair amount of empirical evidence suggesting that many rental housing markets are far from competitive. Cronin (1983), for example, notes that in the Washington, DC, suburbs of Virginia, on average 70 per cent of all units in each submarket are controlled by one owner and that the average number of rental housing firms in each submarket is slightly over four. Mollenkopf and Pynoos (1973) noted that in Cambridge, Massachusetts, 6 per cent of the city's households controlled 70 per cent of the rental housing units. In addition, they estimated that 90 per cent of the apartment owners in the city belong to an association of property owners 700-strong, and that within that organization twenty owners account for 40 per cent of the rental housing stock. Hence even when there are many landlords there is scope for monopoly behavior through collusion. Appelbaum and Glasser (1982; as cited in Gilderbloom 1989) found that in Isla Vista, California, twenty-seven owners controlled 50 per cent of the rental housing stock. Close by in Santa Barbara, Linson (1978; as cited in Gilderbloom 1989) reported that over 50 per cent of the rental housing was owned by only sixty owners and that seven of them could account for 20 per cent of the rental stock. Finally, Gilderbloom and Keating (1982; as cited in Gilderbloom 1989) found that in Orange, New Jersey, just ten owners control close to one-third of the rental housing, and that in Thousand Oaks, California, just one owner controls over 30 per cent of the rentals. In an empirical study of supply side concentration of the rental housing market in Boston, Cherry and Ford (1975) find that housing prices are significantly determined by concentration of within-market segments.

Thus, there are many reasons to believe that housing markets are less than perfectly competitive; the concentration of rental property in the hands of only a handful of major owners and the collusive opportunities arising from the presence of landlord's associations are two. There are theoretical bases for believing in monopoly power as well; for example, Arnott (1989) hypothesizes that the indivisibility and heterogeneity of housing markets leads to monopoly power on the part of landlords.

There is also reason to believe that landlords do not always respond to housing shortages by increasing initial rents and the supply of apartments. For instance, in 1998 the Ontario provincial government relaxed Toronto's rent control laws, giving landlords the right to raise rents to market levels whenever a tenant moved out. However, landlords did not respond with an expected building boom—fewer than 600 new units were built over the subsequent two years while Toronto's population increased by more than 100,000. Vacancy rates in Toronto remain under 1 per cent (Brown 2000).

Again, this does not mean that a model assuming landlords are monopolists is any closer to reality than one assuming that they are perfect competitors. What we are attempting to do in the present chapter is to examine the other extreme that has been neglected in the literature and to point out the aspects that are particular to the monopoly case so that we may better understand the true nature of rent-controlled housing markets.

13.2 THE ALGEBRA OF RENT

We will consider the effects of rent control when the supply of rent-controlled housing is limited. New York, inner-city Mumbai, and Delhi are examples of this. Rent controls can, however, be of many kinds. They can take the form of a rent fixed by a rent control authority or government (see Olsen 1988 for a discussion of the different forms of rent control), or of a law that gives landlords some or full freedom to adjust rents when leasing out property to new tenants but then requiring the rent to be held constant (or adjusted upward only within limits) as long as a tenant remains the lessee (and with the landlord having no right to arbitrarily evict a sitting tenant). This latter form of rent control, called 'tenancy rent control' (see, e.g., Arnott 1995; Basu and Emerson 2000; Börsch-Supan 1986; Nagy 1997), is quite pervasive and is the subject matter of this chapter.

Given tenancy rent control, the presence of even a small positive inflation gives rise to an adverse selection problem. Landlords now prefer short-staying tenants to long-staying tenants (as long-stayers impose greater costs on landlords because of the erosion of real rents during a single tenancy), but they have no way of telling the types apart.[2] Long-staying tenants know their type but have no interest in revealing this information to prospective landlords. Curiously, the relation between rent control and inflation remains a neglected subject. We tried to develop the building blocks of a model for analysing this in Basu and Emerson (2000).

In the present chapter we develop some of the basic theory in a continuous-time model and build into our model some elements of reality—to wit limited supply and monopolistic power on the part of landlords—that have not been modelled thus far.[3]

Each (potential) tenant has an exogenously given duration of tenure t (>0). When we say that a tenant is of 'type t' we mean that the tenant will move from an apartment after t periods. There is a continuum of tenants, and their density function on the tenure duration, t, is given by $f(t)$, with $F(t)$ being the corresponding distribution function. All agents are supposed to have the same discount rate $\delta \in [0, 1)$.

We denote the total number of tenants in the rental market by N. Hence

$$F(\infty) = \int_{t=0}^{\infty} f(t)dt \equiv N.$$

Suppose a landlord leases out to tenants of only type t (i.e., gets a type t tenant, after every t periods), and each time a new tenant comes he fixes the rent so that its real value is \$1. Thereafter the nominal value of the rent remains fixed so long as the tenant does not leave. Let the inflation rate be such that the value of each dollar erodes in each period at the rate of $1 - \beta$, where $\beta \in (0, 1)$. Under these circumstances the present value of the landlord's real income is denoted by $v(t)$. Clearly then,

$$
\begin{aligned}
v(t) &= \int_0^t e^{-(\beta + \delta)x}\, dx + e^{-\delta t}\int_0^t e^{-(\beta + \delta)x}\, dx + e^{-2\delta t}\int_0^t e^{-(\beta + \delta)x}\, dx + \cdots \\
&= \frac{\int_0^t e^{-(\beta + \delta)x}\, dx}{1 - e^{-\delta t}} \\
&= \frac{1 - e^{-(\beta + \delta)t}}{(\beta + \delta)(1 - e^{-\delta t})}
\end{aligned}
\tag{13.1}
$$

The process of adverse selection ensures that, for each rental rate, only tenants of a certain type and above will seek housing in the rent-controlled market. This is due to the fact that short-staying tenants no longer find it worthwhile to rent in this market, as they do not see as much benefit from the erosion of real rents than do long-stayers. (For a technical description of this process, see Step 1 of the proof in Appendix, which demonstrates this result mathematically.)

Hence the central mathematical character in such an analysis is $\hat{v}(t)$—the expected present value of rents (in real terms) earned by a landlord who manages to rent out his house to a tenant selected randomly from a tenant pool with tenure time $x \geq t$, at a rent which is equal to 1 real dollar to start with and thereafter kept fixed nominally (so it erodes each period by $1 - \beta$), and each time a tenant leaves the landlord repeats the above procedure. $\hat{v}(t)$ is given by the following expression:

$$
\hat{v}(t) = \int_t^\infty \frac{f(x)}{N - F(t)}\left[\int_0^x e^{-(\beta + \delta)k}\, dk + e^{-\delta x}\hat{v}(t)\right] dx.
\tag{13.2}
$$

To understand this, observe that $f(x)/[N - F(t)]$ is the probability of picking a type x tenant, conditional on tenants of type t and above being available. The expression in the square bracket is the present value of rents earned when the first tenant is of type x.

Now we are ready to state and prove the one technical result on which we will build our economic analysis.

PROPOSITION 1: *If $t'' > t'$, then $\hat{v}(t'') < \hat{v}(t')$.*

PROOF: See Appendix.

With this technical result in the background, it is now easy to describe a full model of rent control. When tenants make the decision whether to lease a rent-controlled apartment, the alternatives they have to keep in mind are for them to find housing in a non-rent-controlled area or to *buy* a house. Let us assume that an alternative housing arrangement costs C dollars (in present-value terms). For simplicity, we assume that C is independent of the tenant's 'type'. This seems reasonable as well. In buying a house the cost will clearly be independent of whether the person is a long-stayer or a short-stayer. Similarly, in renting an apartment in a non-rent-controlled area, the tenant's type is unlikely to matter because the rent can be inflation-indexed or be made contingent on the length of the tenant's stay.

Now suppose that the rent (per period) in the rent-controlled housing is R. The lifetime rental cost to a tenant of type t is clearly given by $Rv(t)$. Recall that $v(t)$ is the present value of lifetime payment made by a tenant of type t if the real rent at the start of tenure is set each time at 1.

Consider now a monopoly landlord, who sets the real (starting) rent equal to R. Clearly, only those type t tenants for whom $Rv(t) \leq C$ will accept this. Since from Step 1 of Proposition 1, we know $v'(t) < 0$, for all t, it follows that all type t tenants for whom $t \geq v^{-1}(C/R)$ will accept the offer. It follows from the definition of $\hat{v}(\cdot)$ that the landlord's expected present value of rental earned from *each apartment* that is leased out is given by

$$V(R) \equiv R\hat{v}\,[v^{-1}(C/R)].$$

From Step 1 of Proposition 1, we know that as R rises, $v^{-1}(C/R)$ rises. Hence, from Proposition 1 we know that as R rises, $\hat{v}\,[v^{-1}(C/R)]$ falls. It is now transparent that as R rises, $V(R)$ may rise or fall.

Figure 13.1 represents a possible picture of $V(R)$.

Earnings per apartment

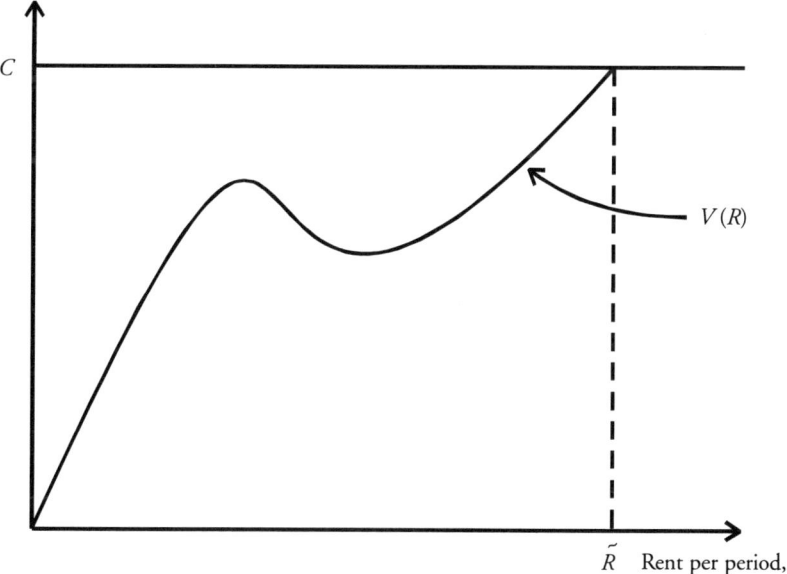

Figure 13.1: Present Value of Rental

Define $\tilde{t}$ to be the supremum of the set $\{t \mid f(t) > 0\}$. In other words, and more informally, $\tilde{t}$ is the upper support of $f(t)$. So $\tilde{t}$, is such that there are no tenants of type $t > \tilde{t}$, and for all $t > 0$, there exists tenants of type $t \in [\tilde{t} - t, \tilde{t}]$.

Now define $\bar{R}$ such that $v^{-1}(C/\bar{R}) = \bar{t}$. Then if rent goes above $\bar{R}$, there are no further takers among the tenants. Hence $V(R)$ is not defined for $R > \bar{R}$. At $\bar{R}$, the only takers are of type $\bar{t}$. Hence $V(\bar{R}) = \bar{R}v(t) = \bar{R}\hat{v}(t) = \bar{R}C/\bar{R} = C$. It is easy to see that, for all $R < \bar{R}$, $V(R) < V(\bar{R})$. This explains the shape of $V(R)$ in Figure 13.1.

It is also evident that $V(R)$ can fall over some stretches. This is especially transparent if tenant types are finite. Then over some increases in R, large numbers of short-stayers can decline the rental offer, leaving the pool of tenants suddenly worse from the landlord's point of view. This is the classic adverse selection problem (Akerlof 1970).

13.3 EXCESS SUPPLY, EXCESS DEMAND, AND EFFICIENCY RENT

The results are the outcome of the landlord's optimization problem when confronted with an earnings curve, $V(R)$. The case of many landlords who drive profits down to zero was analyzed in Basu and Emerson (2000). Here we take on the other polar end: the case of limited supply and monopoly. Rent control applied to a fixed stock of housing, such as in New York, and the evidence supporting the contention that rental housing markets are not competitive, make it worthwhile investigating this polar case.

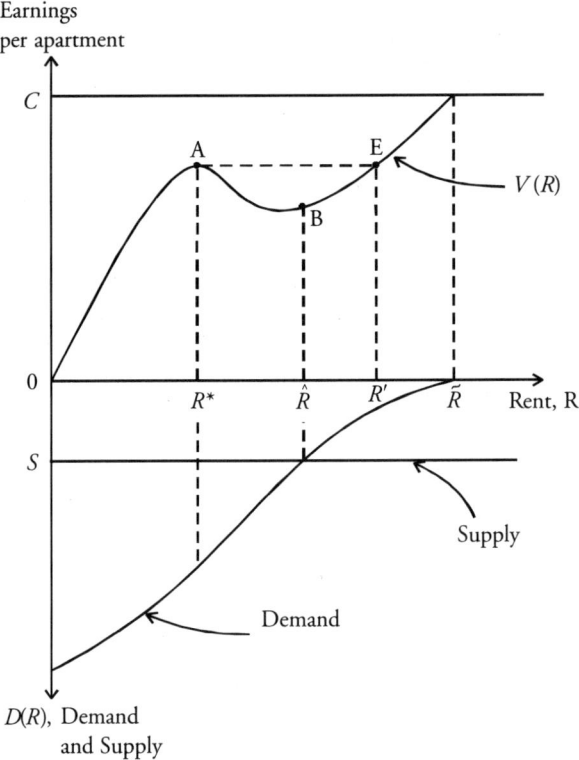

Figure 13.2: Landlord's Optimal Rent

To begin the analysis, let us derive the demand for rent-controlled housing as a function of the (per-period) rent, R. From Section 13.1, we know that, given R, all tenants of type $t \geq v^{-1}(C/R)$ will want to lease rent-controlled housing. Hence, the demand for housing, D, is given by

$$D(R) = \int_{v^{-1}(C/R)}^{\infty} f(t)\,dt \equiv N - F[v^{-1}(C/R)].$$

From Step 1 of Proposition 1, $v^{-1}(C/R)$ rises as R rises. Hence, $D(R)$ declines, as R rises. Such a demand curve is illustrated in the lower panel of Figure 13.2. The upper panel is a reproduction of Figure 13.1.

Next, draw the supply curve in the lower panel. The landlord, it is assumed, owns S units of property. For simplicity, it is assumed that the opportunity cost of leasing out property is zero. Hence the supply curve is perfectly inelastic through the point marked S. Let $\hat{R}$ be the rent that equates demand and supply.

To locate the landlord's optimum rent, consider all rents less than $\hat{R}$, and locate the rent (left of $\hat{R}$) which maximizes $V(R)$. This, in Figure 13.2, is given by R^*. Since $V(R)$ is not necessarily monotonic, there is no reason why R^* will coincide with $\hat{R}$.

Observe that, if the landlords were restricted to selecting a rent less than or equal to $\hat{R}$, they would choose R^*. This is because, for all $R \leq \hat{R}$, they manage to lease out the same number of apartments, to wit, S, and at R^* the per-apartment earnings are maximized. Hence, the total earnings are maximized at R^*.

Next consider rents greater than or equal to $\hat{R}$. As R is raised starting from $\hat{R}$, the earnings of the landlord must eventually (weakly) rise [since $V(\tilde{R}) = C \geq V(R)$, for all R]. However, even if $V(R)$ rises, the total earnings need not rise, because demand falls below S and so more and more apartments remain vacant as R is raised. Let R'' be the rent where total earnings are maximized (subject to $R \geq \hat{R}$).

Let E be the same height as A, and R' the projection of E on the horizontal axis. The landlords' chosen rent will clearly be either R^* or R''. If $R'' \in [\hat{R}, R')$, clearly, their earnings are greater at R^*, because at such an R'', per-apartment earning is smaller *and* fewer apartments are taken. Even if $R'' > R'$, total earnings may be smaller at R'', because at such a rent the landlord will be unable to find tenants for all their apartments.

If the optimum turns out to be at R'', then this is a fairly typical monopoly equilibrium. Monopolists hold back supply in order to push up the price and their earnings.

The interesting case occurs when R^* turns out to be the optimum. Here demand for housing exceeds supply (see lower panel of Figure 13.2). Nevertheless, the landlord prefers not to raise the rent. This is because a higher rent worsens the 'quality' (from the landlord's point of view) of the tenant. This is rather like in models of efficiency wage (e.g., Mirrlees 1975; Stiglitz 1974) or efficiency interest rates (e.g., Stiglitz and Weiss 1981). We shall therefore call R^* the 'efficiency rent.'

Usually, we would expect this kind of a rent to prevail in the market if rent control took the form of an exogenous ceiling on rent. In such a case demand exceeding supply is compatible with equilibrium. What our model illustrates is that, even if there is no ceiling on rents, tenancy rent control can result in behaviour such that the market equilibrium mimics a rent ceiling.

It is interesting to note that the R^* equilibrium is the more likely outcome as S increases. This is because, as S increases, at R'' the landlord's profit is unchanged, because at R'' the number of tenants is constrained by the demand (remember at this point there is excess supply of housing), and so an increase in supply does nothing to the landlord's income. On the other hand, at R^*, the landlord's profit is given by $V(R^*)$ multiplied by S (see Figure 13.2). So an equilibrium with excess demand for housing is more likely when a large portion of the rental stock is under rent control.

13.4 CONCLUSION

Well-meaning urban policymakers of the 1970s and 1980s, attempting to correct the glaring problems of old-style rent controls that placed ceilings on rents (problems that had been illustrated quite vividly by economists), turned to a form of rent control that was more of a tenant's protection legislation than a unit-by-unit rent restriction. This type of 'tenancy rent control' simply restricted landlord's ability to raise rents on sitting tenants and prohibited most side payments and arbitrary eviction. This was seen to be a more flexible program and one that was less susceptible to the inefficiencies of the old rent control laws.

What we showed in Basu and Emerson (2000) is that this new type of rent control brings about different kinds of inefficiencies due to the adverse selection problem brought about by the asymmetric nature of information in these markets. What the present chapter illustrates is that, in the presence of monopolistic landlords, tenancy rent control can cause landlords to operate in such a way that mimics the old-style rent control. To wit, they hold down price, even with excess demand, to attract a better-'quality' tenant (i.e., one that will not stay too long). We call this 'efficiency rent'.

Assuming rental housing markets to be monopolistic may be an abstraction from reality, but, with all of the evidence to suggest large amounts of concentration and possible collusion (through landlords' associations) in many rental housing markets, it is certainly no more of an abstraction than models assuming perfect competition (as is the norm in the literature). We believe that, through the study of both extremes, a clearer and more comprehensive understanding of the reality of rent control will result.

Given the pervasiveness of 'tenancy rent control', it is important to understand fully the nature of the inefficiencies it creates. What the present work illustrates (as does Basu and Emerson 2000) is that certain types of tenants are helped by this policy while other types are hurt. In addition, it illustrates a type of strategic behavior on the part of monopolistic landlords that has not been previously explored in the literature.

A.13 APPENDIX: PROOF OF PROPOSITION 1

To prove Proposition 1, first note that $\hat{v}(t)$ can be simplified, using Eqs. (13.1) and (13.2) to

$$\hat{v}(t) = \frac{\int_t^\infty f(x)(1 - e^{-\delta x})v(x)dx}{N - F(t) - \int_t^\infty f(x)e^{-\delta x}dx}. \tag{A1}$$

From Eq. (A1), it is clear that $\hat{v}(t)$ is the weighted average of the different values of $u(x)$, as x ranges from t to ∞. This is obvious from the fact that, if $u(x)$ is removed (i.e., set equal to 1, for all x) from the right-hand side of (A1), then the right-hand term equals 1. The proposition is now proved in three steps.

STEP 1: We will show that $v(t)$ rises as t falls. In other words, $v'(t) < 0$, for all t. To prove this, note that

$$v(t) = \left(\int_0^t e^{-\beta x}e^{-\delta x}dx\right) + \left(\int_0^t e^{-\beta x}e^{-\delta x}dx\right)e^{-\delta t} + \left(\int_0^t e^{-\beta x}e^{-\delta x}dx\right)(e^{-\delta t})^2 + \cdots$$

$$= \frac{1 - e^{-(\beta + \delta)t}}{(\beta + \delta)(1 - e^{-\delta t})}$$

Then

$$v'(t) = v(t)\left(\frac{(\beta + \delta)e^{-(\beta + \delta)t}}{1 - e^{-(\beta + \delta)t}} - \frac{\delta e^{-\delta t}}{1 - e^{-\delta t}}\right).$$

Note that $v(t)$ is positive. We must now show that the term in parentheses is negative. Rearranging and collecting terms under a common denominator reduces our problem to showing the numerator or

$$(\beta - \delta)e^{\delta t} - \beta - \delta e^{(\beta - \delta)t} \equiv X(t)$$

is negative. We need to prove this for all $t \geq 0$. To see this, note that

$$X'(t) = (\beta + \delta)\delta e^{\delta t}[1 - e^{\beta t}].$$

Hence $X'(t) = 0$ if $t = 0$, and $X'(t) < 0$ if $t > 0$. It follows that, if $X(0) \leq 0$, then $X(t) < 0$, $\forall t > 0$. Finally, note that $X(0) = (\beta + \delta) - \beta - \delta = 0$. This establishes Step 1.

STEP 2: As t falls from t'' to t', the weight on each $v(x)$, for $x \geq t''$, in Eq. (A1) falls. To prove this, suppose $x \geq t''$. The weight on $v(x)$ in Eq. (A1), denoted by $w(x, t)$, is given by

$$w(x,t) \equiv \frac{f(x)(1 - e^{-\delta x})}{N - F(t) - \int_t^\infty f(x)e^{-\delta x}dx}.$$

Note that

$$w(x, t') < w(x, t'') \text{ iff } N - F(t') - \int_{t'}^\infty f(x)e^{-\delta x}dx > N - F(t'') - \int_{t''}^\infty f(x)e^{-\delta x}dx \tag{A2}$$

or $\quad \int_{t'}^{t''} f(x)e^{-\delta x}dx < \int_{t'}^{t''} f(x)dx \tag{A3}$

Inequality (A3) is obviously true. Hence Eq. (A2) is true, which establishes Step 2.

STEP 3: Since the weights on $v(x)$ in (A1) always sum to 1, a decline in the weights on $v(x)$, for all $x \geq t''$, implies positive weights on $v(x)$, for all $x \in [t', t'']$. Hence we know from Step 2 that, as t falls from t'' to t', the weights get transferred from values of $v(x)$ where $x \geq t''$ to values of $v(x)$ where $x \in [t', t'']$. From Step 1, we know $v(x) > v(y)$ for all x, y such that $x \in [t', t'']$ and $y > t''$.

It follows that $\hat{v}(t') < \hat{v}(t'')$. $\hspace{2cm}$ Q.E.D.

NOTES

1. In reality, landlords often are limited only in their ability to raise rents on, and evict, sitting tenants. For the present analysis it is only necessary that the allowable rental increases are insufficient to keep up with inflation, which is generally the case (see Basu and Emerson 2000 for a discussion).

2. This, of course, ignores the possibility that landlords offer tenancy discounts to keep 'good' tenants—those that impose relatively low costs on the landlord. In our model tenants are homogenous except for length of stay. See Hubert (1995) for an excellent analysis of markets with more and less costly tenants.

3. Some monopoly elements were considered in a model of tenancy by Basu (1989), but the focus of that paper was entirely on innovations and the context was that of a backward agrarian economy.

REFERENCES

Akerlof, G., 1970, 'The market for "lemons": quality uncertainty and the market mechanism', *Quarterly Journal of Economics*, 84, 488–500.

Appelbaum, R. P. and T. Glasser, 1982, *Concentration of Ownership in Isla Vista*, Foundation for Natural Progress, San Francisco.

Arnott, R., 1995, 'Time for revisionism on rent control?', *Journal of Economic Perspectives*, 9, 99–120.

—— 1989, 'Housing vacancies, thin markets, and idiosyncratic tastes', *Journal of Real Estate Finance and Economics*, 2, 5–30.

Basu, K., 1989, 'Technological stagnation, tenurial laws, and adverse selection', *American Economic Review*, 79, 251–5.

Basu, K. and P. M. Emerson, 2000, 'The economics of tenancy rent control', *The Economic Journal*, 110, 939–62.

Börsch-Supan, A., 1986, 'On the West German Tenants' Protection Legislation', *Journal of Institutional and Theoretical Economics*, 142, 380–404.

Brown, B., 2000, 'Apartment shortage has Toronto renters searching for space', *Buffalo News*, 1 October, 4A.

Cherry, R. and E. J. Ford, Jr., 1975, 'Concentration of rental housing property and rental housing markets in urban areas', *American Real Estate and Urban Economics Association Journal*, 3, 7–16.

Cronin, F. J., 1983, 'Market structure and the price of housing services', *Urban Studies*, 20, 365–75.

Gilderbloom, J. I., 1989, 'Socioeconomic influences on rentals for US urban housing', *American Journal of Economics and Sociology*, 48, 273–92.

Gilderbloom, J.I. and D. Keating, 1982, *An Evaluation of Rent Control in Orange*, Foundation for Natural Progress, Housing Information Center, San Francisco.

Hubert, F., 1995, 'Contracting with costly tenants', *Regional Science and Urban Economics*, 25, 631–54.

Leibenstein, H., 1957, 'Underemployment in backward economies', *Journal of Political Economy*, 65, 91–103.

Linson, N., 1978, *Concentration of ownership in Santa Barbara*, Santa Barbara Tenants' Union (Mimeo).

Mirrlees, J. A., 1975, 'Pure theory of underdeveloped economies', in L. G. Reynolds (ed.), *Agriculture in Development Theory*, Yale University Press, New Haven.

Mollenkopf, J. and J. Pynoos, 1973, 'Boardwalk and Park Place: Property ownership, political structure and housing policy at the local level', in J. Pynoos, R. Schaffer, and C. Hartman (eds.), *Housing in Urban America* (pp. 56–74), Aldine, Chicago.

Nagy, J., 1997, 'Do vacancy decontrol provisions undo rent control?', *Journal of Urban Economics*, 42, 64–78.

Olsen, E. O., 1988, 'Economics of rent control', *Regional Science and Urban Economics*, 28, 673–8.

Shapiro, C. and J. E. Stiglitz, 1984, 'Equilibrium unemployment as a worker discipline device', *American Economic Review*, 74, 433–44.

Stiglitz, J. E., 1974, 'Alternative theories of wage determination and unemployment in LDC's: The labor turnover model', *Quarterly Journal of Economics*, 88, 194–227.

Stiglitz, J. E., and A. Weiss, 1981, 'Credit rationing in markets with imperfect information', *American Economic Review*, 71, 393–410.

PART III
Globalization and Regulation

14 Global Labour Standards and Local Freedoms

14.1 INTRODUCTION

For some time I have been working on the problem of international labour standards, labour rights, and child labour; and in particular the tensions between global intentions and local aspirations and freedoms. This gives rise to a host of practical problems concerning what the ILO should do, what the World Trade Organization (WTO) could potentially do and what the global policy options are for the US government or the Finnish government. But I plan to dwell relatively little on these practical matters and spend more time on the abstruse theoretical questions that underlie this practical debate. I believe the theoretical debate is important to ensure that our interventions do not go wrong, do not hurt the very constituencies they are meant to help. The impatience that international bureaucrats and policymakers show with abstract debates in their muscular desire to get on with the business of legislating and crafting policy can do much harm. And UNU-WIDER, perched uncomfortably between academe and the world of policy, is a good place to debate some of the abstract principles of economics that underlie global and national-level interventions to uphold minimal labour standards and worker rights.

In the city of Calcutta, a large area called Salt Lake, which was originally a salt marsh, was developed by the local government, with the idea of enabling relatively worse off people to own land and houses. So plots were sold off at a subsidized rate. But it then struck the government that these people to whom the plots were sold could, in the future, lose their land to rich buyers offering to pay a lot more.

The Author thanks Tony Shorrochs for comments, and Alaka Basu, Gayatri Koolwal, Shanti Prasad, and Lorraine Telfer-Taivainen for editorial suggestions.

This was the text of the author's Annual WIDER lecture and was first published in *WIDER Perspectives on Global Development*, Palgrave Macmillan, Basingstoke.

So a law was enacted to disallow the sale of these Salt Lake plots to private buyers. The law was meant to help the disadvantaged people to whom the land was first sold. When I tell economists about this policy, they laugh. Surely, a person who wishes to sell their land will be better off by being able to sell it. So stopping him from doing so can hardly be justified on grounds of *helping* them.[1]

This is part of a larger principle in economics that virtually all economists are brought up on. A contract between two consenting adults, that has no negative fallout on an uninvolved third person, is their business. Government has no reason to intervene and stop such a contract—if anything, government should provide the machinery needed for enforcing such contracts. This 'principle of free contract' is, in turn, derived from a more fundamental axiom of normative economics, the Pareto principle, which asserts that any change, which leaves one or more persons better off and no one worse off, is a desirable change and ought not to be thwarted.

While most economists subscribe to the principle of free contract, many, often unwittingly, support legislative interventions which seem to violate this principle. The same people who laugh at the folly of the government that enacted the Salt Lake legislation frequently support global conventions that disallow workers in poor countries from working at jobs that expose them to significant health hazards or from getting into bonded labour contracts with employers. A 'bonded labour' contract is one in which a worker receives an upfront payment (usually from a big landlord) and promises to pay it back by working for as long as necessary, often lasting several years and maybe even a lifetime, during which time he receives negligible wages (Genicot 2002).

Banning these contracts is often justified by hand-waving references to the ubiquitous 'externality' that economists so often use as an alibi for intervention. I want to argue in this chapter that we need to be more circumspect in justifying bans on such market activity than we have been thus far. The world has gone through a phase of over-vigilant government and over-regulated markets *within nations*. This has given rise to the chorus of demand for economic reform and liberalization. What we are risking now is the same mistake at a *global level*.

Thanks to globalization, it is now easier to intervene in one another's nations and there is a genuine risk of overdoing this. I am not arguing against intervention per se. Governments and international organizations have important roles to play in controlling the market. But interventions have to be evaluated against some well-defined normative principles before they are put into effect.[2] This is necessary to ensure that the agenda of global intervention is not hijacked by lobbies of Northern protectionism, or elite interest groups in the South, and we do not end up with wanton interventionism.

The difficulty of the problem is best illustrated by John Stuart Mill's attempt to grapple with the issue of voluntary slavery. In his *Principles of Political Economy* (1848), after making a spirited case for the principle of free contract, Mill realized that it could take him into troubled waters. In particular, he was uneasy about the fact that some people, driven by their immediate poverty, might be willing to become slaves for life. This dilemma subsequently came to be known as the problem of 'voluntary slavery' or 'waranteeism' (Engerman 1973). While slavery is usually rooted in an initial act of coercion (such as taking people prisoners by force), this is not necessary. There have been times when people, driven by poverty, would opt to become slaves. In the US state of Louisiana, for instance, legislation was passed in 1859 which allowed people the freedom to select masters and become slaves for life.[3] Mill wanted to articulate a principle that would disallow waranteeism and so he argued that very long-term contracts, even if they were voluntary, should not be allowed.[4] This does not seem like a compelling argument to me. It is too *ad hoc*. Besides, it could make it impossible to a get a thirty year mortgage for buying a house.

It is easy for all of us to agree that no worker should have to expose himself or herself to excessive health hazards, and that no one should have to contract away their future labour so as to survive the day. But this does not mean that hazardous work should be banned and bonded labour contracts disallowed. Of course, no one should be poor enough to have to enter into these kinds of contracts. But if people *are* that poor, it is not evident that banning is the right response. There may, nevertheless, be a case for some of these bans but such a case has to be constructed much more carefully and checked against well-articulated principles before they are put into effect. And we may also have, on occasions, to shelve the idea of a ban in favor of more nuanced policy interventions or even no policy intervention at all.

My aim here is to develop criteria against which potential interventions in the market will need to be checked, and to isolate conditions that must be fulfiled before we sanction the use of a legislative intervention in the market. But I want to begin by showing why wanton global intervention is especially risky in today's world.

14.2 GLOBALIZATION AND THE RETREAT OF DEMOCRACY

One matter that has received little attention in the economics or the political science literature, and yet is important for understanding the tensions that can arise between global policy and local freedoms, is the intricate relation between globalization and democracy.[5] This is an important precursor to the analysis of labour standards and so worth recounting, however briefly.

While much has been written in the economics literature on the benefits of globalization (see Bhagwati 1995, 2004; and discussion in Basu 2004), it is important to recognize that one concomitant of globalization is that, *ceteris paribus*, it tends to erode global democracy. This has either not been noticed or has been hushed up by those who did notice it in order not to sully the reputation of globalization. But globalization has many benefits to offer and we would be helping sustain it if we looked at its negative fallout in the eye and worked to counter it, instead of glossing over it.

Democracy entails many things—the existence of a variety of political and legislative institutions, avenues for citizens to participate in the formation of economic policies that affect their lives, and, in the ultimate analysis, a certain mind-set. Yet at its core, democracy requires that people should have the right to choose those who rule them and who have influence over their lives and well-being, and the principle that the vote of one person should count as much as another persons.

Next, note that globalization, almost by definition, means that nations and people can exert a greater influence on other nations and the lives of citizens in other nations. Moreover, it is a fact that the power of one nation to influence another is by no means symmetric. The United States, for instance, can cut off the trade lines of Cuba. It can do so not only by curtailing its own trade with Cuba but by threatening punitive action against those who trade with or invest in Cuba. This is not just a hypothetical possibility, the US Helms-Burton Act is testimony to how it can actually happen. Cuba, on the other hand, can do little to hurt the American economy or polity. Likewise, China can do things to Taiwan, which Taiwan can in no way reciprocate and India to Nepal that Nepal would find hard to counter.

If we now bring the 'axioms' of the two previous paragraphs together, the 'proposition' is obvious: as the world shrinks and powerful governments develop a variety of instruments and ways to influence the lives of citizens in other nations, it is no longer enough for people to be able to choose the leaders of their own nations. Since democracy requires the ability to choose the leaders who have influence over your life, in a globalizing world such as today's citizens, especially those of poorer, weaker nations, need to be able to vote in the elections of the rich and powerful nations. Since such transnational voting does not happen (and even its hypothetical suggestion sounds absurd to us), globalization is bound to cause a diminution of global democracy. This is the 'dismal proposition' that I had tried to establish and warn against in Basu (2002a).

Fortunately or unfortunately, in today's world, to have influence in the affairs of another nation it is no longer necessary to occupy the other nation's land or even to go to war with it. There are, today, a variety of instruments that nations can use to influence outcomes elsewhere. Foremost among these is money. Thanks to the ease of instantaneous electronic links and the improving system of global guarantees, capital has flown across national boundaries as never before. And a rapid withdrawal of such capital can have devastating effects on the debtor nations, as we saw in 1997 when the Asian super-performing economies succumbed to financial crisis (Rakshit 2002).

Like capital, international trade (after a slowdown in the years between the two world wars) has risen steadily. These global linkages have fuelled unprecedented growth rates of national incomes (during the 1990s, China grew at around 8 per cent per annum and India at 6.5 per cent) but they have also created new vulnerabilities. Governments and international organizations can now use the threat of disrupting these flows (or the lure of releasing greater flows of money or goods) to enforce conformity to certain kinds of behaviour. And such threats have been used. International organizations have given money while insisting that the developing countries fulfil certain conditions, many of which have had nothing to do with ensuring repayment. These conditionalities have, at times, even been contradictory, such as requiring the debtor nation to practise democracy and for it to privatize certain key sectors, unmindful of the fact that this was often against the collective wishes of the people.

Some of these conditions have been blatantly in the interest of the donor nation. In 1998, during the Asian crisis, the rescue package put together with money from several industrialized nations, most prominently Japan and the United States, had clauses that required Korea to lift bans on imports of certain Japanese products (which Japan had for long been trying to sell to Korea) and to open up its banking sector to foreign banks (an item that had long been on America's bilateral agenda with Korea).[6]

Given that the benefits of democracy are ample, as modern research has shown, this erosion of global democracy must have negative fallouts. Indeed, it is arguable that the rise in global unrest and political instabilities are a manifestation of this retreat of democracy. And the inchoate demands of the protestors in the streets of Seattle and Washington may be founded in an intuitive but ill-articulated perception of this erosion of democracy. This is a large and important problem, but not one that I am concerned with in this chapter.

Given that cross-country voting is unrealistic and not likely to happen, our best option is to strengthen global governance and make international organizations and global agreements sensitive to the needs of the poor in developing nations. Global conventions signed by ILO member countries, international agreements on non-proliferation of arms, global treaties on the environment, the semblance of global law that is beginning to emerge as the WTO passes judgments on nations' trading practices and the International Criminal Court (ICC) established following the Rome Treaty of 1998, can all have a major role to play in influencing behavior in distant lands and villages.[7] The status of women in an African village and working conditions in a Vietnamese factory can be influenced by a global treaty on the rights of women and a global convention on labour. Hence, these global arrangements can have a major influence on global democracy and local freedoms. In crafting global conventions and agreements we have to keep in mind this larger role that they can and do play. It is with this awareness in the background that I want to enter the arena of global labour standards and local freedoms and well-being.

14.3 THREE EXAMPLES OF LABOUR STANDARDS IN DEVELOPING COUNTRIES

Let us consider three rather controversial matters of international labour standards, where people are often tempted to intervene globally in local practices in developing countries. These are three of several possible contentious matters confronting international policymakers today.

14.3.1 Labour Rights in Export Processing Zones

Many countries have done very well in the export market by creating special export processing zones (EPZ), where firms that produce, exclusively or primarily, for international markets are given special land and benefits by governments.

Mexico's *maquiladoras*, China's special economic zones (SEZs) and Malaysia's numerous EPZs and 'licensed manufacturing warehouses' are examples of this. One of the things that multinationals working in these zones do not want is labour trouble and the disruption of work. To ensure this, some countries require workers in these zones to give up their collective bargaining rights and desist from trade union activity, and suspend the application of minimum wage laws (which are frequently the source of worker–management dispute), which may be effective elsewhere in the nation. That is, workers wanting to work in these zones have to relinquish some of the rights that other workers in other parts of the country can take for granted.

This has led to protests that EPZs that indulge in such practices are anti-labour and should be closed down. I do not think, however, that such a conclusion can be reached quite so easily. This is because no one is compelled to work in an EPZ. If a worker chooses to work in one and is accepted by a firm, then presumably both the worker and the employer like it that way and the principle of free contract seems to kick in. As Judge Robert Sweet, presiding over the case in which McDonald's was sued for causing obesity said, while dismissing the case, 'nobody is forced to eat at McDonald's'.[8]

Moreover, investigations by the US Department of Labour (1990) confirm what most of us suspected—that, overall, workers in EPZs are better off than their counterparts elsewhere. It is true that in some countries workers in EPZs may have fewer rights, but typically they get higher pay and have superior working conditions.[9] It is not surprising that many people would choose to work in an EPZ. Banning the practice of curtailing worker rights in EPZs is thus certainly not axiomatic, especially for the policymaker who values the Pareto criterion. There may be arguments for a ban but, to be compelling, such arguments need a great deal more sophistication than we have seen thus far. Moreover, we may need to distinguish between different kinds of cases.

The US Department of Labour (1990) study on EPZs reports a case in the Guangdong province of China, where some foreign-owned export firms were allegedly offered forced prison labour for work in the SEZs. The prisoners, mainly female prostitutes, would not be paid a salary but earn points towards their release, while the salary on their behalf would be collected by the Provincial Security Bureau. The firms reportedly rejected the offer so the scheme was not put into effect. But even if the firms had accepted it, such a contract quite obviously could not be defended under the principle of free contract. The fact that the workers in this case were being forced into this kind of work meant that the Pareto criterion would not apply here,[10] and so we can rule this kind of practice in an EPZ as wrong without having to look for further justification. But a situation where workers have the option of not working in an EPZ, but nevertheless choose to work in one, is the context where one needs to have more sophisticated arguments for banning the EPZ, and also be prepared for the fact that there may not be such arguments. So we may have no good reason for interfering in the market in these developing nations. In a later section, I shall show how a case for intervention in some free-market transactions can be developed and how we may use it as a litmus test to decide which free-market agreements in labour markets can be construed as not permissible.

Lest it be thought that having special provisions for the export sector is a prerogative of less developed nations, let me emphasize that that is certainly not the case. The idea of export zones with special rights and privileges originated, in all likelihood, in Ireland and quickly took the form of an actual EPZ around Shannon airport in 1960. Moreover, the general phenomenon of having special laws for exporting firms has a long history in industrialized nations. In the United States, typically, there are very stiff anti-trust laws for firms indulging in collusive activity. However, there is another law, the Webb–Pomerene Act of 1918, which exempts firms from the jurisdiction of anti-collusion law if they can show that the bulk of their production is for the export market. In Japan, firms have similar exemptions, and in fact an exporting firm that breaks a cartel can actually be punished by government.

14.3.2 Child Labour

A lot has been written about the reprehensible practice of child labour. But should it be banned? Should we stop the import of products that are made with child labour? I think the answers are not as obvious as may at first sight appear.

Before going into a discussion of these questions, let me briefly recount some of the facts. According to ILO (2002) estimates, there are 186 million child labourers in the world.[11] Child labour is notoriously difficult to measure and any estimate can be subjected to criticism. The ILO estimate tends to overcount in some ways and undercount in some. It counts a child below the age of twelve as a labourer if the child reports having done paid work (or work towards producing a good that is sold on the market) for more than one hour in the previous reference period.[12] What constitutes a 'reference period' can, however, vary—it can be a day or even a week. When it is a week, this amounts to a child doing ten minutes or more work per day as being counted as a child labourer. This would seem to me to be too undemanding a condition: virtually every child in poor regions, if he or she reports truthfully, would be counted as a labourer. On the other hand, domestic work, some of which can be onerous and hazardous, does not count as labour. This leads to serious undercounting of child labour and, especially, 'girl labour'.

Not having any other statistics on hand and given that the above estimate is subject to both an overcount and an undercount, it seems reasonable at this stage to go along with the ILO estimates. It is evident that the incidence of child labour is very high and, in the twenty-first century with technology surging and human wealth reaching for the skies, this cannot but be a source of some embarrassment.

From this it is easy to jump to the conclusion that child labour should be banned. If, however, these children who work do so because they prefer to work and escape acute poverty then it is not obvious that there is a moral case for legislatively banning child labour. I shall show later that there may in some cases be a justification for a legislative ban (Basu and Van 1998) but the argument is a complex one, resting on the possibility of multiple equilibria in the labour market and whether or not this multiple equilibria argument applies in a particular situation is ultimately an empirical matter. I shall return to this later.

As one might suspect, what policy stand we take on child labour depends crucially on what causes it; and the causes of child labour are many. In a region where schools are unavailable or are of a poor quality, the level of child labour is usually found to be high.[13] In regions where schools provide incentives such as midday meals or other subsidies, there tends to be a lower level of child labour (Bourguignon, Ferreira, and Leite, 2003; Dreze and Kingdon 1999; Ravallion and Wodon 2000). Rural households that have more land often have their children working harder than households with no land (Bhalotra and Heady 2003). However, the most compelling reason for child labour is undoubtedly poverty. If a household is sufficiently prosperous, it will not send its children to work, whether or not there is a school in the neighborhood and whether or not it owns land. Barring rare exceptions, poverty seems a necessary condition for child labour, and there is now an enormous amount of micro and macro evidence on this.[14]

A single, elegant exercise that sums the situation up well is a figure that occurs is Humphries (2003), based on an earlier scatter diagram due to Krueger (1997, pp. 281–302). The figure has GDP *per capita* on the horizontal axis and the percentage of children aged ten to fourteen years who work on the vertical axis. On this is plotted a point for each country in 1992. The scatter is a nicely declining one and a line fitted to this is a monotonically downward-sloping one, going from a child labour participation rate of nearly 50 per cent for very poor countries to nearly zero for countries with *per capita* income above US$ 20,000. In the same figure, Humphries inserts points representing the United Kingdom over time, in 1800, 1840 and 1870. During this period the United Kingdom was becoming richer, of course, and we find child labour falling.[15] One can fit a virtual straight line through these points. Once again, we get a monotonically downward-sloping line, suggesting that as people become richer, child labour decreases. But the United Kingdom over time line is significantly above the cross-country line for 1992. What does this tell us? That both lines are downward-sloping suggests that a greater income causes child labour to fall (in other words, poverty is a major cause of child labour) but the fact that in the nineteenth century child labour was greater than in 1992 for comparable *per capita* incomes reveals that income alone is not sufficient to explain child labour. It is possible that social norms matter (Basu 1999; Lopez-Calva 2003) and nineteenth-century Britain had norms that tolerated more child labour. Or that technology matters, and today's world has technology that discourages child labour (Humphries 2003; Tuttle 1999). One can think of many other factors.

However, if the primary reason for child labour is poverty, as seems to be the case, then the use of a legislative ban is harder to justify. One can say that no one should be poor enough to have to make their children work. But the fact is that many people *are* that poor. And then what? If we cannot obliterate poverty, should we tolerate child labour? The answer that I have given is that we should, in that case, try to improve adult wages and employment. This will on its own cause a lot of child labour to go away. But what happens if it turns out impossible to raise adult wage and employment? This is very possible in contexts where poverty itself is stubborn. Should we then legislatively ban child labour? My answer would be, typically, not even then. But the qualification, 'typically', is important because there are significant exceptions.

One possible exception, and therefore a case for a ban, would occur if we had evidence that children were being forced to work by their parents, guardians, or whoever, because the use of force immediately puts such work outside the ambit of the principle of free contract. What constitutes forced labour in the context of children can be a philosophically tricky question (see Satz 2003), since children are treated by us as not always able to judge their own interest.[16] Children often have to be compelled to go to school by their parents—and the parents could be acting entirely in the interest of the children. Should schooling in such cases be described as 'forced schooling'? And even if the answer to that is 'yes', should we treat that as bad just because it is forced? It will be too much of a digression to go into these issues here, and it must suffice to have raised this problem and move on.

The other possible exception where legislative bans are justified is the case where we may have reason to believe that there are multiple equilibria in the child labour market. I shall examine the importance of this argument later.

14.3.3 Forced Labour

A topic where the difficulties of crafting global legislation is amply on display concerns forced labour. Stopping forced labour is part of the four core international labour standards that various global organizations have been campaigning for.[17] But our understanding of 'forced labour' is predicated on our understanding of coercion and voluntariness. This is a topic on which economists have had very little to say, but there is a need to articulate the principles involved more clearly. In economics we make repeated reference to the free choice of agents, on how if people *voluntarily* choose something they must be better off with that thing, and so on. The principle of free contract mentioned at the start of this paper makes critical use of the idea of *voluntary* choice and coercion. But economists have spent very little time on trying to understand what exactly these terms mean. Even in the world of policymaking, these terms crop up. We think poorly of forced labour. The ILO has begun work on counting the number of people who can be thought of as 'forced labourers' and enacting conventions to stop this practice—see ILO (2001) for a discussion. A commendable US law, the so-called Sander's Amendment (1997), has ruled that the United States will not import products that are made using forced child labour.

But to have such policies work effectively and in the interest of the common person, we need to be clear about what the terms mean. Without that, there is a risk that these policies will be hijacked by those who have the best lawyers on their team, and the reference to force and voluntariness will become alibis for powerful lobbies to justify whatever it is that they wish to do.[18] If we are to stop genuine forced labour and not just be an unwitting instrument of protectionism or partisan interest, we need to understand better what forced labour is.

Much of the early Chicago school writings erred on the side of voluntariness. As long as a person's decision was voluntary, most of us agree that the state should keep away. If we could claim that virtually all adult decisions were voluntary, then we can argue against government intervention virtually everywhere. And this fitted well with the Chicago school's market-fundamentalist predilection. Voluntary choice was often thought of by those belonging to this school of thought as equivalent to having choice and being able to reject alternatives. But clearly, if one thinks through a little, one realizes that this cannot be so. When one gives away one's watch to a mugger in response to the mugger having pulled out a gun and asked for the watch, one exercises choice. One did have the option of holding on to the watch and getting a gun shot. But no one would think of a person parting with his watch at gun point as doing so voluntarily, unless one is under the sway of a completely tautological definition (Basu 2000).

Likewise, many have erred on the side of 'force'—finding compulsion and force everywhere. If workers accept a low wage, it is because they are forced to do so. If a woman does not go to work, it must be because she is forced to stay at home. And if she does go to work, this is evidence that she is forced to go to work. To use these all-encompassing categories is to rob this important phenomenon of meaning and significance. If we are to stop coercion and force, we must move away from the traps of tautology on either side. I bring the example of forced labour here not to offer any answers but to impress on the reader that these problems of great legislative urgency can be matters of some analytical intricacy. Policymakers, impatient to get on with the task on hand, may not like to pause over such abstruse matters. But it is incumbent on us to think through this and other topics of labour legislation very carefully before we pass laws or plan interventions.

14.4 WHEN CAN WE INTERVENE?

The topic of international labour standards belongs to the larger subject of the role of intervention in the functioning of markets. In particular, we are concerned here with global interventions in, typically, Third World markets. Should there be interventions and, if so, how should they be enforced? Should trade sanctions be used to punish a country that violates minimal labour standards? Naturally, this has been a very contentious subject[19] and this is not the occasion for me to enter the fray. What I want to do here is simply outline three general principles against which a potential market intervention needs to be tested before it is put into effect.

14.4.1 Irrationality and Behavioral Economics

Human beings are often irrational. We are frequently impatient and willing to make disproportionate sacrifices to make good things happen soon. We lack self-control. We often count wrong, especially where returns *over time* are concerned. In particular, we are atrocious at understanding the implications of interest rates, where compounding is involved. In Indian villages I have noticed that the higher interest rates are usually quoted in per-month terms or per-week terms, whereas the lower interest rates are quoted in per-annum terms. Though I have never put this to the test, my hunch is that stating it in terms of a month or a week allows the lender to camouflage the enormity of the interest rate.[20] Modern behavioral economics has drawn the attention of economists to these systematic 'irrationalities' and opened a welcome chapter in economics. It is a bit dismaying, though, that economists *needed* behavioral economics to realize that not everybody is always rational.

The assumption of total rationality has played a major distorting role in our view of the role of government. One of the major achievements of modern society is that we do not have to continuously to fear physical assault and the loss of our belongings and property. Most of the time when we go out for dinner we do so without having to fear that we will have to evict the people who may have moved into our homes during our absence. When we walk with a briefcase, we typically do not fear that it will be taken away by force. One reason for this great improvement on primitive life is that society recognizes the use of force and physical appropriation of another's belongings as wrong.

What, however, continues to happen nowadays is that people get duped out of their wealth. I am not using the word 'duped' here to mean obvious cheating, such as selling adulterated goods or defective products or misinforming the consumer. Against these, in principle, there is recourse. I am referring here to deals in which there is no cheating involved as such, and all information is put on the table. The duping here occurs in more subtle ways, by using the inability of agents fully to comprehend the information that is made available to them or the complications of the contract about to be signed, and also because of some common irrationalities. For instance, complex financial deals and sophisticated contracts, which are difficult for the common person to understand, make lots of people routinely lose out. The main reason society does not provide enough safeguards against this, in the same way that it prevents the physical appropriation of property, is the success of economists in persuading the policymaker that no one ever *does* get duped because human being are always rational.

When a villager regularly takes loans at the interest rate of 10 per cent or more per month, as I saw happen regularly during my field work in Jharkhand (see note 20), economists explain this in terms of the alleged fact of high default rate or monopoly pricing or fragmented credit markets. What we do not do is to allow for the fact that the borrower may not understand what '10 per cent per month' means in terms of the enormity of the repayment burden. Until recently, a paper that made such an assumption would be unlikely to get past the journal refereeing process. Fortunately, this is beginning to change now.

In reality, a failure to understand the meaning of interest rates is one of the major factors behind the prevalence of high rates in rural areas. I also believe that some of the poorest people are poor because of lopsided deals they have made over the years. And it is possible that the people who become moneylenders in rural areas are, maybe through a process of natural selection, individuals who are better at understanding the details of intertemporal deals.

The US Federal Trade Commission has brought to our attention the phenomenon of 'predatory lending'. This refers to the practice of (predatory) lenders locating an attractive property, typically with a single, elderly person living in it who, in their opinion, will be unable to pay back a large loan, and then offering them a large loan for home improvement at seemingly attractive terms. The aim here is to work into the contract exorbitant charges in the event of a default and even attempt to foreclose on the property when the person fails to pay back the loan.[21] If borrowers assessed their ability to repay loans realistically then there would be no scope for this phenomenon. Hence, the existence of predatory lending suggests a less than full rationality on the borrower's part. Once this is recognized, there arises scope for legislative action that could not have been justified otherwise. Consider, for instance, hazardous work. If there is reason to believe that workers do not fully comprehend the long-run harm done to their health through certain kinds of hazardous work, then there could be a case for legally banning certain kinds of such work or fixing, legislatively, the amount of compensation that a worker who is involved in this kind of work must be paid. Presumably this will be an amount greater than what would occur naturally through the market, since the latter relies on the worker's perception of the hazard.

If we could show that parents do not understand the loss of human capital or damage to health that occurs in a child who works, then again this could be reason to ban child labour. We must be careful, however, to recognize that it is not enough to show that a child's health is hurt through labour or that there is loss of human capital (even if this were huge). This argument is predicated on our being able to show that parents underestimate these costs of child labour, and adults in hazardous work underestimate the long-run costs of such work.

The recognition of less than full rationality and the inability to calculate right among ordinary people also opens the door for better redress for those who are tricked into accepting deals that hurt them. This is, however, a very difficult and risky area for government to get involved in, for it can easily lead to excessive intrusion in free- market decisions. Hence, this is a future policy option for us to keep in mind and discuss but not one on the basis of which I would encourage interventions in labour markets.

14.4.2 Multiple Equilibria

A general case where a legislative intervention in free-market contracts may be justified without having to disregard the Pareto principle is one where we have multiple equilibria. This idea has now been used in several areas of development economics, from the theory of child labour to the analysis of policies for releasing a nation caught in the vicious circle of poverty.[22] The idea is simplest to outline in a game-theoretic setting. Consider the problem of hazardous labour (though this could equally apply to child labour, or statutory limits on hours of work, or the worker's right to be part of a union). Let us consider a case where there are no government restrictions on hazardous work. Each worker looks around the labour market, at the wages that prevail for 'safe' work (S) and 'hazardous' work (H), and decides which one to opt for. The worker, in other words, has a right to a safe environment but this is not an inalienable right. If she wishes to divest herself of this right in order to earn some more money (since, presumably, hazardous work will command a higher wage than safe work), he has the freedom to do so.

Now, without going into the details of the economy, it seems reasonable to posit that what a person wishes to do could depend on what others do, since what others do could affect the wages that hazardous work and safe work command. Improvising a bit on standard game-theory notation, let me denote by $B(H, H)$ the payoff that a worker gets if she does hazardous work and all other workers do hazardous work. The first H in $B(H, H)$ denotes what she does and the second H denotes what everybody else does. Likewise, I shall use $B(H, S)$ to denote the payoff that a worker expects to earn if she does hazardous work while everybody else does safe work. $B(S, H)$ and $B(S, S)$ are interpreted similarly. These are all net payoffs, and so take account of the pleasure of the income that such work brings and also the cost of leisure and hazard associated with this work. Let us now consider a case where the following assumptions are true:

$$B(H, H) > B(S, H) \qquad (14.1)$$

and

$$B(S, S) > B(H, S). \qquad (14.2)$$

It is true that I have not specified the full game because it has not been stated what the payoff to an individual would be if half the remaining population does safe work and half does hazardous work, and so on. But even without having to go into this, it is evident that the assumptions of the payoffs just specified give rise to at least two Nash equilibria—one in which everybody chooses hazardous work and another in which no one does. For simplicity, let us assume that these are the only Nash equilibria of this game. Assumptions (14.1) and (14.2) in themselves do not tell us anything about the relative values of $B(H, H)$ and $B(S, S)$. Let us now assume that

$$B(S, S) > B(H, H). \qquad (14.3)$$

Now, let us now consider an equilibrium in which everybody chooses to do hazardous work. Note that each individual, if asked separately what she prefers, hazardous work or safe work, will say 'hazardous' [this follows from (14.1)]. Nevertheless, there is a case here for banning hazardous work. Such a ban would give rise to the outcome in which everybody does safe work. And since this is a Nash equilibrium, if people were *now* given the freedom to choose between safe and hazardous work, they would choose safe work. Moreover, by (14.3), they are all better off in this new equilibrium than the old one. Disallowing free contract has, far from leading to a Pareto suboptimal outcome, taken the economy to a Pareto superior outcome. Hence, if some economy has the characteristics just described, a legislative ban on hazardous work will be justified and, in fact, be required under the Pareto principle.

Another interesting possibility occurs if some people satisfy (14.3) but others satisfy the opposite inequality. Assume, for example, that there are two kinds of people in society: type R and type P. We may think of R as 'rich' and P as 'poor'. Suppose now that everybody satisfies (14.1) and (14.2) but types R and P have differences over how they evaluate outcomes (S, S) and (H, H). Assume, in particular, that

$$B_P(S, S) > B_P(H, H) \tag{14.4}$$

and

$$B_R(S, S) < B_R(H, H). \tag{14.5}$$

In this economy, once again, there are two Nash equilibria—one where everybody chooses S and another everybody chooses H. But neither of these two states Pareto dominates the other. Hence, if beginning from a state where everybody chooses H, the state bans H (and so deflects society to the equilibrium where all choose S), this will not be a policy in violation of the Pareto principle. In the context of child labour (see Basu and Van 1998; Basu 1999) it can be shown, by constructing a full-fledged economic model, that multiple equilibria (namely, payoffs that have the above characteristics)[23] could well occur under realistic conditions. It will be interesting to attempt the same for hazardous work.[24] Only after one has done so can one make a case for placing legislative restrictions on workplace hazard.

At times we find some economists expressing skepticism about these kinds of multiple equilibria results, and policies that arise from them, on the ground that we do not as yet have hard empirical evidence of there being multiple equilibria. This, however, is a fallacy because, just as we may not have evidence of multiple equilibria, we do not have hard evidence of the non-existence of multiple equilibria either. Even if we observe an economy in one equilibrium, from this we cannot assume that the economy has a unique equilibrium. Hence, at least for the time being, the actual crafting of policy has to depend on the demonstration of theoretical possibilities and the marshalling of circumstantial empirical evidence, because empirically we have no reason for favouring one assumption over the other.[25]

I want to end this section with two observations about law. Legal intervention in a multiple equilibrium model is very different from what we see in standard models of law and economics. In the standard model, when a law is needed to change behavior, one needs the persistent presence of the law. For instance, if you declare parking on a busy road illegal and subject to a fine, you will have to have the law perennially there if you wish to prevent cars from being parked on that road. However, in the above model, once hazardous work has been banned for a while, even if the law is revoked, the economy does not return to the original situation, since the new outcome is an equilibrium in its own right. In Basu and Van (1998), we had called this a *benign* legal intervention. The advantage of economies with appropriate multiple equilibria is that we may not need perennial legal or government intervention. A benign intervention may be enough to 'cure' the problem once and for all.

The second point that is important to keep in mind, especially in the context of this chapter, is that the efficacy of a nation's law depends critically on the extent of globalization. I have glossed over this in the model specified here, but in a more fully specified model, with globally mobile capital, the wage that workers in a nation can command will depend on the extent of capital attracted into the country.

The basic point is intuitively obvious. Consider a closed economy that bans hazardous work. This will have a certain effect on the wage rate that prevails for safe work, because presumably capital that supported firms involved in hazardous work will move over to nonhazardous work and likewise labour will move from one sector to another. Presumably something like this occurs behind the sparse game model described above. Now suppose the same ban occurs in an open and globally integrated economy. It is entirely possible that capital will now shift not just to another sector but out of the country to another nation where there is no such law. And given that labour is not so mobile between nations, this will cause the 'safe' wage in this country (post ban on hazardous work) to be lower than what it would be in a closed economy. Hence, a government concerned about labour welfare will find it harder to enact certain laws, effectively, in a more globalized environment. This is a theme that will be picked up later.

14.4.3 The Large Numbers Problem

Let me finally turn to a theoretically more intricate route for justifying interventions to stop *certain kinds* of free-market contracts in the labour market. In economics, we often go from reasoning about a single contract or a limited number of contracts to taking a position on such contracts, in general. As we have seen earlier, economists often argue that if a rational person wants to voluntarily sell his house and another rational person wants to buy it, then these two persons must be better off by the exchange, and since no one else has any obvious negative externality from this, this constitutes a Pareto improvement and so the state should not intervene. From this they jump to concluding that the state should not stop individuals from selling their homes. An intervention that disallows such sales, they argue, is wrong.

This deductive jump is however not as innocuous as may seem at first sight. Can we always go from arguing about the moral status of each single transaction to the moral status of a *class* of such transactions? This is a philosophically difficult question that economists tend to gloss over by answering or implicitly assuming the answer to be 'yes'. The serious investigation into this was undertaken by a philosopher. Derek Parfit, in his celebrated book on moral reasoning (Parfit 1984), answers this question in the negative. That is, he argues that there are certain actions such that, each one of them may be morally acceptable, but the totality of these actions may not be.[26]

At one level, Parfit's answer seems in easy accord with general equilibrium theory, in particular, the concept of pecuniary externalities. Each single person signing an agreement to do hazardous work or to give up his right to join a trade union or her right to be protected against sexual harassment in the workplace may have no externalities, because, by assumption, wages and prices are not affected by a single person's behaviour. So a single person signing such a contract voluntarily must lead to a Pareto improvement (that person and his or her employer are better off and there is no effect on anybody else). But if lots of people sign such contracts, wages can get affected and this could hurt the welfare of uninvolved individuals, thereby rendering the new outcome Pareto non-comparable to the old outcome.

Let me call this standard general equilibrium (GE) proposition the 'GE reversal claim' (GERC). The paradoxical conclusion that GERC enables us to reach is this: If we are committed Paretians (and I maintain that we ought to be), then we should not object to individual workers agreeing to sign away their right not to be sexually harassed in the workplace, or their right not to expose themselves to large health hazards at work, or their right to collective bargaining in order to get some benefit in exchange—for instance, be able to work in an EPZ, where work conditions may be particularly nice or simply to earn a higher wage. But we may nevertheless, without having to face the charge of inconsistency, enact a law that prohibits sexual harassment in the workplace or excessive health hazard at work or forbids everyone from giving up his or her right to bargain collectively, whether or not an individual worker and an employer find such a contract worthwhile. This is because the enactment of a law amounts to banning a whole *class* of actions or contracts, and we can, by GERC, take a different normative stand on a class of actions and on each action in the class.

It is interesting to note that the justification for this kind of a legislative ban, is very different from the ones commonly proposed. Here, the case for the ban is founded on the harm it does to 'others'—those who suffer the pecuniary externalities of many people doing some kind of trade and exchange.[27] Hence, if a firm offers job contracts which involve high wages but requires workers to forego the right not to be sexually harassed at work (and this is made clear to the worker at the time of her taking up employment), the reason why this should be stopped is not to protect individual workers from being harassed by this entrepreneur, but because of what firms offering such contracts do to wages and the well-being of workers—and, in particular, to the well-being of those who are especially strongly averse to harassment. This principle sounds inimical to our commonsense, simply because we have reasoned so poorly for so long in these difficult areas of labour rights. But this is a very sensible way to understand why we may wish to legislate against EPZs where workers have to give up their rights or against contractual sexual harassment in the workplace, or against hazardous work.

The discerning reader would have noted that just because allowing workers the freedom to forego a certain right does not lead to a Pareto improvement—this does not mean that workers should *not* be allowed to forego the right. All that we have shown is that, when the GERC principle holds, we do not, on purely Paretian grounds, have a case for either upholding the right as inalienable or permitting workers to trade the right away. And, equivalently, upholding the right and not upholding it are both compatible with Paretianism. From this to go to a definite prescription we need further moral axioms.

One route to such a further axiom is to note that we, quite naturally, think of certain human preferences as (morally) fine—or what will be here called 'morally maintainable'—and others as not. A person's racist preference for giving jobs only to Whites would be considered by most people as 'wrong' or not morally maintainable. On the other hand, we would consider a person's propensity not to work four days a week as fine—that is, morally maintainable. We would agree that this is likely to be harmful for the person himself, but not a preference that most of us would consider morally wrong.

Now among preferences that we consider morally maintainable, it seems possible to make some further distinctions. I have argued in (Basu 2003) that there are certain morally maintainable preferences, which are special in the sense that most of us would agree that not only do people have the right to have these preferences but, in addition, no one should have to pay a penalty for having such a preference. These may be called 'inviolable preferences'. Thus while we agree that it is fine for Rip to want to sleep all day, four days a week, we at the same time see nothing wrong in the fact that Rip will have to pay the price of poverty for having this preference. On the other hand, most of us would agree that no one should have to pay a price for being averse to being sexually harassed or bullied. Hence, the preference not to be sexually harassed is an *inviolable preference*.

What exactly is considered an inviolable preference is of course not given *a priori*. Being a normative matter there may also be disagreement between different societies about what should figure under the description of inviolable preference. Is a father's aversion to let his child do hard labour an inviolable preference or merely a morally maintainable preference? What about the preference not to be exposed to huge health hazards at work?

These could of course be contentious matters.[28] But most of us would agree that there exist preferences that are inviolable. And this enables us to reach clear conclusions in choosing between certain Pareto noncomparable alternatives. Suppose states x and y are Pareto noncomparable but in x some people who are especially averse to being sexually harassed has to pay a special price for having this preference. We should in that case consider y superior.

The use of the GERC, along with the normative criterion of marking certain preferences as morally inviolable, could allow us to reach policy prescriptions about labour rights—in particular, to treat certain rights as inalienable.

It is worth noting that the moral criterion developed here belongs to neither welfarism nor deontological ethics. It involves a blending of welfarism and nonwelfarist considerations.[29] I shall here call it a 'miscible moral'. What I am arguing is that we should be welfarists in applying the Pareto criterion (that is, use individual welfare data to check if between states and y, everybody is at least as well off in x as in y and there exists one person who is better off in x than in y and, if so, to declare x to be socially superior to y) but among alternatives that are Pareto noncomparable we should be prepared go beyond people's welfare information. We should look at the basis of, or what *underlies*, a person's preference or welfare. If in state x, person i is unable to go to the cinema and so i prefers state y to x, and in state y, person j is exposed to large workplace health hazard or faces sexual harassment at work and so j prefers x to y, then in deciding between x and y, society has to look beyond i's and j's welfare intensities. Society should choose x because y results in a violation of what we consider (to the extent that we do) to be someone's inviolable preference. The reason why this argument is partly welfarist is because it fully heeds the Pareto principle. But it is not entirely welfarist because nonwelfare considerations kick in when the Pareto principle gives us no verdict. Observe that under this miscible moral system if an adult says that she hates to be harassed but is willing to put up with it (because of the higher wage promised by the sleazy employer),[30] society has no authority to stop this transaction by saying that the harassment hurts her.

In the existing literature there is some discussion on unacceptable preferences and this has led to the suggestion of using preference-based welfare criteria but only after the 'purification' of individual preferences (see Sen 1997, for discussion on this). It seems to me that such purification should be allowed only in the event of Paretian noncomparability. Suppose a person gets pleasure (as an end in itself) from hurting others and suppose this person plans to hurt another person. Should this be permitted? Or, more specifically, should this be allowed if the pleasure the 'hurter' gets is very large compared to the pain the 'hurtee' feels?[31] My response will be: No, this should not be allowed, no matter how big the pleasure and how small the hurt.[32] But now consider another case where the hurter offers to pay money to a free person in order to inflict pain on him and suppose the latter, after properly thinking through, considers the offer acceptable. Should this transaction be allowed? Under the miscible moral system, that I have proposed, the answer is 'yes'.

There is also an important theoretical agenda that this inquiry opens up, which I want to mention here only in passing. The possibility of there being transactions which by themselves are Pareto improving but in their collectivity may lead to a Pareto non-comparable or Pareto inferior state arises by the GERC. The claim is of course standard and is to be found in our textbooks. But that is no reason for accepting it. Is it really a reasonable principle? If each of a class of actions lead to a welfare increase, can all the actions together cause welfare to decline in any meaningful sense? The fact that this is assumed to happen in GE theory is no real consolation. Can such a theory be founded in a consistent logic? We know from the works of Aumann (1964) and Hildenbrand and Kirman (1988) that this is possible in a society with uncountably many individuals. I tried to show in Basu (1994) that in certain game-theoretic situations we can get such results with a countably infinite number of individuals. But evidently this is a topic that deserves further investigation.[33]

14.5 NOTES ON INTERNATIONAL LABOUR STANDARDS

It is time to return to the mundane. Given the above problems and arguments, what is to be done? How can we have global labour standards that do not tramp on local freedoms? This needs to be approached with an open mind. We should be prepared to do whatever is necessary, including doing nothing. The human mind has a natural propensity to prompt us, wherever we see a problem, to do something. In reality, there are lots of problems about which nothing can be done, or at least nothing can be done without making matters worse. The recognition that we live in a necessarily second-best world is important for realistic and successful policymaking.

Some of the problems of labour standards are a concomitant of poverty. The only way to fight these is to fight poverty and we will be successful to the extent that we can be successful in mitigating global poverty.

Also worth keeping in mind is what mainstream economics keeps telling us— that much of what workers have achieved over the last century have been achieved by virtue of the greater demand for labour and the consequent, automatic empowering of labourers. The power of what can be achieved by creating new legal rights may be small compared to what can be achieved by ensuring that labour demand keeps rising, which would enable workers on their own to ask for more and get it. This implies that trade channels ought to be left open. The use of trade sanctions to achieve certain ends should be discouraged. A direct implication of this is that labour standards should be the charge of the ILO and not the WTO.

Further, the culture of contracts—whereby consenting adults can in general make agreements among themselves and have reasonable expectations that the agreements will be fulfiled because the state will help uphold such contracts or, more minimally, not negate them—can play an important role in improving living standards. The significance of this is routinely underestimated by policymakers.

Nevertheless and despite all the above caveats, as we have already seen, there can be contexts where we may legitimately intervene in free market contracts and exchange and do so without abandoning the Pareto axiom. The Pareto reversal proposition explains why we may want to legislate against certain classes of actions that may, on their own, seem to be Pareto improving. We have also seen that the existence of multiple equilibria provides a powerful case for banning certain voluntary, free-market exchanges. But all this applies to any legitimate governing authority. There is nothing special here about global interventions. Interestingly, the multiple equilibria argument developed above can be taken further to provide a new justification for and meaning to the idea of *international* labour standards. As has already been pointed out in the above discussion on hazardous work, in a world with globally mobile capital, a ban on certain labour market practices may not succeed in deflecting the economy to the 'good equilibrium' the way it would have done in an economy where firms were country-specific. This is because every new, unilateral labour legislation would typically be expected to cause some flight of capital.

The only way to get around this is to have all countries—or, at least, all similarly placed countries—legislate simultaneously. This would limit the flight of capital associated with new legislation or government intervention. This is the principle reason why we need *international* labour standards as opposed to purely idiosyncratic and country-specific laws. But this also means that, when talking of international labour standards in poor countries, the initiative and the specific details of what goes into a package of international standards must come from developing countries and not from the governments or lobbies of industrialized nations, no matter how well-meaning they are.

I also hesitate about the use of consumer sanction in industrialized countries to improve labour standards in poor nations. Moral monitoring through consumer activism has the risk of playing into the hands of big businesses that have the power to direct opinion. Also, there is the genuine risk of witch-hunting—the criticism of some companies as violating labour codes that takes the form of hysteria and has very little actual basis. Given that labour standards in poor nations will be lower than in rich nations (not to allow for this is to hurt the poor nations), it is easy for the laity in industrialized nations to get misled into thinking that a firm that pays much lower wages in Vietnam compared to the United States is violating a moral standard.

Once the initiative gets passed onto poor countries and gives voice to workers and the dispossessed in those nations, we may end up getting a very different package of labour standards than the ones that are currently favoured. We may reach the conclusion that in very poor countries, such as Ethiopia and Nepal, where more than 40 per cent of children in the age group 10–14 years are labourers, child labour should not be banned, because this could cause starvation deaths or drive children to other more dangerous livelihoods. We may reach the conclusion that employers must provide bathrooms for women close to the workplace, because people close to the grassroots know that this is a major reason why women cannot take on certain kinds of jobs. It may be agreed upon that there should be no minimum wage laws in the poorest nations.

Giving voice to the Third World means risking that the agenda of international labour standards that we come up with will be significantly different from what would have emerged under the old order. But that is what global democracy is all about.

NOTES

1. I have written about this Salt Lake phenomenon in Basu (2003).
2. This becomes especially important if we view development as essentially an expansion in the freedoms of the people concerned (Sen 1999). Then the curbing of free transactions in markets in the name of encouraging development has to be carefully crafted and justified so as not to be a self-contradictory policy.
3. It is interesting to note, however, that the Louisiana legislators were rather partial and granted this freedom only to persons of colour.
4. In later work Mill (1859) took a somewhat more sophisticated position on this, but never really managed to outline a compelling general principle for the exceptions.
5. I have discussed this at length in Basu (2002a).
6. The pressures on poor nations are not always in the deliberate self-interest of the donors and the industrialized nations. Many of the issues championed in various international fora are (as I discuss below) meant to be for the good of the poor nations. So my concern is not with the specific demands but with the mechanism through which these are brought to bear on the world. Once we create an effective mechanism, this can become a conduit for exploiting poor nations and bending them against their political will. The global mechanisms that are made available need to have built-in safeguards to prevent anti-democratic uses.
7. Shadows were cast on this process as the current US administration has sought to withdraw from some major international treaties. On 6 May 2002, for instance, the US administration announced its intention to withdraw from the Rome Treaty and has actually worked to undermine the ICC by striking bilateral deals with nations to bypass the court, such as that signed with India on 26 December 2003.
8. See *USAToday*, 22 January 2003: www.usatoday.com/money/industries/food/2003-01-22-mcdonalds-lawsuit x.htm
9. Kabeer (2000) for an excellent analysis in the context of Bangladeshi workers.

10. That is, the signing of this contract between the prison authority and a firm in the EPZ would not automatically lead to a Pareto improvement, since there is no reason to believe that the workers would be better off by this.

11. A 'child' for this purpose is someone below the age of 15 years.

12. For someone between the ages of 12 and 15 years, 14 hours or more work in the reference period has him/her classified as a child labourer.

13. For analyses of the relation between schooling and child labour, see Levison, Moe, and Knaul (2001) and Rosati and Rossi (2003).

14. See, for instance, Edmonds (2004), Humphries (2003), Krueger (1997), Ray (2000). For surveys of some of this evidence, see Basu and Tzannatos (2003) and Brown, Deardorff, and Stern (2003).

15. The claim is not without controversy, since child labour data, prior to 1951, was very uneven in Britain. Humphries' (2003) numbers are based on her own computations.

16. Note that we typically speak about the rights of 'consenting *adults*' and not '*people*'.

17. The other three being the stoppage of child labour, prevention of discrimination in the workplace and upholding of workers' right to form unions and bargain collectively.

18. There is some evidence that this happening, as, for instance, when the Sanders' Amendment was cited in the charge that was brought in the United States against Brazil's juice exporter, Sucocitrico Cutrale Ltd.

19. See, for example, Bhagwati (1995, 2004), Chau and Kanbur (2002), Kanbur (2003), Maskus (1997), Satz (2004), Singh (2003), and Winters (2003).

20. In the village of Nawadih, in the state of (now) Jharkhand, in eastern India, where I did four field trips in the early 1990s, I found that whenever interest rates crossed 100 per cent per annum, it would typically be stated in per month terms. Lots of poor farmers told me, for instance, that they had got loans from local landlords or money lenders at 10 per cent per month. I have my doubts if they realized that this amounted to an interest rate of over 200 per cent per annum.

21. 'Predatory lending' has been the subject of very good analysis by the US Federal Trade Commission. See, for instance, www.ftc.gov/os/2000/05/predatorytestimony.htm

22. See, for instance, Basu and Van (1998); Edmonds and Pavnick (2005); Emerson and Souza (2002); Hoff and Stiglitz (2001); Lopez-Calva (2003); Matsuyama (1992); Murphy, Shleifer, and Vishny (1989); Nurkse (1953); Rosenstein-Rodan (1948).

23. That model corresponds to the case described by Eqs. (14.1), (14.2), (14.4), and (14.5).

24. If we build a competitive model with hazardous work and multiple equilibria, one important distinction will be that no equilibrium will Pareto dominate another equilibrium. This is because by the first fundamental theorem of welfare economics we know that each equilibrium must be Pareto optimal. However, our main claim would remain valid despite this. If, starting from an equilibrium where some workers do hazardous work, a ban is imposed on hazardous work and the economy moves to another equilibrium, we can be sure that this ban does not lead to a Pareto inferior outcome, because the new equilibrium must also be Pareto optimal.

25. In the context of child labour, we now know that both these are available. In fact, the theoretical possibility of multiple equilibria occurs not only in models with exogenous law but even in models where child labour regulation is an endogenous choice of the citizenry (Doepke and Zilibotti 2004).

26. See also Basu (2002b), Genicot (2002), Neeman (1999).

27. Needless to say, this is only for freely entered upon, contractual exchange. In coercive situations, one can appeal directly to the welfare loss of the aggrieved party.

28. This should not surprise us. We are trying to develop a normative policy principle. From Hume's law we know that we can never reach such a principle from propositions of pure facts and logic. When, on occasion, we feel we have done so, it must be that we have unwittingly slipped a normative axiom into the discourse. All I am doing here is confronting the unavoidable normative axiom directly.

29. Welfarism has been critiqued effectively in the literature and in many different ways (see Sen 1997, 2003). The critique being presented here is, however, different because it is based on rejecting welfarism only in the event of Paretian reticence.

30. The argument that she may be opting for this because of poverty is clearly no reason to stop the transaction. I would consider it a case for doing some thing to eradicate poverty so that no one has to make awful choices like this. But if someone is so poor as to wish to make this choice, clearly she would be even worse off if she were not allowed to do it.

31. I am grateful to Abhijit Banerjee for drawing my attention to this moral quandary.

32. Note that in making this statement we have gone beyond welfarism, because if the same welfare profile were generated through another underlying story—for instance, the first person wanting to listen to music which the second person does not like—we may have reached a different prescription.

33. Following Parfit's (1984) lead I have, in Basu (2002b), explored some possibilities of such reversal results in finite societies, but this has to be viewed as no more than a beginning.

REFERENCES

Aumann, R., 1964, 'Markets with a continuum of traders', *Econometrica*, 32, 39–50.

Basu, K., 1994, 'Group rationality, utilitarianism and Escher's waterfall', *Games and Economic Behavior*, 7, 1–9.

—— 1999, 'Child labor: Cause, consequence and cure with remarks on international labor standards', *Journal of Economic Literature*, 37, 1083–119.

—— 2000, *Prelude to Political Economy: A Study of the Social and Political Foundations of Economics*, Oxford University Press, Oxford.

—— 2002a, 'The retreat of global democracy', *Indicators*, 1, 1–10.

—— 2002b, *Sexual Harassment in the Workplace: An Economic Analysis with Implications for Worker Rights and Labor Standards Policy*, Department of Economics Working Papers 02–11, MIT, Cambridge, MA.

—— 2003, 'The economics and law of sexual harassment in the workplace', *Journal of Economic Perspectives*, 17, 141–57.

—— 2004, 'Globalization and development: A re-examination of development policy', in A. Kohsaka (ed.), *New Development Strategies: Beyond the Washington Consensus*, Palgrave Macmillan, Basingstoke.

Basu, K. and Z. Tzannatos, 2003, 'The global child labour problem: What do we know and what can we do?', *World Bank Economic Review*, 17(2), 147–73.

Basu, K. and P. H. Van, 1998, 'The economics of child labour', *American Economic Review*, 88(3), 412–27.

Bhagwati, J., 1995, 'Trade liberalization and "fair trade" demands: Addressing the environmental and labor standards issues', *World Economy*, 18, 745–59.

—— 2004, *In Defense of Globalization*, Oxford University Press, New York.

Bhalotra, S. and C. Heady, 2003, 'Child farm labour: The wealth paradox', *World Bank Economic Review*, 17(2), 197–228.

Bourguignon, F., F. H. G. Ferreira, and P.G. Leite, 2003, 'Conditional cash transfers, schooling, and child labor: Microsimulatng Brazil's Bolsa Escola program', *World Bank Economic Review*, 17(2), 229–54.

Brown, D., A. Deardorff, and R. M. Stern, 2003, 'Child labour: Theory, evidence and policy', in K. Basu, H. Horn, L. Roman, and J. Shapiro (eds.), *International Labour Standards*, Blackwell, Malden, MA.

Chau, N. and R. Kanbur, 2002, 'The adoption of international labour standards: Who, when, and why', *Brookings Trade Forum 2001*, Brookings Institution, Washington, DC.

Doepke, M. and F. Zilibotti, 2004, 'The macroeconomics of child labour regulation', UCLA, Los Angeles (Mimeo).

Dreze, J., and G. Kingdon, 1999, 'School participation in rural India,' *Development Economics Discussion Paper 18*, LSE, London.

Edmonds, E., 2005, 'Does child labour decline with improving economic status?', *Journal of Human Resources*, 40(1), 77–99.

Edmonds, E. and N. Pavnick, 2005, 'Child labour in the global economy', *Journal of Economic Perspectives*, 19(1), 199–200.

Emerson, P. M. and A. P. Souza, 2002, 'Is there a child labour trap? Intergenerational persistence of child labour in Brazil', *Economic Development and Cultural Change*, 51(2), 375–98.

Engerman, S., 1973, 'Some considerations relating to property rights in man', *Journal of Economic History*, 33, 43–65.

Genicot, G., 2002, 'Bonded labor and serfdom: A paradox of voluntary choice', *Journal of Development Economics*, 67(1), 101–28.

Hildenbrand, W. and A. Kirman, 1988, *Equilibrium Analysis*, North-Holland, Amsterdam.

Hoff, K. and J. Stiglitz, 2001, 'Modern economic theory and development', in G. Meier, and J. Stiglitz (eds.), *Frontiers of Development Economics*, Oxford University Press, Oxford.

Humphries, J., 2003, 'The parallels between the past and the present', in K. Basu, H. Horn, L. Roman, and J. Shapiro (eds.), *International Labour Standards*, Blackwell, Malden, MA.

ILO, 2001, *Stopping Forced Labour*, ILO, Geneva.

—— 2002, *Every Child Counts: New Global Estimates on Child Labour*, ILO, Geneva.

Kabeer, N., 2000, *The Power to Choose: Bangladeshi Women and Labour Market Decisions in London and Dhaka*, Verso, London.

Kanbur, R., 2003, 'On obnoxious markets', in P.K. Pattanaik and S. Cullenberg (eds.), *Globalization, Culture, and the Limits of the Market: Essays in Economics and Philosophy*, Oxford University Press, New Delhi.

Krueger, A., 1997, 'International labor standards and trade', *Annual World Bank Conference on Development Economics 1996*, World Bank, Washington, DC.

Levison, D., K. Moe, and F. Knaul, 2001, 'Youth education and work in Mexico', *World Development*, 29, 167–88.

Lopez-Calva, L.-F., 2003, 'Social norms, coordination, and policy issues in the fight against child labour', in K. Basu, H. Horn, L. Roman, and J. Shapiro (eds.), *International Labour Standards*, Blackwell, Malden, MA.

Maskus, K., 1997, *Should core labour standards be imposed through international trade policy*, International Trade Division, World Bank, Washington, DC (Mimeo).

Matsuyama, K., 1992, 'The market size, entrepreneurship and the big push', *Journal of Japanese and International Economics*, 6, 347–64.

Mill, J. S., 1848, *Principles of Political Economy* (1970 ed.), Penguin, Harmondsworth.

——— 1859, *On Liberty*, Parker, London.

Murphy, K. M., A. Shleifer, and R. Vishny, 1989, 'Industrialization and the big push', *Journal of Political Economy*, 97(5), 1003–26.

Neeman, Z., 1999, 'The freedom to contract and the free-rider problem', *Journal of Law, Economics and Organization*, 15 (3), 685–703.

Nurkse, R., 1953, *Problems of Capital Formation in Underdeveloped Countries*, Oxford University Press, New York.

Parfit, D., 1984, *Persons and Reasons*, Clarendon Press, Oxford.

Rakshit, M., 2002, *The East Asian Currency Crisis*, Oxford University Press, New Delhi.

Ravallion, M. and Q. Wodon, 2000, 'Does child labor displace schooling? Evidence on behavioral responses to an enrollment study', *Economic Journal*, 110, 158–76.

Ray, R., 2000, 'Analysis of child labor in Peru and Pakistan: A comparative study', *Journal of Population Economics*, 13(1), 3–19.

Rosati, F.C. and M. Rossi, 2003, 'Children's working hours and school enrollment: Evidence from Pakistan and Nicaragua', *World Bank Economic Review*, 17, 283–96.

Rosenstein-Rodan, P. N., 1948, 'Problems of industrialization in Eastern and South Eastern Europe', *Economic Journal*, 53, 202–11.

Satz, D., 2003, 'Child labor: A normative perspective', *World Bank Economic Review*, 17, 297–310.

——— 2004, 'Noxious markets: Why should some things not be for sale', in P.K. Pattanaik and S. Cullenberg (eds.), *Globalization, Culture, and the Limits of the Market: Essays in Economics and Philosophy*, Oxford University Press, New Delhi.

Sen, A., 1997, 'Individual preference as the basis of social choice', in K.J. Arrow, A. Sen, and K. Suzumura (eds.), *Social Choice Re-examined*, Macmillan, London.

——— 1999, *Development as Freedom*, Alfred Knopf, New York.

——— 2003, 'Processes, liberty and rights', in A. Sen (ed.), *Rationality and Freedom*, Harvard University Press, Cambridge, MA.

Singh, N., 2003, 'The theory of international labour standards', in K. Basu, H. Horn, L. Roman, and J. Shapiro (eds.), *International Labour Standards*, Malden, MA: Blackwell.

Tuttle, C., 1999, *Hard at Work in Factories and Mines: The Economics of Child Labour during the British Industrial Revolution*, Westview Press, Boulder, CO.

U.S. Department of Labour, 1990, 'Worker rights in export processing zones', *Foreign Labour Trends*, 90–32, US Department of Labour, Washington, DC.

Winters, L. A., 2003, 'Trade and labour standards: To link or not to link?', in K. Basu, H. Horn, L. Roman, and J. Shapiro (eds.), *International Labour Standards*, Blackwell, Malden, MA.

15 Globalization and Development
A Re-examination of Development Policy

15.1 INTRODUCTION

The nature of policymaking in developing countries has been undergoing a sea change in recent times. This is due in part to the increasing maturity of the discipline of development economics and in part to the changing nature of the global economy. Development economics has advanced rapidly on both the theoretical and empirical fronts. Better interaction with mainstream economic theory, and the increasing availability of data sets that enable us to analyse aspects of the economy that were previously beyond scrutiny, have deeply influenced the study of development. As far as the real world goes, technological advancement and globalization have had a huge impact on the nature of policy-making in developing countries and, more generally, policy-making for development.

This chapter not only investigates the changing face of development policy, but also goes further by raising new analytical issues and urging action in areas that have thus far seen little policy action. The two main themes are labour market policies in a globalizing world and the scope for policy intervention to curb poverty and inequality in a world with increasing mobility of capital and professional labour.

The former has been the subject of very good analysis and heated debate in various fora, such as the ILO and the WTO, so the aim here is to shed some new light with the help of modern theory. Although the problem of inequality and poverty in the context of globalization has been studied, there is still scope for good analysis. This chapter does some spade work on the subject.

The paper was presented as plenary lecture at the regional meeting of Econometric Society in Lahore (28 December 2002) and as keynote address to a conference in Johannesburg (22 October 2002). It has benefited from the comments of Takashi Kurosaki, Koji Nishikmi, Jelf Nugent, and Hiroshi Sato.

First published in A. Kohsaka (ed.), *New Development Strategies: Beyond the Washington Consensus*, Basingstoke: Palgrave Macmillan.

15.2 THE CHANGING GLOBAL SCENARIO

This section discusses the global backdrop against which the subsequent analysis of development policy will be conducted.

15.2.1. Colonizing the Future

Two remarkable developments from the point of view of real-world economics are the recent advances in information technology and globalization. The best historical equivalent to the rise of the information technology (IT) industry was the invention of the wheel in about 3500 BC in Mesopotamia where depictions of the wheel on clay have been discovered and dated at just after that time. While the wheel is at times useful as an end in itself—in fact there is evidence that soon after its invention there were ancillary inventions of toys and games in which the wheel was the central feature—its main value is that it raises the productivity of other activities. Likewise while computers and other IT innovations can serve as ends in themselves, their main advantage is that they facilitate other activities, be it trade, communication, or the simulation of nuclear bombs. The IT industry is currently one of the most profitable industries in the world and will probably remain so for some time. But it is conceivable that eventually it will be just one more industry, its value to the world being that it has made virtually all other industries more efficient and profitable.

Globalization is a close concomitant of the IT industry. It has been facilitated by the cheap and easy modes of communication and trade made possible by the rise of the IT industry. The two main components of globalization are international trade and global capital flows. Both these have grown rapidly. Tables 15.1 and 15.2 provide an overview of the historical trends.

Table 15.1: Merchandise Exports as Percentage of GDP, 1870–1995

	1870	1913	1950	1995
Western Europe	8.8	14.1	8.7	35.8
Asia	1.7	3.4	4.2	12.6
Latin America	9.7	9.0	6.0	9.7
Africa	5.8	20.0	15.1	14.8
World	4.6	7.9	5.5	17.2

Source: Maddison (2001).

Table 15.2: Value of Foreign Capital Stock in Developing Countries, 1870–1998

	1870	1914	1950	1998
Total ($ billion in 1990 prices)	40.1	235.4	63.2	3030.7
Stock as percentage of GDP	8.6	32.4	4.4	21.7

Source: Maddison (2001).

It is evident that the period between the two World Wars marked a retreat from globalization, but barring this period of aberration the movement has been forward. Exports as a percentage of gross domestic product (GDP), taken as the average of all countries of the world, rose from 4.6 per cent in 1870 to 17.2 per cent in 1998, with a brief reversal between 1914 and 1950. The same is true of the value of foreign capital stock, both on its own and as a percentage of GDP. Today, however, foreign capital stock as a percentage of GDP is below the level in 1914, but two factors make this less significant than it appears to be. From 1914 to 1998, world GDP experienced enormous growth, so it is not surprising that the stock of foreign capital lagged behind in relative terms, despite its own immense growth. In fact the total stock of foreign capital in all countries in 1998 stood at an astonishing $3030 billion, which was far higher than ever before. Second, it seems reasonable to presume that, until the early twentieth century, most of the foreign capital flows were from the imperial powers to their respective colonies. In those days it was necessary to establish political control before sending one's money somewhere. Now that this is no longer necessary the world has become much more of a market place. Of course, investing countries still use subtle forms of political control and checks on recipient countries, but that is very different from the control of a colony. One of the main features of globalization is the ability to invest in distant countries with little direct control.

These trends have continued in recent years, as is illustrated in Table 15.3. In the ten-year period covered in the table, all regions saw a rise in capital flows, the overall average rise being approximately 50 per cent. Trade in goods as a percentage of GDP also rose substantially in all regions.

Table 15.3: Globalization Indicators, 1989–99

	Trade in goods as a percentage of goods GDP[1]		Gross private capital flows[2] as a percentage of PPP GDP	
	1989	1999	1989	1999
Low-income countries	41.3	60.0	0.8	1.2
Middle-income countries	69.0	81.5	1.9	4.9
High-income countries	93.5	123.5	2.1	4.9
South Asia	25.6	38.1	0.3	0.6
Sub-Saharan Africa	78.1	95.6	2.1	4.9
Latin America and Caribbean	49.8	74.6	2.2	7.3

Notes: [1] Excludes services.

[2] The sum of the capital that flowed into and out of the countries.

Source: World Bank (2001).

Of course, not all regions have merged with the global markets and financial economy with equal rapidity. The fast integration has occurred in most of Asia and parts of Latin America (Kohsaka 2002); Africa has also done well as a percentage of its own GDP. Also, the composition of capital has changed, with a sharper increase in foreign direct investment and portfolio investment than in bank credit and loans. For some countries, such as India, the dominant form of capital market integration has been the flow of foreign capital into the stock markets.

While political instability or war can reverse the trend, it seems reasonable to conclude that the process of globalization, after a period of vacillation caused by the shock of the two world wars, is firmly on course. Ideas, goods, and money now flow almost instantaneously between distant countries and cities. One implication of this is that inventions spread rapidly, generating interest in faraway places and facilitating further inventions. A ten-year old computer now looks ancient, it seems hard to believe that 10 years ago email did not exist.

Contrast this with innovations in the design of the wheel. Initially (that is, around 3500 BC) it was made of solid disks. It took 1500 years for human beings to realize that it would be more efficient to carve the disk into a ring stretched by spokes. This made it lighter and better able to absorb shocks. It then took another 1400 years to realize that roller bearings would minimize friction and enable the wheel to turn more easily. Not only did inventions occur at great intervals, but ideas took a very long time to spread from one region to another. The Mayans built some of the world's most magnificent stone pyramids but were unaware of the machinery or wheeled vehicles that would have made their work considerably easier.

All these processes went on for hundreds and even thousands of years. The pace has picked up astonishingly in the past few decades, thus changing the nature of the global game faster than most of us can comprehend, creating new opportunities and new tensions, and rapidly altering the efficacy of policy instruments.

In today's world, the struggle is no longer to colonize and control new lands but to 'colonize the future', that is, to lay claims on tomorrow's output (Basu 2000a). Two factors have made this feasible: the ability to take out patents and copyrights and enforce them, and the widespread availability of stocks, shares, and other financial assets. The colonizers of today try to secure a large number of patents and hold huge amounts of shares. When tomorrow comes and the output emerges from factories and offices, part of this output will already have claimants from today (those holding patents and those holding shares) and the remainder will be there to be split between the providers of tomorrow's inputs—labour, raw materials, and so on.

It is worth noting that this colonization of the future is not happening equitably across countries. In 1995, for instance, 235,440 patent applications were filed in the United States, whereas in some of the poorer countries fewer than 100 applications were filed. Hence, the global inequalities of today are likely to be reinforced tomorrow.

15.2.2 Erosion of Global Democracy

Another aspect of globalization and the rise of modern technology is that it has a tendency to erode global democracy. To understand this, observe that it is much easier today for one country to interfere in the affairs of another. In ancient times to influence the policy of another nation, the only option would be to muster an army or sail the high seas to attack. Today, not only have military actions become much more arms-length and effective,[1] but a variety of economic reprisals are possible by the click of a mouse. Of course, coordinating these economic actions is not always easy, because it may involve the participation of firms and corporations that in principle are free agents. Nevertheless, countries have successfully used the threat of cessation of trade or the withholding of capital flows to influence policy. For example, the US Helms-Burton Act has been used to apply pressure on Cuba not only by curbing US business and trade with that country but also by threatening to cut off business with countries that trade with and invest in it. For instance, the Act has been used to dissuade Mexicans, Italians, and Canadians from doing business with Cuba, which is something they would not have contemplated of their own accord.

Now, if we use a rudimentary definition of democracy, namely, a political system where ordinary citizens have the ability to influence the choice of leaders who influence their lives by exercising their vote, it should be immediately transparent that globalization has a tendency to erode democracy.

Even though it is not possible for people in one country to influence events in another through the democratic electoral process, the leaders of some countries have developed more and more instruments to influence the lives of people elsewhere in the world (a definitional implication of globalization), with a consequent diminution of global democracy (Basu 2002a). This phenomenon has major political and economic implications and calls for thought and institutional innovation.

15.2.3 Globalization and Marginalization

While, on the whole, globalization creates more opportunities than it destroys, it can have the negative effect of marginalizing some people. If this is left unchecked it could lead to political instability and social decay. Not only are marginalization and the consequent rise in poverty undesirable in themselves, but we now have the ability to deal with them effectively. We are at a point in history where it is possible to talk—without inviting the label of idealistic crank—of rooting out poverty from the world altogether. And unlike smallpox there will be no need to keep a small supply of the germs of poverty in stock to counter possible terrorist attacks.

To understand how globalization can lead to marginalization, consider a poor person in a Third World nation, say, India whose livelihood is off-shore fishing. If India were to modernize and become more integrated into the world economy, it might well be that technologically advanced fishing companies would go out to the high seas and bring in larger catches than ever before. The exportation of these catches would make India better off as a whole, but it could diminish the available stock of fish closer to the coast line, resulting in smaller catches by the poor fisherman thereby leaving him worse off. This is an obvious 'resource route' which can result in some people getting marginalized and made worse off by the processes of technological advancement and globalization.

But there is another, more complex, route. Suppose now that the fisherman catches all his fish from a lake. He consumes some of the fish and sells the remainder on the market, which enables him to buy other essentials such as salt, sugar, other foods, and clothing (no one in today's world is totally self-sufficient). At first sight, it may appear that the activities of the deep-sea fishing companies would have no effect on his standard of living, but this is actually unlikely. It is entirely possible that the larger hauls of ocean fish would cause the price of all fish to drop. Hence the fisherman would receive less money for his fish and could buy fewer essentials.[2] Hence he would become poorer even though the resources to which he had access remained unchanged. The extent of such 'market-route' marginalization could be very significant, but as economists and statisticians have shown little awareness of the phenomenon, no data is available.

Together, the market and resource routes could create very large constituencies of losers in the globalization process. Apart from its innate unfairness, this could cause large-scale disillusionment, dissent, and ultimately political instability. It is arguable that some of the myriad forms of global dissent that we are seeing today, ranging from terrorism to roadside protests, have their roots in such marginalization. To describe this as a cause of today's global dissent is not to deny that there may be other causes. Economists and social scientists tend to concentrate overly on the proximate causes of global dissent, not realizing that unless the deeper underlying causes are recognized and dealt with we shall perennially be putting out little fires.

Markets can be very good instruments for generating greater productivity and efficiency but they do not have an in-built mechanism to ensure better distribution of the fruits of progress. Hence it is essential to establish institutions to improve the distribution of goods and services and to obliterate poverty.

With this global scenario as back-drop I shall now consider some concrete themes in development policy. The focus will not be just on developing countries but also on the poorest sections of these countries, particularly in terms of the well-being of workers and inequality in general.

15.3 LABOUR MARKET POLICY

Labour market policy is important both because it can influence the performance of the whole economy and because workers are typically the poorest constituents of the economy[3] and therefore deserve special attention. Moreover, this is the area in which globalization has changed the terms of debate more dramatically than anywhere else.

Labour markets have always been more closed than the market for goods and services. For reasons mired in politics and sociology, people have generally preferred to stay where they have been born and raised and where they have their cultural roots, which means that the economic incentive for moving had to be substantial for those who decide to migrate. Moreover, nations have tended to erect barriers against immigration that are more formidable than those erected to curtail the importing of goods and services. Hence labour markets are one area in which countries have felt relatively free to have their own laws and regulations, designed according to their own tastes, politics, and cultural prerogatives. Of course, through the import and export of goods, which are ultimately made by labourers, this freedom had its limits. But what has happened in recent times is that this freedom to craft one's own laws has been more severely curbed by the global mobility of capital. Even if workers from country x cannot move to the factories of country y, globally mobile capital means that the factories of country y can now come to the workers of country x.

This *de facto* labour mobility has two implications. First, when a developing country now drafts a new law for its workers, for instance to enhance some workplace right, it has to be sure that this will not drive capital away to another country. Hence countries' legislative freedom is much more limited in today's globalized world.[4] Second, this *de facto* labour mobility means that industrialized countries are now paying greater attention to working conditions and the labour market in developing countries. This is based on the fear, often misplaced, that the outward flow of capital will result in a loss of jobs in industrialized countries. Child labour is a matter of moral concern to most people, but opposition to it can be used as an instrument of protectionism. In effect, the blending of economics and politics is making the crafting of labour market policy a much contested and intricate matter. To illustrate these points I shall construct a simple model of statutory working hours.

15.3.1 Statutory Limit on Working Hours and International Labour Standards

Statutory limits on working hours is an old topic. It was hotly debated in the United States in the nineteenth and twentieth centuries and in Britain during the industrial revolution, when it was routine for workers and even child labourers to work for fourteen hours a day. As early as 1825, skilled workers in Boston were unionizing and holding strikes to have their working day limited to ten hours. In 1842, the Ten Hour Republican Association distributed campaign leaflets for a statutory limit on working hours and the movement soon gained considerable momentum (Murphy 1992).

The standard argument against a statutory limit proceeded by appealing to the Pareto principle. If a worker voluntarily accepts a work offer that entails fourteen hours of work a day, then, while this may seem unbearably long to us, it must be the case that the money earned by the worker more than compensates him for the hard work. And since the employer makes the offer voluntarily, (s)he must be gaining from this as well. It seems we have a Pareto-improving deal here, so why should we object, especially if the agents involved are adults and therefore able to judge what is in their own best interests? So widely accepted was this argument that, when the first Factories Act came into force in Britain in 1802, it was still impossible to set limits on working hours for adult males, who were supposed to know what was in their own best interests. So the Factories Act of 1802, displaying a rare gender bias which probably helped rather than hurt women, set limits on the number of hours that women and children could work. The limit was twelve hours within the time range of 6 am and 9 pm. An upper limit of twelve hours today would be seen as enabling employers to extract an unreasonable amount of work from individuals, but in the early nineteenth century there were those who worried that the new interventionist Factories Act would encourage sloth in the working class and hurt Britain *vis-à-vis* its trading partners.

The general question of whether all voluntary contracts among adults should be allowed without government intervention has been a hotly debated subject, going back to at least John Stuart Mill's classic works of 1848 and 1856.[5] I shall here pursue a line that contests this by recognizing the possibility of multiple equilibria, which is an important feature of developing countries (Hoff and Stiglitz 2001).

Consider a very poor country where many people's incomes are close to the subsistence level. In such an economy, workers' job decisions will reflect their concern for survival. A simple way to model this is to think of workers making their job decision in the way that workers do in developed countries but with an additional eye to subsistence or survival. If their incomes tend to fall below the subsistence level they will work as much as is feasible to ensure that they stay above subsistence. This can be captured by specifying the workers' utility function as follows:

$$U(x, 1 - e) = \begin{cases} x - u(e), \text{ if } x \geq s \\ x - u(1), \text{ if } (x < s), \end{cases} \tag{15.1}$$

where x is the amount of consumption by the workers, e is the amount of work done by them, and s is the subsistence level of consumption. As usual we shall assume that the cost of work, $u(e)$, increases with the amount of work and at an increasing rate. That is, $u'(e) > 0$ and $u''(e) > 0$. Effort, e, is supposed to be an element of $[0,1]$ and we normalize by setting $u(0) = 0$. That is, there is an upper limit to the amount of work that individuals can possibly do, and this is by definition equal to 1. Thus 1 represents a very large amount of work, say, fifteen hours a day. According to (15.1), until individual workers reach a consumption level of s (the subsistence consumption), consumption is their sole objective—note that $u(1)$ is a constant. They will work as hard as necessary to reach this target, but once they have done so the first line of the utility function takes over and they take an interest in increasing their consumption and their leisure time.

To work out an individual's supply function, suppose that w is the market wage rate and the price of the good being consumed is 1. For the moment, we shall ignore the subsistence factor (i.e., pretend that $s = 0$) and work out the worker's supply. Since the person has no other source of income, x must be equal to ew. Making this substitution in the first line of (15.1) and working out the first-order condition we get

$$w = u'(e) \tag{15.2}$$

If w rises, for Eq. (15.2) to hold, e must rise as well. This follows from the assumption that $u''(e) > 0$. Hence Eq. (15.2) describes an upward-sloping curve. Let us describe the inverse of Eq. (15.2) by $e = e(w)$. What we just proved is that $e'(w) > 0$. Now let us bring in the subsistence requirement, where $s > 0$. Define w^* to be such that $w^*e(w^*) = s$. Therefore for all $w \leq w^*$, $we(w) < s$. Now, solving the full maximization problem of the labourer, that is, with the subsistence constraint taken into account, for $w \quad w^*$ the supply is given by $e(w)$, but for all $w < w^*$ the supply is given by $\min\{1, s/w\}$. The particular form $\min\{1, s/w\}$ simply takes account of the fact that 1 is the technically feasible maximum amount of work. Summing up this in a single equation, we have the following labour supply function, $E(w)$, of the labourer:

$$E(w) = \begin{cases} e(w), \text{ if } w \geq w^* \\ \min(1, s/w), \text{ if } w < w^*. \end{cases} \tag{15.3}$$

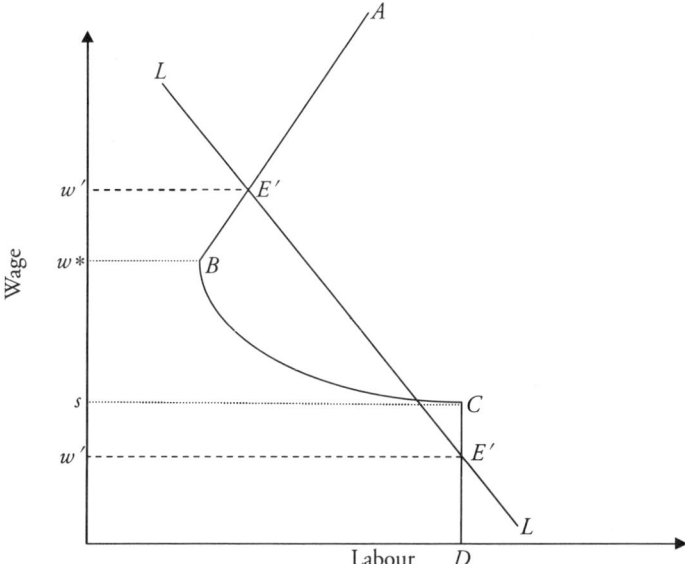

Figure 15.1: Labour Market Equilibrium

This supply curve is illustrated in Figure 15.1 by the line *ABCD*. Since the supply curve has a backward-bending section, clearly there can be multiple labour market equilibria. To complete the story, suppose that there are m identical workers. In this case the aggregate supply curve of labour is an m-fold horizontal aggregation of an individual's supply curve. Without loss of generality, let *ABCD* in Figure 15.1 represent such a supply curve.

Let us suppose that the country in question has n firms, each endowed with the production function $f(L)$, where L is the amount of labour used by the firm and $f(L)$ is the output produced by it, and for all L, $f'(L) > 0$, and $f''(L) < 0$. Assuming that the price of the good is 1 and the wage is w, a firm will demand L units of labour, where $f'(L) = w$. Using d to denote the inverse of this function, a single firm's demand function for labour is $L = d(w)$. It is easy to check that this is a downward-sloping curve. Aggregating the demand curves of all n firms we have the market demand curve for labour. This is illustrated by the line LL in Figure 15.1. Equilibrium is defined in the usual way—as a wage rate at which aggregate demand equals aggregate supply: $mE(w) = nd(w)$.

In the case illustrated in Figure 15.1 there are three possible equilibria, of which the points marked by E' and E'' are the two stable ones. If the wage rate is low, at w', workers are forced to work many hours just to survive, and the increased supply of labour in turn keeps the wage low. If wage is high, at w'', workers are happy to work less and this 'limited' supply reinforces the wage high.

Clearly, workers are better off at the high wage equilibrium, that is, at E''. Since all agents are price-takers, both equilibria will be Pareto efficient. This is illustrated formally, in the context of child labour, by Basu (2002b). So the difference between the two equilibria is that at E'' workers are better off and profit-earners are worse off; and vice versa at E', in a way akin to what Swinnerton and Rogers (1999) have illustrated in the context of child labour.

Suppose now that this economy is stuck at the equilibrium E'. Since workers usually comprise the poorest class in developing countries there is reason for the government to try to move the economy to the superior equilibrium, which in this economy is easy to achieve. The government simply has to announce a statutory limit on the number of hours that a worker can work. This will shift the CD segment of the supply curve to the left. If this shift is sufficient there will be only one remaining point of intersection between the supply and demand curves, namely E''. Hence, the wage will rise to w'' and the market will settle into that equilibrium.

In this model, the statutory equilibrium law has a very interesting feature. Once it has been imposed for a while it ceases to be necessary, and it can be removed without the economy reverting back to the old equilibrium. Basu and Van (1998) refer to such a legal intervention as a 'benign intervention'. It is a law that is meant to deflect the economy from one equilibrium to another preexisting equilibrium. The law does not hold the economy where the market could not have held it, rather it simply helps to select from the various positions where the market could potentially hold it. Moreover, once the new equilibrium is established the law plays no further role. Since multiple equilibria are germane to developing economies, benign legal intervention has a large role to play in development economics.[6]

Before moving on to the subject of globalization, let us briefly digress to the subject of free contracting and Pareto improvements. It is a staple of economics that, if two consenting adults agree to a contract or an exchange, which has no negative fall-outs on any uninvolved third party, then the contract or exchange results in a Pareto improvement. Hence, economists typically believe that government and politicians should not intervene in such contracts and exchanges. Laws such as rent control legislation are often frowned upon by economists precisely for this reason. If a person is willing to take up a tenancy that requires him or her to pay a low rent but to vacate the dwelling at a day's notice, and if the landlord wants to have such a tenant for the rent agreed upon, there seems to be no obvious reason why the government should disallow such a contract since it will lead to a Pareto improvement. This, however, is a matter of considerable contention and misunderstanding, the debate going back at least to the writings of John Stuart Mill in the early 1800s.

There are three ways in which one can justify a legal ban on certain kinds of voluntary contract among adults while continuing to adhere to the Pareto criterion.[7] The first is to argue that there is a difference between single acts of exchange or contract and a class of such acts. In particular, it may be the case that each such act can be justified on Paretian grounds but a class of such acts cannot be justified. Parfit (1984) laid the philosophical foundations of this argument and I have tried to formalize this in Basu (2003; see also Genicot 2002).

The second argument is based on the recognition that human beings are often irrational, and systematically so, as the new literature on behavioral economics has made obvious to economists (others already knew it). If people are irrational when making choices over time, it is possible for some to be systematic losers in certain kinds of market transaction, such as when taking credit. If in today's society a weak person is deprived of his property by someone stronger than him, this is not considered acceptable. But if someone makes a borrower part with her or his property because the latter has failed to repay a loan taken out on terms that reflected the borrower's irrationality and miscalculation we do not raise an eyebrow. The reason for this is the presumption, deeply embedded in traditional economics, that no one miscalculates, no one is irrational. But once we recognize that people are often irrational we can legislate against certain transactions, for instance by putting an interest rate ceiling on credit agreements. It is true that this could prevent some efficiency-raising transactions between smart borrowers and lenders, but it could also prevent some irrational borrowers from being duped. Good policymaking entails intelligently balancing the potential gains and losses.

The third argument based on the proposition that some economies are characterized by multiple equilibria, where each equilibrium can be Pareto optimal. In such a situation, if we are in the vicinity of one equilibrium and we disallow a particular transaction, we may shift the economy to a Pareto suboptimal outcome. But if all transactions in a certain class of transactions are disallowed, we may move to a new Pareto optimal outcome. Hence the new outcome is not Pareto-dominated by the old equilibrium. The model constructed above belongs to this third category. If individual workers and a firm are prohibited from entering a contract in which each individual is required to work, say, fourteen hours a day, this will be Pareto suboptimal. However, if no worker is allowed to work more than a certain amount of hours per day, say ten, it is entirely possible that all workers will be better off by such legislation.[8] Hence we cannot use the Pareto criterion to rule against such legislation. This argument was not available to those who debated statutory limits on working hours in the nineteenth century or even in the early twentieth century. It is only the advance of economic theory that has enabled us properly to understand the role and consequences of such legislation.

We shall now analyse how globalization can render certain benign laws ineffective. In general, under globalization the labour market legislation adopted by a country can lose much of its force due to the fact that capital is able to escape to another country. This also applies within a country if that country has a federal structure. In the United States there was once a considerable degree of interstate competition in terms of relaxing or not enforcing labour laws in order to attract or retain capital (see Kelley 1905), and this eventually prompted the nationwide imposition of the Fair Labour Standards Act in 1938 (see Bhagwati [1995] and Engerman [2002] for a lucid account of the history of labour standards). India, given its large size and growing regional freedom in terms of the law, including labour laws (Besley and Burgess 2002), can learn lessons from this experience.

To understand the problem that globalization creates in terms of development policy, suppose now that there are many countries, say t, just like the one described above. So what was described in Figure 15.1 refers to country 1 and there are identical countries, 2, 3, …, t. Let us suppose that each country is caught in the 'bad' equilibrium, namely at E'. In each country there are n firms. But now let us suppose that these firms are mobile across countries. Each can pick up its capital and move to another country should the need arise. To keep the analysis general, suppose that a firm has to incur a fixed cost of C to shift its operations to another country.[9]

If $C = \infty$, we have the case of a closed economy; and if we have $C = 0$ we have a fully globalized world in which capital can move costlessly among economies. C can be a product of nature and governmental nurture. Some transactions' cost of movement are in the nature of economic life. Certain kinds of capital are typically sunk, and they cannot be uprooted and moved without loss; and even when they can be moved or sold off, transportation costs or advertisement and selling costs may have to be incurred. In addition to such natural costs, the government can enact laws and impose taxes that make movement of capital costly. Hence C can take different values depending on the policy followed, which may in turn be the product of the attitude towards globalization. In reality, C is probably never zero or infinite, but polar cases can shed light on and help us understand the kinds of response we can expect from the market. I have already discussed the case of $C = \infty$, so now turn to $C = 0$.

Suppose that all the economies of the world are caught in the 'bad' equilibrium and therefore wages are equal to w' in each country. Now suppose that the government of a single country wants to nudge its economy towards the better equilibrium. In the case of $C = \infty$, as we have already seen, it could simply impose a statutory limit on working hours—say, ten hours a day. If the limit is severe enough there will be only one point of intersection between the demand and supply curves for labour. Such a case is illustrated in Figure 15.2. Each worker is

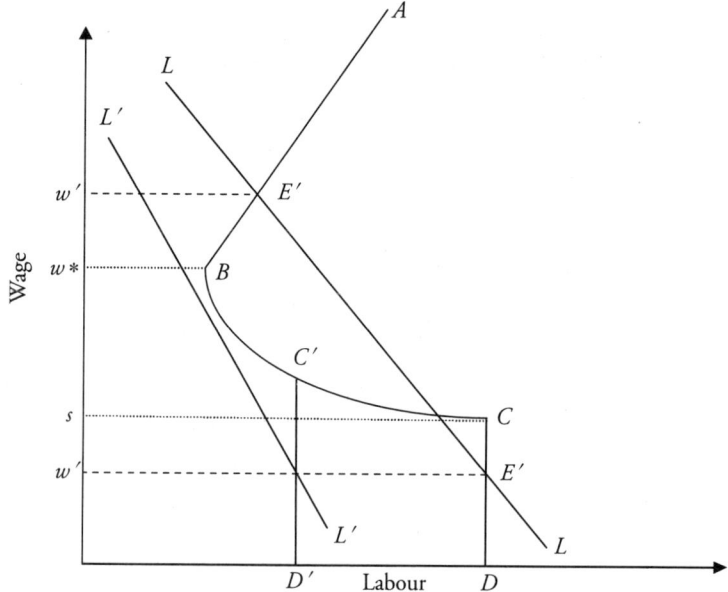

Figure 15.2: Labour Market Equilibrium, Small Nation Case

allowed to work for a maximum of D'/m hours, where D' is as described in Figure 15.2. Then the new supply curve for labour is $A'A''BC'D'$, and if C is so high that no firms leave the country the goal of raising labour standard will be achieved. If, however, C is very low—say, the extreme case of $C = 0$—as soon as the new law comes into effect and the wage rises, firms will begin to pack up their capital and leave for greener pastures. This will cause the demand curve of labour to move leftwards, which will continue as long as wages elsewhere are lower. Clearly, some of the aims of the government intervention will be defeated. The extent of this will depend on market parameters and how large the country is in comparison with the rest of the world, because as firms move out of this country to other countries, wages elsewhere will rise.

In the extreme case of this country being so small that the flight of firms does not raise wages in the rest of the world, wages here must fall to w' (Figure 15.2). So the new law to improve labour standards in country 1 causes firms to leave, and demand for labour keeps falling in country 1 until it reaches the position of $L'L'$, wages fall to w' and nothing of value is achieved by the country. In reality, $C > 0$, so we would not expect such extreme response, but the theoretical result is nevertheless suggestive. It also explains why international labour standards is so politically charged and highly contested.

The reason why the demand for international standards has arisen is obvious from this model. When capital is free to move between a set of countries, each country's power to impose unilateral labour standards tends to be impaired. One way of restoring this power is for the cluster of relevant nations to coordinate their labour market policies, which is exactly what international labour standards is all about.[10]

Before moving on to consider actual policy questions it is important to offer some words of caution about the use of this model. Clearly the model applies to similarly placed countries and we have considered an extreme version of this—a set of identical countries. The model cannot be applied to a set of countries at very different levels of industrialization. The sort of work carried out by children in Ethiopia will be something that no worker in the United States or Japan would have to do. Hence there will be very little movement of firms between Ethiopia and these countries and the banning of child labour in Ethiopia is unlikely to impact seriously on them. The reason why this is important to caution against is that the misunderstanding of this is one of the major factors that has given rise to the chorus for international labour standards in developed nations and created a platform for Northern protectionism.

15.3.2 Labour Market Policy in a Globalized World

When thinking about labour market policy it is worth reminding ourselves that legislative intervention is not the only way to ensure workers' well-being and rights. Ultimately the biggest guarantor of labour standards and well-being is labour demand. If the demand for labour is sufficiently high so that employers have to compete with one another to obtain workers, the workers will be able to get themselves decent wages and have their basic rights ensured. Hence open trade channels that boost demand for labour in developing countries is good policy.

Developed nations can do more to boost workers' welfare by opening their doors to products from the Third World than they can by campaigning for 'social clauses' and chanting slogans. However, at times the market cannot do much to help workers' interests. This can happen when there are multiple equilibria or when the moral status of a set of actions happens to be different from the moral status of individual acts. For certain kinds of labour standard, such as the requirements not to engage in sexual harassment or expose workers to excessive hazards, legislative action and government intervention may be needed (Basu 2003).

While in these matters countries could once enact laws unilaterally, the scope for such action has diminished with globalization. One can also find examples of this within large countries. As mentioned earlier, the United States experienced ruinous interstate competition until 1938, when the enactment of the Fair Labour Standards Act brought all states under a common labour code. In 1904, for instance, the glass industry of New Jersey declared that if it were prevented from the night-time employment of children under the age of sixteen it would shift to Delaware or West Virginia, which imposed no such restrictions. In India, individual states have often competed over labour market policy, which has often led to flights of capital away from states that attempted to implement pro-poor labour market policies (Basu 2002c; Besley and Burgess 2002). Besley and Burgess classify the amendments made by Indian states to the Industrial Disputes Act of 1947 as pro-poor, neutral, or anti-poor. They show that states with more pro-poor amendments have ended up discouraging investment, often hurting the very constituency they meant to help. One lesson of this is that a pro-poor policy may cease to be so when capital is mobile, as is the case in India.

Returning to the subject of international labour policy, as we saw in the previous section, there is a case for concerted international action. However, this does not mean that international labour standards should be enacted and enforced in the way that they are currently being enforced. Ideally they should be formulated, designed and executed by developing countries. The current suggestion that they be designed and enforced by the WTO fails on this score. Despite the fact that the WTO operates on the principle of one-country one-vote and, wherever possible, on consensus, much has been written about the 'greenroom effect' and behind-the-scene attempts by rich countries to set the agenda in advance. As long as power is vested disproportionately in the hands of the industrialized nations, the risk of trade sanctions being used for protectionist purpose cannot be ruled out, nor can the risk of international labour standards being used as an instrument for Northern protection (Bhagwati 1995; Bhagwati and Hudec 1996).

Hence, for now, the matter of labour standards is best left to the ILO, which is unlikely to interfere with trade. The main method used by the ILO is to draft a convention and then to encourage countries to sign it—in this case signing means a commitment on the part of a government to enforce the terms of the convention. Convention 138, for instance, entails a commitment not to allow children below the age of fifteen years to do regular work. A similar effort has been made by the United Nations. Its 'Global Compact' is a voluntary agreement to uphold minimal labour standards, but unlike the ILO's conventions the signatories are not countries but corporations and multinational companies. Corporations that sign the Global Compact are essentially committing themselves to abjure certain labour market practices that are deemed harmful to workers. Unlike in the case of the WTO—which, if it were to introduce a social clause in its agreements, would use trade sanctions and other forms of punishment as retribution for countries that violated the specified standards—the ILO and the UN work on the basis of self-enforcement by the signatories and rely on the power of publicity and social disapproval.

15.4 GLOBAL INEQUALITY AND POVERTY

Another area where there is growing need for the global coordination of policies is the mitigation of inequality and poverty. As noted above, globalization has a tendency to marginalize sections of the population. Hence, it has created a concomitant need for policies to control inequality. Ironically, however, globalization often makes it hard for countries to control inequality and poverty unilaterally.

Analysts have claimed that inequality is higher today than in medieval times. This is hard to substantiate because there is little historical data on inequality and because the products we consume today are very different from what our forebears consumed. Ghenghiz Khan may have been very rich but he had no way of taking a holiday in Hawaii or visiting the French Riviera for a quick weekend break. If he had a debilitating headache, he would not have been able to swallow an aspirin and get straight back to the business of conquering others. How, then, can Ghenghiz Khan's wealth be compared with that of a billionaire of today? While it would take the best of cliometrics and historical research, not to mention intelligent guesses, to amass inter-temporal inequality data and a lot of abstract theorizing to make sense of it, what we can assert on the basis of cursory research is that (1) the plight of the very poor have remained more or less the way it was during the time of Ghenghiz Khan, (2) inequality today, irrespective of whether or not it is greater than in the past, is astronomically high[11] and regional inequality, by all accounts, is higher than ever before, and (3) inequality and poverty are unnecessary today as we have the technology required to provide all human beings with food, housing, basic health facilities, and most other necessities of life.

The reason for the first assertion is simply the knowledge that the very poor today, as in the past, live barely above the subsistence level, that is, they are barely alive. This is because death truncates income distribution at the bottom end. It did so thousands of years ago and it continues to do so today in large parts of Sub-Saharan Africa and some parts of Asia and Latin America. Evidence on regional inequality emerged quite clearly from Maddison's (1979, 2001) research on economic progress over the last 1000 years. Tables 15.4 and 15.5 summarize some of the relevant data.

In the year 1000 AD, while within regions there would have been grave inequalities, since each region had kings and subalterns, across regions there was immense homogeneity, with per capita income standing at just above $400. The reason for this could be a Malthusian one. With very little in the way of technology, each region supported the size of population that food production allowed. By 1998, disparity had soared, with the income ratio between the richest and poorest regions being 19 to 1.

Table 15.4: Per capita GDP by Region, 1000–1998 (1990 dollars)

	1000	1820	1950	1998
Western Europe	400	1232	4594	17921
Japan	425	669	1926	26146
Asia (excl. Japan)	450	575	635	2936
Latin America	400	665	2554	5795
Africa	416	418	852	1368
Interregional spread	1.1:1	3:1	15:1	19:1

Source: Madison (2001).

Table 15.5: Per capita GDP, Asian Countries, 1820–1998 (1990 dollars)

	1820	1950	1998
China	600	439	3117
India	533	619	1746
Japan	669	1926	26146
South Korea	–	770	12152
Vietnam	546	658	1677

Source: Madison (2001).

Another interesting feature of the growing regional disparity is that the disparity within regions has grown over time. From Table 15.4 it is clear that a certain amount of global inequality had already emerged by 1820, when the interregional per capita income spread was 3 to 1. However, if we take a region such as Asia (Table 15.5), we find that at that time the disparity within the region was not very high (a spread of 1.3 to 1). By 1998 huge inequalities had emerged, so Asia is now much more heterogeneous than it was in 1820.

In India, regional inequality has been on the rise, since at least the 1960s (see, e.g., Rao, Govinda, and Kalirajan 1999), and one suspects that this process goes quite deep and probably indicates a rise in overall inequality.

Finally, if we compare countries for which we have recent data, we find that inequality is still on the rise. For instance, the income ratio of the richest 10 per cent of the world to the poorest 10 per cent has risen very sharply, from 52 to 1 in 1988 to 64 to 1 in 1993. During this same period the Gini coefficient of the world income distribution, based on household survey data from 91 countries, deteriorated from 62.8 to 66.0 (Milanovic 1996; Thorbecke and Chutatong 2002).

While all this does not tell us unequivocally whether interpersonal inequality is greater or lesser today as part of a secular trend, it does suggest that global inequality is growing. Moreover, because the world as a whole is now a much richer place. It is this that makes today's poverty, even if it were comparable to the poverty of a hundred or two hundred years ago, so intolerable. It is shocking that there are regions in the world where 25 per cent of children aged ten to fourteen are still toiling away as full-time labourers.[12]

If we could skim off a little from the richest segment of the population and make this available to the poorest, there would be no acute poverty in the world. The rich might not even feel the burden, but the poor would certainly feel the benefit. It is possible that many of the rich would be willing to participate in such a project if there were sure ways of directing the funds to the poorest and ensuring that they received adequate medical facilities, food, clothing, and housing.

The problem of global inequality and poverty is not so much an intellectual problem as a problem of determination and commitment, of finding ways to transfer basic necessities, which are now available in abundance, to the needy. This problem relates closely to some of the issues considered earlier in this chapter. Consider, for instance, the subject of international labour standards and environmental standards. These would not have been such major issues if it were not for the fact that the nations of the world have such dramatically different living standards. Hence a level of labour well-being that may appear tolerable to Ethiopians would seem downright degrading to the Swiss, and the level of pollution that Mexicans are required to endure would be totally unacceptable to the Japanese. If there were less global inequality, there would also be less variation in labour and environmental standards across nations. Hence some of the controversies touched on above would be mitigated automatically if we had a more equitable world.

The degree of inequality between, say, the poorest in Burundi and the richest in Switzerland would be considered unacceptable if it were to occur within the same country. If the poorest people of Burundi were to reside in the United States, the latter would find it impossible to ignore their plight or dismiss it as the fault of the Burundians themselves. This is where the subject of global democracy, developed early in this chapter, comes into play. If global democracy were sufficiently developed, then in the world as a whole the inequalities that exist today would be questioned, debated, and no doubt considered intolerable.

There is another sense in which globalization and inequality are intimately related. In today's globalized world, even if a country wants to reduce the extent of inequality in its territory it may not be able to do so because of the ability of professional workers and capital to cross boundaries. Let us formalize this idea a little. Suppose there are two identical countries, 1 and 2. In each country, in the absence of government intervention, there is a high-productivity person (in short, rich) with an income of x (>0) and a low-productivity person (poor) with an income of 0. Now suppose that the government introduces an income tax, with tax rate t that is used to transfer money from the rich person to the poor person. In other words a fraction, t, of the rich person's income is transferred to the poor person. It seems reasonable to argue that the rich person's incentive to work will be affected by the tax. Hence let us assume that the rich person's income, given a tax rate of t, is given by $x(t)$, where

$$x(t) = x - 16t. \qquad (15.4)$$

Of course, an expression such as this can only be true within some bounds of t. Instead of complicating the algebra with a formal specification of bounds, we shall simply be careful to remain within reasonable limits when we consider the examples below.

Since the poor person receives whatever is collected from the rich person as an income subsidy, the poor person's marginal income is unaffected by his or her labour, so I shall assume that the poor person's level of work is unchanged by the subsidy. Hence after the tax system is put in place the rich person's income is $(1-t)x(t)$ and the poor person's income is $tx(t)$. Let us suppose that each country's social welfare, W, is positively related to its per capita income, m, and negatively related to the income gap, g, between the richest and the poorest persons. Since all these variables depend on t, assuming that there is no international migration we can derive these terms as

$$m(t) = [x(t)]/2 \qquad (15.5)$$

and

$$g(t) = x(t)(1-t) - x(t)t. \qquad (15.6)$$

As this is meant to be a simple illustrative exercise let us assume that social welfare, W, consists of four times the per capita income minus the income gap. Hence

$$W(t) = 4m(t) - g(t) = (1+2t)x(t). \qquad (15.7)$$

Assuming from here on that $x = 16$, substituting Eq. (15.4) into the Eq. (15.7) and maximizing it, it is easy to check that the optimal value of t is 1/4 and $W(1/4) = 18$. Check also that $W(0) = 16$.

So in this model we have a clear policy prescription. If each country could be sure that its tax policy would not cause an out or in migration, then it would fix the tax rate at 25 per cent. This would cause social welfare to rise to 18 from a base line of 16, which occurs when there is no tax.

Now, let us bring in the global aspect of the problem. We shall assume that people will want to work in the country that offers them the higher income, that if both countries offer the same income they will prefer to remain in their own country; and (this is only for simplicity) that both countries only allow high-productivity persons to migrate to their country. In other words, the only people who are able to migrate are high-productivity people. Since both countries are innately identical, taken together the above assumptions mean that high-productivity people will migrate from their own country if and only if the other country charges a lower tax rate.

With these assumptions it is clear that the global optimal occurs when each country sets t at 1/4. Both countries then achieve a welfare level of 18. Let us now suppose that country 1 sets t at 1/4 and country 2 sets it at 0. The latter's welfare level is then given by 26 + (2/3). To see how this is calculated, note that country 2 will now have two rich people and one poor person, because of migration. So its per capita income will be [2(16) + 0]/3 and the income gap will be 16. Applying the welfare function to this we get welfare level as 26 + (2/3).

If, on the other hand, country 1 sets t at zero and country 2 sets it at more than zero (for instance $t = 1/4$) then all productive workers will leave country 2 and its welfare will be given by zero. Clearly it would then be better off to lower t to 0. In short, if we view this as a game between countries, setting t at 0 is the dominant strategy for both countries, although both would be better off if they could set t at 1/4. In other words, what we have is a classic prisoners' dilemma. The payoff matrix Figure 15.3—in which for simplicity we assume that each country has to choose between setting t_i at 0 or 1/4, where the subscript i refers to the country in question and the payoffs are the welfare levels of the two countries—sums up this model. It is obvious that no matter what the other country does, it is better for both of them not to tax the rich and to leave income distribution untouched, even though both would be better off if they taxed the rich and transferred some money to the poor.

The model just described, though highly simplistic, drives home the need for global cooperation if countries are interested in improving their income distribution. Globalization means that independent countries do not have quite the independence that countries had in earlier times in terms of exercising their individual policy prerogatives.

While we talk about the need for global coordination for labour market policies, for environmental policies, for trade norms, we seldom talk of global coordination when it comes to discussing countries' policy for achieving greater equity. This chapter, however, has indicated that some of the same problems arise here. The fact that capital and professional worker, migrate from one country to another in response to economic incentives has created a need for international coordination.

There is, however, a need to sound two words of caution. First, the model must not be taken as justification for closing borders against the movement of people or capital. The advantages of the free movement of resources, physical and human, in and out of countries are immense, and the pressure should be kept up on governments to keep the corridors open rather than set up barriers. Second, in the above model there is no conflict between poverty and inequality—the policy that curbs inequality also curtails poverty. But that does not always happen in reality. Some policies can curb poverty only by increasing inequality. At times countries can be so overzealous in controlling inequality that they are unmindful of the fact that this may lead to an increase in poverty.

What should one do if such a conflict arises? I have taken the view elsewhere and would endorse it again, that poverty alleviation should be the priority

Country 1	Country 2	
	$t_2 = 0$	$t_2 = 1/4$
$t_1 = 0$	16, 16	26 + (2/3), 0
$t_1 = 1/4$	0, 26 + (2/3)	18, 18

Figure 15.3: The International Inequality Game

objective. So if there is a conflict between poverty removal and inequality mitigation, we should go for poverty removal. However, there are many situations in which the alleviation of inequality and of poverty are compatible objectives and may even be complementary.

With regard to the objective of reducing poverty, one way of encouraging countries to do so would be to present their economic performance not in terms of their overall per capita incomes, as is the common practice, but according to the per capita incomes of the poorest 20 per cent of each country, or what may be called its 'quintile income'.[13] The objective of raising quintile incomes would be a good way of combining poverty alleviation with growth. The focus on quintile income would make for a natural Rawlsian focus on the weakest sections of the population. At the same time, a focus on the *relatively* poor, rather than on those below an exogenously defined poverty line, means that one would eventually be interested in everybody. The target of raising poor people above an exogenously given poverty line can be reached, but the aim of raising the living standard of the poorest 20 per cent of a society can never be fulfiled totally, because there will always be a category of the poorest 20 per cent of a population. In a perfectly equitable society, for instance, the objective of raising quintile income would coincide with the objective of raising per capita income. This is the strength of the quintile income target proposed here.

One could legitimately ask whether a focus on quintile income, rather than the more common per capita income, would make that much of a difference. The cross-country view of these measures in Table 15.6 shows the substantial difference between the two. As can be seen, there is not only a large difference in absolute terms (in the 1990s the bottom 20 per cent of people in Peru had a per capita income of less than $1000, whereas the figure for the country as a whole was more than $4000; most people in Sierra Leone were poor, but those in the quintile income group were unbelievably poor), but also reference to the quintile measure changes the rankings of countries quite sharply. For example, the United States drops from first place in per capita terms to a position under Norway, Japan, and Sweden. Likewise Bolivia, which ranks higher than India, drops to below India when the quintile measure is used.

Of course, the quintile income is less inclusive than, say, the human development Index, compiled by the UNDP, which takes account of life expectancy and literacy as well as income, but it has the advantage of a sharper focus. It is not claimed here that other measures of standards of living do not matter; rather we merely stress that when looking at a country's economic (or, more narrowly, income) performance we should focus our attention on the bottom of income distribution.

Despite the advantages of, and the moral case for, focussing on quintile incomes, there is a considerable problem with lack of data. In Table 15.6 the data on the share of income that the bottom 20 per cent of each country commanded were collected in different years (shown in parentheses), which reflects the fact that such data are compiled only occasionally. Moreover, it is arguable that the poorest 20 per cent people have a very different consumption pattern from the average person. Hence, ideally, we should use index numbers that are specific to this class when computing their per capita incomes. Clearly these are data that we will have to strive to collect if we are seriously interested in focussing on the poor.

Table 15.6: Per capita Incomes and Quintile Incomes, Selected Countries

	Per capita GNP, 1999 (US dollars at PPP)	Percentage share of income of bottom quintile (survey years in parenthesis)	Per capita income of poorest 20 % (or quintile income), 1999 (US dollars at PPP)
Sierra Leone	414	1.1 (1989)	23
India	2149	8.1 (1997)	870
Bolivia	2193	5.6 (1990)	614
China	3291	5.9 (1998)	971
Peru	4387	4.4 (1996)	965
Thailand	5599	6.4 (1998)	1792
Mexico	7719	3.6 (1995)	1389
Malaysia	7963	4.5 (1995)	1792
Korea	14637	7.5 (1993)	5489
Sweden	20824	9.6 (1992)	9996
France	21897	7.2 (1995)	7883
Japan	24041	10.6 (1993)	12742
Norway	26522	9.7 (1995)	12863
United States	30600	5.2 (1997)	7956

Source: Compiled from World Bank (2001).

To reiterate what was said above, concern for poverty alleviation should take priority over concern for income inequality. Inequality is something we should strive to minimize, but not be jeopardizing improvement of the income of those in the lower quintile group. To understand this, consider the policy question of whether to extend intellectual property rights over a long duration, say fifteen years. If this were judged to be against the lower-quintile–income-maximization criterion, then we should have to work out precisely what such a policy was likely to do to the poorest 20 per cent. It is, for instance, entirely conceivable that it would make a few people very wealthy, namely those who patent commercially useful ideas. This should not be considered a reason for or against extending the rights to intellectual property, but if such a policy could somehow help to raise the living standard of the poorest people, then it should be considered worthwhile.

Similar issues arise in respect of tax policy. If an overly aggressive tax system were used to divert money from the rich to the poor, it is arguable that this could damage richer people's incentives to such an extent that the poor would end up being hurt when all effects were taken into account, despite the income subsidy they received. This is because their incomes could drop so much (because there was not enough capital to work with or because firms would shut down, causing a drop in demand for labour) that despite the subsidy they would be below the pre-policy intervention level. In such a case the tax would obviously be counter productive. Note that tax policy is not being evaluated here in terms of general efficiency, as mainstream economics normally does, but in terms of a kind of truncated efficiency where efficiency is judged by its effect on the poorest.

A large number of policies, however, have no obvious effect on the poorest people, and from these we should choose the ones that could minimize inequality. And since the effectiveness of many such policies will crucially depend on what is being practiced in other countries, we return to the subject of the global coordination of equity policies. In practical terms this will require a lot of institutional spade work. As mentioned earlier, for discussions of intercountry trade policy we have the WTO, and for crafting intercountry labour market policy we have the ILO, but there is no forum for coordinating the effort to alleviate poverty and inequality. These issues have been written about and discussed by the World Bank, but inadequate attention has been paid to the intercountry dimension of this problem, which is an increasingly significant omission in a globalizing world.

15.5 CONCLUSION

This chapter has addressed the changing nature of development economics, and in particular the changes brought about by the process of globalization. The focus has been on curtailment of the freedom of individual countries to draft policy. In the ancient world, where the cost of crossing borders and the uncertainties awaiting in distant lands caused most countries to isolate themselves, national governments could pursue the policies they wanted with impunity and with little concern for what other countries were doing. This is no longer the case, and goods, capital and people are flowing in and out of countries at unprecedented levels and at a much lower cost than ever before.

This chapter has argued that one consequence of globalization is the erosion of global democracy. Because powerful nations and the activities of multinationals can have huge effects on the well-being of people in other countries, a retreat of democracy is a natural concomitant of globalization. This has led to the increasing marginalization of some groups and contributed to global political instability. One way of controlling this problem would be to strengthen the democratic working of international organizations, giving much more voice to poorer countries than they currently have.

As an illustration of one area in which globalization has had a major impact on the nature and effectiveness of development policy, this chapter has considered the subject of international labour standards. Models have been constructed to show the need to coordinate policies among developing countries, which will require a democratic forum to help with the coordination. We further argue that global inequality is reaching an intolerable level, and once again this is an area where policies will have to be coordinated across countries if they are to be effective. To this end there is need for international organizations to take the initiative.

Advances in development economics, the availability of large data sets, and the growing sophistication of economic theory have enabled us to understand economic underdevelopment and global poverty much better than ever before. The challenge is to combine this understanding with political will in order to do away with poverty, which is unacceptable and unnecessary in this generally prosperous world.

NOTES

1. Dreze (2000), drawing on the research of Sivard (1996), reports that the ratio of civilian casualties to military casualties in armed conflicts rose from 1:1 to 5:1 between the beginning and end of the twentieth century.

2. This argument of course hinges on there being more technological advancement in the fishing sector than in other sectors, but this causes only a minimal loss of generality. As long as there is not a perfectly balanced innovation in all sectors, some sectors will see more technological advances than others. One such sector could be 'fishing'.

3. In making this remark I am aware that there are other, more acutely dispossessed people in society. There are the jobless and the homeless, and within households there may be women and girls who are intolerably poor. Needless to say, we need policy instruments to reach out to them. But here, I am not trying to be comprehensive but to consider some major illustrative examples involving the poor in places where globalization is changing the nature of development policy. And it seems reasonable to focus on the case of the working class in poor nations.

4. This is also noted by Kimura (2002), who rightly points out that the competition is not just about wages and workers' compensation but also about the productivity of workers, their education, language skills, and so on.

5. See Mill (1970, 1971). For more contemporary references see Kanbur (2001), Neeman (1999), Satz (2001) and Basu (2002b), Trebilcock (1993). The model constructed here is similar to those in Raynauld and Vidal (1998) and Singh (2002).

6. Once we recognize that social norms are an ingredient in human choice, the role of benign legislation becomes even more important, since a law can help shape human preference and norms; and once these have been formed—even if the law were to be removed—behaviour might well remain unaltered.

7. It could be argued that the Pareto criterion should not be treated as a sacred cow, and that if we eschew the Pareto criterion we can develop justifications for intervention. Such an argument has been advanced by Kanbur (2001), Sen (1970), and others. Kanbur actually looks into a variety of Paretian and non-Paretian justifications for intervening in what he describes as 'obnoxious markets'. Here we shall remain within the Paretian framework.

8. While this sounds obvious, it involves some tricky problems concerning the number of agents involved in an economy. Conventional economics makes very strong assumptions about this, and the only reason why such assumptions are accepted by economists is that they are so used to them (Basu 2002c).

9. A similar analysis pertaining to child labour was conducted in Basu (1999).

10. While the consequences of such coordinated action have not been studied empirically, a recent empirical study of which countries are likely to ratify labour market conventions sheds interesting light on what prompts countries to ratify standards (Chau and Kanbur 2001). There seems to be evidence of a peer-group effect—that is, if the other countries in a country's 'peer group' ratify a convention it is more likely that the latter country will ratify it.

11. Based on *Fortune* magazine data and world development indicators, it seems that in 1998 the 50 richest Hollywood personalities earned as much as the entire population of Burundi (11 million), and that the rise in the value of stocks owned by Bill Gates of Microsoft was equal to the combined income of the entire 60 million people of Ethiopia. Turning to more serious data, the income ratio between the richest 10 per cent of the world and the poorest 10 per cent was 64:1 in 1993 (Thorbecke and Chutatong 2002, based on Milanovic 1996).

12. According to the Census of England and Wales of 1861, 36.9 per cent of boys and 20.5 per cent of girls aged ten to fourteen were regular labourers.

13. The normative basis of using quintile incomes to rank countries and evaluate economic policy is discussed by Michael Lipton, Paul Streeten, and myself in Meier and Stiglitz (2001).

REFERENCES

Basu, K., 1999, 'Child labor: Cause, Consequence and Cure with Remarks on International Labor Standards', *Journal of Economic Literature*, 37, 1083–119.

—— 2000a, 'Whither India? The prospect of prosperity', in R. Thapar (ed.), *India: Another Millennium*, Penguin, New Delhi.

—— 2000b, *Prelude to Political Economy: A Study of the Social and Political Foundations of Economics*, Oxford University Press, New York.

—— 2002a, 'The retreat of global democracy', *Indicators*, 1, 77–87.

—— 2002b, 'A note on multiple general equilibria with child labor', *Economics Letters*, 74, 301–8.

—— 2002c, *Labor laws and labor welfare, in the context of the Indian experience*, Cornwell University (Mimeo).

—— 2003, 'The economics and law of sexual harassment in the workplace', *Journal of Economic Perspectives*, 17, 141–57.

Basu, K. and P. H. Van, 1998, 'The economics of child labor', *American Economic Review*, 88, 412–27.

Besley, T. and R. Burgess, 2002, *Can labor regulation hinder economic performance? Evidence from India*, London School of Economics (Mimeo).

Bhagwati, J., 1995, 'Trade liberalization and "fair trade" demands: Addressing the environmental and labor standards issues', *World Economy*, 18, 745–59.

Bhagwati, J. and R. Hudec, 1996, *Fair Trade and Harmonization* (Vols. 1 and 2), MIT Press, Cambridge, MA.

Chau, N. and R. Kanbur, 2001, *The adoption of international labor standards conventions: Who, when and why?*, Cornell University, Ithaca, NY (Mimeo).

Dreze, J., 2000, 'Militarism, development and democracy', *Economic and Political Weekly*, 37, 1171–83.

Engerman, S., 2002, 'The history and political economy of international labor standards', in K. Basu, H. Horn, L. Roman, and J. Shapiro (eds.), *International Labor Standards: Issues, Theories and Policy Options*, Blackwell, Oxford.

Genicot, G., 2002, 'Bonded labor and serfdom: A paradox of voluntary choice', *Journal of Development Economics*, 67, 101–28.

Hoff, K. and J. E. Stiglitz, 2001, 'Modern economic theory and development', in G. M. Meier and J. E. Stiglitz (eds.), *Frontiers of Development Economics*, Oxford University Press, Oxford.

Kanbur, R., 2001, *On obnoxious markets*, Cornell University, Ithaca, NY (Mimeo).

Kelley, F., 1905, *Some Ethical Gains through Legislation*, Macmillan, London.

Kimura, F., 2002, *Development strategies for economics under globalization: Southeast Asia as a new development model*, Keio University, Japan (Mimeo).

Kohsaka, A., 2002, *National economies under globalization: A quest for new development strategies*, Osaka University, Japan (Mimeo).

Maddison, A., 1979, 'Per capita output in the long run', *Kyklos*, 32, 412–29.

—— 2001, *The World Economy: A Millennial Prospective*, OECD, Paris.

Meier, G. and J. Stiglitz, 2001, *Frontiers of Development Economics: The Future in Perspective*, Oxford University Press, Oxford and New York.

Milanovic, B., 1996, *True world income distribution, 1988 and 1993: First calculation based on household surveys alone*, World Bank, Washington, DC (Mimeo).

Mill, J. S., 1970, *Principles of Political Economy*, Penguin, Harmondsworth.

—— 1971, *On Liberty*, reproduced in Mill, *Utilitarianism, Liberty and Representative Government*, Dent and Sons, London.

Murphy, T. A., 1992, *Ten Hours' Labor: Religion, Reform, and Gender in Early New England*, Cornell University Press, Ithaca, NY.

Neeman, Z., 1999, 'The freedom to contract and the free rider problem', *Journal of Law Economics and Organization*, 15, 685–703.

Parfit, D., 1984, *Reasons and Persons*, Clarendon Press, Oxford.

Rao, M. S., R. T. Govinda, and K.P. Kalirajan, 1999, 'Convergence of incomes across Indian states: A divergent view', *Economic and Political Weekly*, 34, 769–78.

Raynauld, A. and J.-P. Vidal, 1998, *Labor Standards and International Competitiveness: A Comparative Analysis of Developing and Industrialized Countries*, Edward Elgar, Northampton, MA.

Satz, D., 2001, *Why shouldn't some things be for sale*, Stanford University, CA (Mimeo).

Sen, A., 1970, 'The impossibility of a Paretian liberal', *Journal of Political Economy*, 78, 152–7.

Singh, N., 2002, 'The impact of international labor standards: A survey of economic theory', in K. Basu, H. Horn, L. Roman, and J. Shapiro (eds.), *International Labor Standards: Issues, Theories and Policy Options*, Blackwell, Oxford.

Sivard, R., 1996, *World Military and Social Expenditures 1996*, World Priorities, Washington DC.

Swinnerton, K. and C. A. Rogers, 1999, 'The economics of child labor: Comment', *American Economic Review*, 89, 1382–5.

Thorbecke, E. and C. Chutatong, 2002, *Economic inequality and its socio-economic impact*, Cornell University, Ithaca, NY (Mimeo).

Trebilcock, M., 1993, *The Limits of Freedom of Contract*, Harvard University Press, Cambridge, MA.

World Bank, 2001, *World Development Report 2000-01: Attacking Poverty*, Oxford University Press, Oxford and New York.

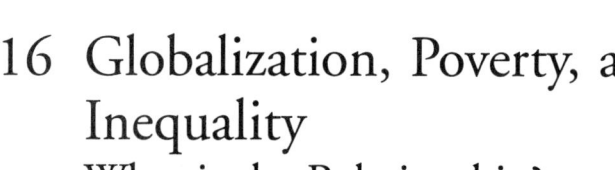

16 Globalization, Poverty, and Inequality
What is the Relationship?
What can be done?

16.1 THE QUESTIONS

Forbes Online of 27 February 2003,[1] offers some information about the world's ten richest people. Much of the information would cause little surprise. The list shows that big money comes from software innovation, retailing scale economies, the business of oil, investment luck, and inheritance. What is, however, really striking—more so as one ponders the matter—is just *how* rich these ten people are.[2] Together they had, in 2002, a net worth of $217 billion, ranging from Bill Gates in the lead with $40.7 billion to John Walton (son of Sam Walton, founder of Wal-mart) at the rear with $16.5 billion.

To understand how staggering this is, let us look at Tanzania in the same year, 2002. In that year Tanzania, with a population of 35 million, had a GDP of $10.15 billion (World Bank 2004). In other words, if one assumes that the ten richest people earn a return of 5 per cent on their assets,[3] their earning in one year would be roughly equal to the total annual earnings of the entire population of Tanzania. And, of course, Tanzania has its own share of the very wealthy. If we leave them out—say 1 per cent of the richest Tanzanians—and look at the poorer end of the spectrum, we will get a gap between the world's richest and poorest that is difficult to comprehend.

I have benefited from the comments of Tony Aiddison, Carol Graham, Rhys Jenkins, Ethan Ligon, Machiko Nissanke, Omar Robles, Elisabeth Sadvovlet, Alice Sindzingre, Eric Thorbecke, Rolph Van der Hoeven, David Zilberman, and especially Tony Shorrochs and an anonymous referee of *World Development*.

First published in *World Development*, Vol. 34 (2006), 1361–73.

If we leave out individuals and turn to nations, the gaps of course shrink but are still striking. Take the richest and the poorest countries (in terms of per capita income) in the list of 152 countries[4] for which detailed data are provided in the *World Development Indicators 2005* (World Bank 2005). These are, respectively, Norway and (tying at the bottom rank) Burundi and Ethiopia. Ethiopia and Burundi have a per capita income of $90 and Norway $43,400. If we make purchasing power parity (PPP) corrections on these, they get a bit closer but still the gap is huge. A person picked at random in Norway is expected to be sixty times as rich as a person chosen randomly in Burundi, even with a PPP correction.

I do not present these numbers to advocate any obvious normative proposition, such as how bad governments are in the Third World to leave their citizens so poor or how mean governments are in the industrialized countries not to divert more money to poor countries. Once one takes account the realities and constraints within which policy makers and politicians in poor and rich nations function, none of these propositions survive—at least not in any obvious way. There are many changes that each of us may want but not one of us may be empowered to do anything about.

The reason I present these statistics is to draw attention to the fact that, even though the debate on whether global inequality has risen or fallen in recent times may be unresolved, the *amount* of inequality is staggering; the *hiatus* between the richest and the poorest people is *too large* and the *extent* of poverty on earth (whether or not it has risen in recent times) is unacceptable. I like to believe that there will come a time when, looking back at today's world, human beings will wonder how primitive we were that we tolerated this.

From this observation to proceed to answering the question, 'What should be done?' turns out to be much harder than what persons of action commonly suppose. That is the reason why, despite having so many persons of action, inequities have persisted from the time of the pharaohs, and in fact recorded history, to present times. What has to be recognized is that the intellectual design problem of how to mitigate poverty is a difficult one, and that could be so even if all of us were single-minded in wanting to remove poverty and we had the science and technology at our disposal (as we probably already do) to remove everybody's poverty. This is because, in contrast to a single individual, for a group of persons to translate their preferences into actions can be a very difficult problem, as rudimentary game theory teaches us.

The aim of this chapter is to study the relation between globalization, inequality, and marginalization and to ask policy questions about what we should do. I shall briefly review the empirical literature on globalization, inequality, and poverty, and the possible interconnections between these (Sections 16.2 and 16.3), and argue that such analysis ought to be combined with theoretical analysis (Sections 16.4 and 16.5), which allows us to explore the realm of the possible— of things that may not have happened as yet but could happen. I shall argue that even if our empirical verdict remains ambiguous, we can think constructively about policy and agency (Section 16.6). While there is a considerable literature on the tradeoff between inequality and growth,[5] what is unusual in this chapter is its attention to the tradeoff between poverty and inequality. This allows us to formulate some clear rules about how much inequality ought to be tolerated in society. The paper formalizes the concept of 'poverty-minimizing level of inequality'.

16.2 THE FACTS

Has globalization led to greater inequality or less? This question has greatly exercised the minds of many analysts. The reason why this question has loomed so large in our debates is that, for many ideologues, how we answer this question amounts to a verdict on globalization. I shall however take the view that seeking a verdict on globalization is a hopeless project. First of all, it is too catch-all a term and therefore it can be good and bad, depending on what aspect of it we are looking at, in which period, and at which location. When the Spaniards came into contact with the Incas in the early sixteenth century, that was a step toward globalization. And judging by the fact that the native population of the new world rapidly declined under the combined might of the sword and new bacteria, clearly this globalization was not good for the native population. And even if it could be argued that the natives are better off *today* than they would have been had they remained 'undiscovered', it could still be argued that (barring the case where their discount factor were indistinguishably close to one), their welfare, aggregated over the last few centuries, has been adversely affected. On the other hand, when the British came in contact with the Chinese of Hong Kong, that was also a step towards globalization; and it is maintainable that on this occasion globalization benefitted all parties involved.

This diversity of experience suggests two things—that a single answer for the effect of globalization is too much to expect and that globalization is *potentially* beneficial for all.[6] The latter suggests the need for policy design that can convert the potential benefit into actual benefit and that will indeed be the driving motive behind the policy analysis in this chapter.

But let me begin with the facts. Has inequality in the world increased? We will see that the answer is mired in debate. If we take a very long run view, the answer is fairly transparent. Over the last five centuries, the world has become more globalized and much more prosperous, and, if we consider inter-regional inequality (in contrast to interpersonal inequality), it is clear that inequality has grown.

The concept of globalization, as measured by trade volumes and capital flows, has been written about a lot (Basu 2004a; Bhagwati 2004; Wolf 2004). The total value of exports all over the world in the year 2002 was $6,455 billion, up from $3,452 billion in 1990; and the total amount of foreign direct investment globally in 2002 was $631 billion, while it was $202 billion in 1992 (World Bank 2004).

As far as prosperity and inequality go, though there is scope for debate about whether global regional inequality has increased or decreased over the last two or three decades,[7] the trend, viewed over a long stretch of time and measured as the ratio between the richest and the poorest, seems to be an unequivocal deterioration. According to the calculations of Angus Madison (2001), displayed in

Table 16.1, if we track per capita GDP of large regions of the world, the growing disparity is obvious. The richest region was 1.8 times richer than the poorest region half a millennium ago, whereas, currently, the richest region has a per capita income that is 20 times the income of the poorest region.

Table 16.1: Levels of GDP per capita, 1500–1998, (1990 PPP dollars)

	1500	1700	1913	1998
U.S.A.	400	527	5,301	27,331
Sweden	695	977	3,096	18,685
U.K.	714	1,250	4,921	18,714
Japan	500	570	1,387	20,413
India	550	550	673	1,746
China	600	600	552	3,117
Africa	400	400	585	1,368
Ratio of Richest to Poorest	1.8:1	3.1:1	9.4:1	20:1

Source: Maddison (2001).

What has happened in recent times remains more controversial (see, for instance, Bourguignon and Morrison 2002; Galbraith 2002; Heshmati 2004; Melchior 2001; Milanovich 2002; Naschold 2004).[8] A comprehensive way to measure inequality is to compute the Gini coefficient. If we do this for countries, what do we find? The answer depends critically on whether we use population-weighted or -unweighted data, and a part of the controversy is caused by this difference. If we use population-weighted data, this means that we pretend that all Chinese earn the per capita income of China and all Indians earn the per capita income of India and so on, and then compute the Gini coefficient of the world. The use of unweighted data means that each country is treated as one person earning the per capita income of that country. So evidently both methods have their shortcomings. It should be recognized that this problem is encountered in economics at various levels. Even within the household there is often a lot of inequality, and this is especially significant for households that have internal conflicts of interest (Basu 2006). But thanks to the inadequacy of data we often are compelled to treat the household as a single decision-making unit.

If we go via the route of using unweighted data for each country, then we find that the Gini coefficient of intercountry inequality has grown over the last few decades (Milanovich 2002). On the other hand, if we use population-weighted data, we find that the Gini coefficient has been declining slowly but almost monotonically since the late 1960s, with the pace of decline picking up a bit in the 1990s (Melchior 2001; Melchior, Telle and Wiig 2000). The latter is driven in large measure by the strong economic growth in China since the late 1970s and India since the early 1990s, since population weights of these countries are very high.

It should now be clear that, depending on exactly what one chooses to use as the measure, one can find almost any evidence that one seeks. Is one measure clearly superior to another measure? If we are interested in *individual* well-being, as much of economics is, it may seem right that we use population-weighted data.

To treat China and Canada as comparable units does not seem right. But there are two possible responses to this. Given the significance of the nation state as a political unit, and given that our political perceptions are shaped by awareness of intercountry situations, there may be a case for trying to find out what is happening to intercountry incomes. Second, if we are interested, ultimately, in the individual, we should be looking at neither the population-unweighted nor the population-weighted intercountry inequality, but global interpersonal inequality. This is because counting all the people of China as one person is to lose vital information and to treat all the people of China as if they each earn the per capita income of China is also to lose important information, especially since inequality in China has been growing. The same is true of India. Fortunately, how this debate is resolved is not critical to what I want to argue here.

If I were to try to associate global inequality to globalization, I would take the longer run view of what has happened, since globalization is a process that has been with us for centuries. It has gone through some brief periods of retreat (Williamson 2002), but the long-run process has been a slow and steady one of the globe coming together. The long-run regional inequality (and I am not equating this to interpersonal inequality and poverty, though interpersonal inequality has probably moved in tandem with regional inequality) seems also to have increased over the very long run. But no matter what view we take of the trends, it seems easy to argue that there is reason for concern. First, while the Gini coefficient is important, the gap between the richest and the poorest is important as well. If a sizable population feels increasingly marginalized because they find themselves becoming poor relative to global wealth, this is bound to stoke political volatility and even if that did not happen, this would seem normatively unacceptable to me. And, as we saw, the gap between the poorest and the richest is rising if we take a long-run view of this. Secondly, no matter what has been the trajectory and no matter what its connection to globalization, the level of inequality that we see today, as cited at the start of this chapter, is far too large for complacency.

16.3 THE POSITIVE AND NEGATIVE FALLOUTS OF GLOBALIZATION

To understand how globalization can have the negative fallout of marginalizing people, consider the case where the world markets for goods and services are suddenly and fully opened up. Given that a disproportionately large share of the world's GDP comes from the industrialized countries, it seems reasonable to predict that the prices of goods in poor nations will converge more rapidly towards prices in industrialized nations than the latter converge towards the former. In other words, international prices of goods and services will move to somewhere between prices in industrialized nations and prices in developing countries but closer to the former.

Labour being less mobile than goods and services, it seems reasonable that for sections of the labour force in poor nations, and especially for the illiterate and unskilled, who are unable to take advantage of the new technology, wages will lag behind prices.[9] Hence, for some of the poorest people there can be a period of increased hardship before the benefits of opening up trickle down. This is one of the important problems of rapid globalization. To a certain extent, the reported increase in inequality within poor countries (see Banerjee and Piketty 2005 for India) is a consequence of this.

Conversely, it is natural to expect that, with globalization, the skilled end of the labour market in poor countries will benefit disproportionately. Their access to modern technology will increase their pay. Also as their compatriots find jobs in developed countries and move out, the shortage of their skill in the home country will push up the price for their work and make them rich. Banerjee and Piketty's study shows that the group that has gained disproportionately in India over the last decade is the richest 0.01 per cent of the population. It is not hard to show that as income stretches out in this manner for some, the poorer people are not just poorer compared to the richest, but their absolute welfare may decline because of the rise in the price of goods or by their getting excluded from the 'market'.[10]

During a field visit to the village of Jakotra, in a remote corner of Gujarat, bordering on Pakistan, I found a palpable concern among the poor villagers about what globalization might do to them (Basu 2004b). The villagers of Jakotra earn their livelihood largely from handicrafts and mainly embroidery work on textiles. The villagers were concerned that their meager livelihood could get wiped out by competition from some international producer who decides to manufacture embroidered clothing in large factories and export to India. Talking to the villagers I realized what a double-edged sword globalization is. On the one hand, they have benefitted in the last decade because of globalization and their ability to sell their product in faraway lands and cities.[11] On the other hand, they rightly feared that this prosperity may not last. Moreover, these people are still poor enough that end of prosperity for them could mean acute poverty, destitution and even starvation. When that happens it would clearly not be good enough to point these people to the *potential* benefits of globalization. The right policy is to craft government interventions that provide a safety net for the poorest people at times of transition.

Something analogous is true for developed countries concerned with the problem of outsourcing. The overall benefits of outsourcing are clear enough. When the US automobile industry beg eroding because of competition from Japan, if the US government thwarted competition by blocking Japanese cars from coming into the country, it is likely that there would be many more automobile workers in the United States today, but the country would also be poorer for this. In the early 1990s it had looked as if the Japanese economy would overtake that of the United States. But it was the openness of the IT sector in the United States, drawing talent from all over the world, which prevented this from happening.

Something similar is true for the current outsourcing problem. To block outsourcing will mean more people in the United States doing call center jobs, data filing work, and rudimentary software work, but it will almost certainly mean the loss of competitive advantage for the United States and overall loss for the country. But this is not to deny that there are people who are being hurt, certainly in the short run, by outsourcing. The right policy here, as in the case of poor countries facing competition, is not to stop outsourcing but devising policies to soften the consequences of competition for the population that is hurt by it. This policy question is addressed in Section 16.6.

I construct a simple model in Section 16.5 to illustrate some of the policy dilemmas mentioned in this chapter and the risks of globalization. But I should emphasize that the message of this must not be read as one against globalization. The potential benefits created by the easier flow of goods, services, software products, and labour are enormous and to stop these would be a gross error. At the same time, the fear of these getting stopped must not lead us to praise all aspects of globalization. By pointing to its negative fallout, this chapter hopes to encourage policies to counter them and to distribute better the spoils of globalization. Not only should this be viewed as a moral imperative, but to ignore the marginalizing groups is to risk political instability and war in the long run.

16.4 THE QUINTILE AXIOM

In designing policy it is important to try to spell out clearly what our ultimate objectives are. A new tax, a subsidy or a new restriction on trade is seldom good *in itself.* The goodness or badness of such action depends on what it does to what we value ultimately for society. There may indeed be philosophical difficulties in spelling out, once and for all, ultimate or basic value judgements, as Sen (1970) argued. New situations, new policy conundrums may compel us to abandon some judgement that we had earlier held as fundamental.[12] But keeping in mind that new situations and new choices may make us want to mould our objectives, we must ask what is it that the policymaker should try to maximize.

I have elsewhere (Basu 2001) suggested a simple normative rule, which has attractive properties, not least of which is simplicity. Where traditionally we associate each country's main objective with its per capita income, the normative criteria that I have proposed elsewhere and am going to maintain here would require us to associate it with the per capita income of the poorest 20 per cent of the population. I call this the 'quintile income' of a country.

More formally, let the income profile of a country with n people be given by $(x_1, x_2, \cdots, x_n)$ and assume, without loss of generality, that individuals are so named that

$$x_1 \leq x_2 \leq \cdots \leq x_n.$$

Clearly, this country's per capita income is given by

$$y = (x_1 + x_2 + \cdots + x_n)/n.$$

On the other hand, the country's quintile income is given by

$$q = (x_1 + x_2 + \cdots + x_t)/t,$$

where $t = n/5$.

What is being suggested is that in evaluating a country's well-being we should focus on the country's quintile income. Henceforth, this normative principle will be referred to as the 'quintile axiom'.

The quintile measure should not be confused with a poverty measure (or inverse of a poverty measure) of a society. Hence, the objective of raising the quintile income of a country need not coincide with the objective of lowering poverty. This will certainly be so if we use an absolute measure of poverty (which can become zero and so leave no further target unfulfilled, whereas that can never happen with the target of maximizing quintile income) and may not be true even for most relative poverty measures. The quintile axiom I am recommending is a much more *overall* normative target with which policymakers should be concerned.

At first sight this indicator may seem arbitrary; but, as a rule, any single indicator for measuring a nation's well-being is arbitrary till we get used to it.

There are ways in which the quintile axiom or the general idea behind it can be generalized. We could, for instance, give weights to the incomes of people at different levels of poverty with the poorest people getting the highest weights and then look at the weighted per capita income of the society (some of these variants are discussed in Basu [2001]). But I am here interested in suggesting a measure that is simple and easy to understand. The quintile axiom is a suggestion in that spirit.

It is worth seeing how evaluating an economy using the quintile income not only makes a large difference to the absolute numbers, as is only to be expected, but can change the rankings. Table 16.2 gives the per capita incomes and quintile incomes of a selection of countries. As expected Norway and Japan move up the ranking ladder sharply, the United States moves down. At the poorer end Romania,

Table 16.2: Quintile Incomes of Nations, 2002

Country	Per capita income US #, PPP	% of income accruing to poorest 20 %	Quintile income US $, PPP
Norway	36,690	9.6	17,611
USA	36,110	5.4	9,750
Switzerland	31,840	6.9	10,985
Japan	27,380	10.6	14,511
Finland	26,160	9.6	12,557
Sweden	25,820	9.1	11,748
Korea, South	16,960	7.9	6,699
South Africa	9,810	2.0	981
Trinidad & Tobago	9,000	5.5	2,475
Malaysia	8,500	4.4	1,870
Russian Federation	8,080	4.9	1,980
Romania	6,490	8.2	2,661
Peru	4,880	2.9	708
China	4,520	4.7	1,062
Guatemala	4,030	2.6	524
India	2,650	8.9	1,179
Bangladesh	1,770	9.0	797
Sierra Leone	500	1.1	28

Source: Computed from World Development Indicators 2004.

India, and Bangladesh make relative gains, whereas China, somewhat surprisingly, loses out. The sharpest losses caused by shifting attention from per capita income to quintile income occur in Peru, Guatemala, and Sierra Leone.

The quintile income measure, viewed as an equity-conscious measure of welfare, has several normative advantages. Unlike a policy that tries to minimize poverty or minimize inequality, the objective of maximizing the quintile income has a natural dynamism, because it is a moving target. In a country with gross inequalities, this measure will suggest that we focus on the conditions of the poorest people. But if the better off people are ignored totally and for too long, they will soon be a part of the bottom quintile of the society and so deserve attention. If there is full equality in society, this measure does not allow the policymaker to sit back. Since in such a society the quintile income coincides with the per capita income, the aim now will be to raise the per capita income.

Also, a focus on the quintile income does not mean that the growth rate is to be ignored. It is simply that the growth rate should be measured in terms of the growth rate of the per capita income of the bottom quintile of society. And there is the advantage of directness in this new measure. Instead of saying or claiming that we should aim to increase income growth and expect the benefits to reach the poorest sections, this measure says we should aim to increase the growth rate of the quintile incomes.

It is true that, unlike the United Nations Development Programme's (UNDP) human development index, the quintile income ignores non-income aspects of development. But my defense against this criticism is twofold: First, what I am recommending is not that we ignore non-income aspects of development but that, where we would have focussed on per capita income, we focus on quintile income, instead. Second, I would conjecture that, in general, quintile incomes will have a closer a relation to a counry's various standard-of-living indicators, like infant mortality, life expectancy, literacy, and so on, than per capita incomes. This is something that will in fact be interesting to investigate later.

The focus on quintile income also suggests how we should view inequality. In general, I would view inequality as undesirable, but poverty as the greater evil. So, the amount of inequality that we should tolerate is the amount 'necessary' to minimize poverty, which will here be equated with maximizing quintile income.[13] It is, for instance, arguable that a society of perfect equality (at least given our contemporary values and preferences) would be crushingly poor. Hence, the focus on quintile income will steer us away from attempting perfect equality. It should be evident that the welfare criteria being suggested here is different from the well-known one in which welfare is equated with $\mu(1 - G)$, where μ is the per capita income of the society being evaluated and G its Gini coefficient (Sen 1976). In this measure, welfare is deflated according to the amount of inequality in the country, whereas in my measure, welfare is deflated by the poverty of the poorest quintile of society.

In the next section, a model is developed which illustrates the notion of the 'right' amount of inequality. The model will also show how this may depend on the level of globalization. This naturally gives way to the idea of having to coordinate policies across countries, which is what the last section of this chapter will be concerned with.

16.5 AN ILLUSTRATIVE MODEL

I shall in this section develop a simple, highly stylized model to illustrate some of the principles discussed thus far. In particular, the model will illustrate (1) how the 'quintile axiom' may imply that we have to tolerate a modicum of inequality and (2) how globalization weakens each nation's ability to control poverty and thus directs our attention to the need for inter-country coordination of policy.

Consider a world with 'many' identical countries. Each country has a population of n. And of these n people, p are 'productive' and u are 'unproductive'. Here,

$$n = p + u, \quad p, u \geq 0, \, n > 0.$$

Output in a country occurs because of the work done by productive people. The unproductive people live off the externality of other people's work.

The amount of work, $h \in [0,1]$, that a productive person does is negatively related to the (proportional) income tax rate, t, that prevails in the country where he resides. To keep the analysis simple, I shall assume

$$h = 1 - t, \tag{16.1}$$

where $t \in [0,1]$ is chosen by the government and is treated by citizens as exogenous.

The (pre-tax) income, Y, that accrues to a productive person who puts in h units of work is given by

$$Y = Ah, \quad A > 0. \tag{16.2}$$

If every productive person does h units of work, every unproductive person gets an income, y, given by

$$y = ah, \tag{16.3}$$

where $A > a > 0$. This captures the externality assumption.

The assumption of linearity, namely, $Y = Ah$ and $y = ah$, is purely for algebraic simplicity. I could just as well have assumed $Y = f(h)$, where $f'(h) > 0$. What is unusual here, and at variance from textbook models of the economy, is the assumption of externality. I am assuming that when productive people in a country work hard, they benefit of course, but also the (nonworking) unproductive people of that country benefit, however little. In a more realistic model, the benefit accruing to the unproductive peole would depend on *how many* productive people there are, but that will not make any significant change to my model and so will be ignored here.

Government's sole activity in this model is to transfer income, through the choice of a tax rate from the rich to the poor. If the tax rate is t, the post-tax incomes of the productive and unproductive people, denoted by, respectively, $\bar{Y}(t)$ and $\bar{y}(t)$ are given by

$$\bar{Y}(t) = (1 - t)Y, \tag{16.4}$$

$$\bar{y}(t) = y + \frac{ptY}{u}. \qquad (16.5)$$

Since each unproductive person receives an equal share of the total amount of tax revenue collected by the government, his total post-tax income is a sum of the externality, y, and the tax subsidy, ptY/u.

Using Eqs. (16.1)–(16.3) to substitute for Y and y, Eqs. (16.4) and (16.5) can be rewritten as

$$\bar{Y}(t) = (1 - t)^2 A, \qquad (16.6)$$

$$\bar{y}(t) = (1 - t)\left(a + \frac{pAt}{u}\right). \qquad (16.7)$$

A typical picture of how individual (post-tax) incomes vary with the tax rate is illustrated in Figure 16.1. We use $\hat{t}$ to denote the tax rate t, where $\bar{Y}(t)$ and $\bar{y}(t)$.

A government that is Rawlsian would be focussed entirely on the unproductive people as long as $t \leq \hat{t}$. It would focus on the welfare of the productive people if $t > \hat{t}$. Suppose now that the government is not exactly Rawlsian but follows the more pragmatic quintile axiom outlined above. If $u/n \geq 1/5$ and $p/n \geq 1/5$ then it would behave like a Rawlsian. Up to $\hat{t}$, it would equate this society's welfare with the welfare of the unproductive people and, beyond $\hat{t}$, it would equate society's welfare with the welfare of the productive people (who are now poorer).

Let us, for now, assume $u/n, p/n \geq 1/5$ and also assume that

$$t^* = \arg \max \bar{y}(t) < \hat{t}. \qquad (16.8)$$

Consider now a government that is committed to the quintile axiom trying to decide what tax rate it should choose. Clearly this government's problem is as follows:

$$\text{Max}_t \text{ min } \{ \bar{y}(t), \bar{Y}(t) \}.$$

Given assumption (16.8), we know that the solution to this will coincide with $\arg \max y(\hat{t})$. From the first-order condition of maximizing $y(\hat{t})$ as described in Eq. (16.7) we get

$$t^* = \frac{1}{2} - \frac{au}{2Ap}. \qquad (16.9)$$

It is easy to see that

$$\hat{t} = \frac{u(A - a)}{A(u + p)}.$$

It is already evident that being concerned about poverty necessitates tolerating a certain amount of inequality. But to see this more clearly, let us focus on a special case. Assume $a = 1$, $A = 4$, and $u/p = 2$. This implies

$$t^* = 1/4 \text{ and } \hat{t} = 1/2;$$

$$\bar{Y}(t^*) = 9/4 \text{ and } \bar{y}(t^*) = 9/8.$$

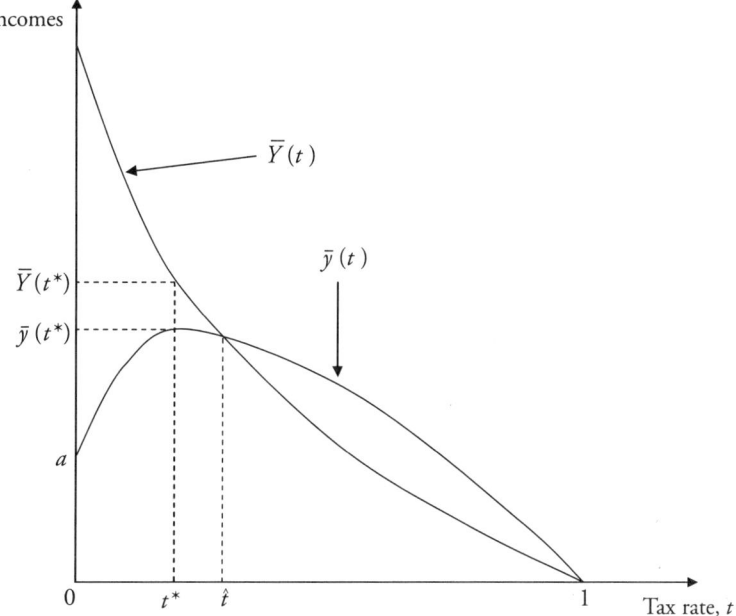

Figure 16.1: The Tradeoff between Inequality and Poverty

That is, a government totally focussed on the poor would choose a tax rate of 25 per cent. This would mean that some people would be twice as rich as some other people. This is an inequality that has to be tolerated in order to help the poor. This is the 'poverty-minimizing level of inequality'.

If, instead, government was committed to eradicating inequality, it would set the tax rate at 50 per cent. In that case incomes would be

$$\bar{Y}(\hat{t}) = \bar{y}(\hat{t}) = 1.$$

In other words the poorest people would find their incomes reduced if total equality was to be achieved.

To complete the discussion, let us see how a government committed to maximizing per capita income would behave. Such a government's aim would be to

$$\underset{t}{\text{Max}} \ \frac{\bar{Y}(t)p + y(t)u}{p + u}.$$

Taking population to be constant, this reduces to the following problem:

$$\text{Max} \ (1 - t)^2 Ap + (1 - t)(au + Atp).$$

It is easy to see that as t decreases, per capita income rises. Hence, such a government would set $t = 0$ and the incomes of the productive and unproductive people would be 4 and 1, respectively.

Up to now the entire analysis has been done by assuming that there is no movement of workers possible from one country to another. In other words, the economies were treated as if they were closed. To see how globalization complicates the picture let us now assume that economies are open. Since in this simple model there is only one good and no capital, the only way to model globalization is to allow labour to be mobile across national boundaries.

I shall consider basically a model of 'real tax competition' (Atkinson 1999). Workers will want to move to a country where post-tax income is the highest, thereby setting off tax competition between governments.

Let us assume that workers will study the tax (and subsidy) structure of different nations and try to migrate to countries where they have the highest (post-tax subsidy) income. Each government sets its tax rate and can decide whom (among all those who so desire) to allow into the country. Let us also assume that, if all countries have the same tax/subsidy rates, then each person stays in his or her home country.

The problems of domestic policy in the event of globalization of the kind just described can be illustrated in many different ways. Let me here consider the case where each country aims to maximize its quintile income. We have seen that if the boundaries of nations were exogenously closed, then each nation would set $t = 1/4$. Now, let globalization remove the exogenous hindrance to labour movements.

Note that each country setting $t = 1/4$ is no longer in equilibrium. Suppose one country lowers t, clearly all productive people from other nations will want to migrate to this country. If the government now decides that it will (1) allow some of the productive people to come in and (2) not allow any unproductive person to come in, it will clearly be able to increase the income subsidy per capita that it gives to its poorest people. Given the government's aim to maximize the income of its poorest people, clearly this government will be better off.

From the above analysis it should be evident that there is no $t > 0$ such that if all governments choose that t, we have a Nash equilibrium. It is easy to see that in equilibrium every country will set $t = 0$. Real tax competition will result in an erosion of taxation and in equilibrium we will have all productive people earning $A (= 4)$ and all unproductive people earning $a(=1)$. Each country ends up behaving *as if* it were interested in maximizing per capita income with no concern for poverty or equity. Globalization erodes each national government's power to have equity-conscious policy. The mobility of labour and, in a more realistic model the mobility of capital compromises a nation's policy efficacy.

Since, from the point of view of governments, the equilibrium outcome is suboptimal (all governments prefer $t = 1/4$ to $t = 0$), there is evidently need for the international coordination of anti-poverty policies. I agree with Atkinson (1999) that redistributive policies by individual governments are possible and one must not turn a blind eye to this. But, at the same time, as globalization progresses, there is increasing need for the coordination of policies across countries.

When we see the enormous poverty in, Ethiopia for instance, we tend to blame it on its government. While most governments have room to improve their performance and the Ethiopian government may have more than its share to do, it would be wrong to overlook that how much control Ethiopia has over Ethiopian poverty depends in part on what happens in Kenya, Tanzania, India, China, and the United States.

16.6 THE POLICY OPTIONS

From the theoretical construction in the last section to move to real-world policy is not an easy task. Countries are at different levels of development and policy instruments available to a government are more varied than choosing tax rates and immigration rules. How can countries coordinate policies in such a world? Do we need a central coordinating organization, like we have ILO for labour policies and WTO for trade policies, for crafting and coordinating anti-poverty and greater-equity policies? These are matters about which we can only speculate, marshalling the insights gained from abstract theoretical models and wisdom from empirical studies and combining them with commonsense, intuition, and guess work.

Much has been written about the nature of pro-poor growth in developing countries (see, for instance, Klasen 2004) and also about the specific problem of pro-poor growth in the context of globalization.[14] Instead of going over the same ground, I want to concentrate here on two policy suggestions which seem to have few antecedents in the literature.

16.6.1 Equity for Workers

I had briefly suggested in Basu (2004b) that one way to counter the problem of some workers losing out because of globalization, whether they be workers in developed countries losing work to outsourcing or labourers in poor countries losing jobs to low-cost high tech imports, is to give workers claims to a fraction of corporate equity income.[15] I do not mean profit sharing in the firm where the worker works but, more radically, that a fraction of equity earnings from all firms should be given to workers in all firms and even to labourers who are currently without work. The full details of this will be complex and will have to be worked out carefully, but the broad idea is that a fraction of equity in firms should be owned by government or some governmental organization on behalf of people in the poorest category, for instance, the bottom quintile. Presumably, workers belong to this category and so will be able to partake in the profits earned by firms.

So, when work is outsourced and some workers lose their jobs, a part of the extra profit generated by the outsourcing should be earned by the workers by virtue of their owning equity. This can be an important policy that guards against excessive marginalization of workers. Moreover, it can help diminish some of the antagonism that exists among workers in developed and poor countries to globalization.

Moreover, if it is true that, over time, the share of labour income will decline (see Basu 2004c, for discussion), then this scheme will have the advantage of automatically softening some of the impact of this on workers, because a part of what they lose out because of dwindling employment and labour income, they will get back in terms of higher equity income.

Among the difficult questions that an actual plan will have to sort out is that of intercountry transfers. The discussion in the above paragraphs is conducted under the implicit assumption that this policy will be implemented by each country, separately. Maybe that is how we have to start. But in today's globalizing world, especially given the huge amount of inter-regional inequality, there is a moral case for extending this, however minimally, to the world as a whole. This will entail developing rules for some intercountry transfer of equity income. In the absence of this, the above economic policy could have the adverse side-effect of heightening nationalism. But our institutions of global governance are so under-developed that the details of how intercountry transfers can be worked out will need some radical innovation in our international organizations. This relates closely to the subject matter of the next section.

16.6.2 A New International Organization for Coordinating Equitable Development

My second suggestion is to urge the need for a new international organization or a new division of an existing international organization that helps coordinate intercountry anti-poverty policies. As we have seen above, achieving greater global equality and reducing global poverty may require the use of policy interventions that are *coordinated across countries.* Unilateral effort by a country is likely to cause the flight of capital and skilled labour from the country and impoverish those who stay behind. Hence, we may get into a Prisoner's Dilemma type of situation where each country would like to take steps to curb inequality or to help the poorest but not be able to do so.

The theoretical possibility of this happening was illustrated in the last section. This is also a very real problem in today's globalized world. Inequality *within* China, India and several other developing countries is on the rise. As argued above, this is closely connected to globalization and this probably explains why China and India—two of the fastest globalizers—are most affected by this problem. Yet there is no institutional arrangement or even infrastructure for countering this. The fact that the income gap between the richest and the poorest people in the world as a whole is way larger than the gap that occurs inside any country is a reflection of the fact that we have no global political institution to address this. No government would be able to tolerate this kind of hiatus within its region of control.

That there may be coordination problems in trade is well-recognized and we have the WTO to help mitigate such problems. That labour market policies need coordination is known and we have the ILO to address this. For environmental problems we have the UNEP or the GEF. But there is nothing comparable to these for anti-poverty and anti-inequality policies. Yet, as demonstrated in the previous sections, this is an area where the coordination problem may be no less acute. Hence, there is clearly a perceived need for a coordinating agency.

This ties up with the objective of giving workers an equity stake discussed in the previous section. In an ideal world these stakes should cut across national barriers. This will once again create a need for a global coordination agency. The same agency that coordinates anti-poverty programmes could also have this as a part of its mandate for the future.

To work out the details of this will not be an easy task. My aim here was to float the idea and place it in the public domain.

NOTES

1. See http://www.forbes.com/lists/2003/02/26/billionaireland.html
2. Another striking commonality among these people that should especially interest academics is that three of these ten are university dropouts (Bill Gates, Harvard; Paul Allen, Washington State University; Lawrence Ellison, University of Illinois).
3. In all likelihood they earn much more—they would not be among the ten richest if they invested their wealth as poorly as most of us do.
4. The list is comprehensive if one is interested in countries that have populations of more than one million. The list, therefore, omits some really small nations, like Lichtenstein.
5. There is, for instance, a considerable empirical literature that shows how inequality can hamper growth (see, e.g., Birdsall, Ross, and Sabot 1995; and Deininger and Squire 1998).
6. A potential benefit for all does not seem to me to be reason for celebration. If it is the case that we expect that the potential will be realized, then of course we should celebrate, but the reason for the celebration is not the potential gain but the fact that we expect an actual Pareto improvement. If, on the other hand, we do not expect the potential to be realized, it is not clear why we should be happy that there has been a potential gain.
7. And debate there has been aplenty: see, for instance, Atkinson (1999), Melchior (2001), Milanovich (2002), and Wade (2004).
8. Some of these controversies on global inequality are mirrored in the discussion on global poverty (see Chen and Ravallion 2001; Reddy and Pogge 2003; Reddy and Minoiu 2005).
9. There can also be increased unemployment among the unskilled. This is possible to explain theoretically once we recognize that employing each person entails some cost on the part of the employer (supervising, conflict mitigation with other employees, breakage of instruments of work) and so, unless the productivity of the worker is above a certain cut-off level, it is not worth employing the person even for a very low wage.
10. A simple adaptation of Atkinson (1995) could illustrate this.
11. Some recent studies seem to confirm, at the level of India, what I saw among the artisans of rural Gujarat. India's opening up in the 1990s, far from hurting the handicrafts sector, seem to have benefitted it. Through the 1990s the share of handicrafts exports in the overall manufacturing exports of India has risen from 2 per cent to 5 per cent (Leibl and Roy 2003).
12. We may maintain that 'one must not kill (a human being)' is a basic value judgement. Then, seeing a friend in terminal condition and suffering from acute pain, we may legitimately revise the basic value judgement to say that 'one must not kill except to relieve a person in pain and in terminal condition'. Sen had suggested that the possibility of having to revise what we think is a basic value judgement will always be there.
13. I put the word 'necessary' within quotes to show awareness that this may itself be malleable. As societal organization changes and our norms and preferences change, the inequality necessary to minimize poverty may itself change. And in a very long run policy exercise one may try to change this parameter. For a recent discussion of the twin objectives of poverty mitigation and the control of inequality, see Dagdeviren, van der Hoeven, and Weeks (2004).
14. Many of the references already cited in this chapter deal with this subject.

15. This is derived from a recognition that what is popularly posed as a conflict between labourers in the developing nations and labourers in industrialized countries should, more accurately, be construed as a problem of global capital versus labour (Basu 2004b; Chau and Kanbur 2003).

REFERENCES

Atkinson, A. B., 1995, 'Capabilities, exclusion and the supply of goods', in K. Basu, P.K. Pattanaik, and K. Suzumura (eds.), *Choice, Welfare and Development*, Oxford University Press, Oxford.

—— 1999, 'Is rising inequality inevitable? A critique of the transatlantic consensus', *WIDER Annual Lecture 3*, UNU-WIDER, Helsinki.

Banerjee, A. V. and T. Piketty, 2005, 'Top Indian incomes, 1956–2000', *World Bank Economic Review*, 19.

Basu, K., 2001, 'On the Goals of Development', in G.M. Meier and J. E. Stiglitz (eds.), *Frontiers of Development Economics: The Future in Perspective*, Oxford University Press, New York.

—— 2004a, 'Globalization and Development: A Re-examination of Development Policy', in A. Kohsaka (ed.), New Development Strategies: Beyond the Washington Consensus, Palgrave Macmillan, Basingstoke.

—— 2004b, 'Globalization and babool gum', *The Little Magazine*, 5.

—— 2004c, 'The Indian economy: Up to 1991 and since', in K. Basu (ed.), *India's Emerging Economy: Problems and Prospects in the 1990s and Beyond*, MIT Press, Cambridge, MA.

—— 2006, 'Gender and say: A model of household behavior with endogenously- determined balance of power', *Economic Journal*, 116.

Bhagwati, J. N., 2004, *In Defence of Globalization*, Oxford University Press, Oxford.

Birdsall, N., D. Ross, and R. Sabot, 1995, 'Inequality and growth reconsidered: Lessons from East Asia', *World Bank Economic Review*, 9.

Bourguignon, F. and C. Morrisson, 2002, 'Inequality among world citizens: 1820–1992', *American Economic Review*, 92.

Chau, N. and R. Kanbur, 2003, *On Footloose Industries. Asymmetric Information and Wage Bargaining*, Cornell University (Mimeo).

Chen, S. and M. Ravallion, 2001, 'How did the world's poorest fare in the 1990s?' *Review of Income and Wealth*, 47.

Dagdeviren, H., R. van der Hoeven, and J. Weeks, 2004, 'Redistribution does matter: Growth and redistribution for poverty reduction', in A. Shorrocks and R. van der Hoeven (eds.), *Growth, Inequality and Poverty: Prospects for Pro-poor Economic Development*, Oxford University Press, Oxford.

Deininger, K. and L. Squire, 1998, 'New ways of looking at old issues: Inequality and growth', *Journal of Development Economics*, 57.

Galbraith, J. K., 2002, 'A perfect crime: Inequality in the age of globalization', *Daedalus*, 131.

Heshmati, A., 2004, *The World Distribution of Income and Income Inequality*, IZA Discussion Paper No. 1267, Bonn.

Klasen, S., 2004, *In Search of the Holy Grail: How to Achieve Pro-poor Growth?*, University of Munich (Mimeo).

Leibl, M. and T. Roy, 2003, 'Handmade in India: Preliminary analysis of crafts producers and crafts production', *Economic and Political Weekly*, 38, 27 December.

Madison, A., 2001, *The World Economy: A Millennial Perspective*, OECD, Paris.

Melchior, A., 2001, 'Global income inequality: Beliefs facts and unresolved issues', *World Economics*, 2.

Melchior, A., K. Telle, and H. Wiig, 2000, *Globalization and Inequality—World Income Distribution and Living Standards 1960–1998*, The Royal Norwegian Ministry of Foreign Affairs, Studies on Foreign Policy Issues, No. 6b, Oslo.

Milanovich, B., 2002, 'True world income distribution, 1988 and 1993: First calculation based on household surveys alone', *Economic Journal*, 112.

Naschold, F., 2004, 'Growth, redistribution and poverty reduction: LDCs are falling behind', in A. Shorrocks and R. van der Hoeven (eds.), *Growth, Inequality and Poverty: Prospects for Pro-poor Economic Development*, Oxford University Press, Oxford.

Reddy, S. and C. Minoiu, 2005, '*Has world poverty really fallen during the 1990s?*', Princeton University (Mimeo).

Reddy, S. and T. Pogge, 2003, *How Not to Count the Poor*, Columbia University (Mimeo).

Sen, A., 1970, *Collective Choice and Social Welfare*, Holden-Day and Edinburgh: Oliver and Boyd, San Francisco.

—— 1976, 'Real national income', *Review of Economic Studies*, 43.

Wade, R. H., 2004, 'Is globalization reducing poverty and inequality?', *World Development*, 32(4), 567–89.

Williamson, J. G., 2002, *Winners and Losers over Two Centuries of Globalization*, WIDER Annual Lecture 6, UNU-WIDER, Helsinki.

Wolf, M., 2004, *Why Globalization Works*, Yale University Press, New Haven.

World Bank, 2004, *World Economic Indicators, 2004*, World Bank, Washington, D.C.

World Bank, 2005, *World Economic Indicators, 2005*, World Bank, Washington, D.C.

17 International Credit and Welfare
A Paradoxical Theorem and Its Policy Implications

with Hodaka Morita

17.1 INTRODUCTION

There is a small literature that argues that the benefits of international credit do not accrue to the recipient developing country, ending up, instead, benefitting the donors or the coffers of large corporations that sell goods to the developing countries.[1] The aim of this chapter is to subject this claim to careful theoretical scrutiny. What we find is that, while this hypothesis need not always be true, there do exist parametric configurations under which it is valid. This is interesting because of its paradoxical nature. At first sight it seems that the availability of credit (or, more generally, availability of credit at better terms) cannot make the recipient, whether it be an individual or a nation, worse off because the recipient has the option not to take the credit or to pay a higher interest than what the donor demands (by, for instance, burning money). However, such simple logic runs into difficulty, especially in the domain of strategic international finance.

We construct a formal model and show that, when a nation buys goods from large corporations with monopolistic power, the availability of cheaper credit may actually leave the recipient worse off. In particular, a poor developing country that is currently borrowing money from a profit-maximizing international bank or financial institution may become worse off if some 'benevolent' organization steps in, in place of the profit-maximizing bank, and begins to lend hard currency at

The authors are grateful to Abhijit Banerjee, Jonathan Eaton, Raquel Fernandez, Arvind Panagariya, Priya Ranjan, Debraj Roy, an Associate Editor, and an anonymous referee of *European Economic Review* for comments and suggestions.

First published in *European Economic Review*, Vol. 50 (2006), 1507–1528.

zero or a subsidized interest rate. Since public foreign lending is usually motivated by altruism and the need to fill in for market failures (Eaton 1989, p. 1308) it seems quite surprising to find that there are situations where the recipient nation does better when it gets its foreign capital from private sources.

To see the logic, note that if a poor country has to borrow money from a profit-maximizing lender at an interest i and pay a price p to a manufacturer for the good, then (assuming the exchange rate is 1) the *effective* price on the margin is $(1 + i)p$. Here the lender and the manufacturer compete in an interesting manner with the lender controlling i and the manufacturer controlling p. Now suppose that a benevolent lender steps in and sets the interest rate equal to zero, which reduces the effective price from $(1 + i)p$ to p. The manufacturer takes advantage of this by raising its price from p to p'. On one hand, this price rise results in a welfare loss, because a part of the country's purchase is financed by its own foreign exchange. On the other hand, for the other part of the purchase that is financed by the lender, the country is still benefitted from the benevolent lending because the effective price p' turns out to be still lower than $(1 + i)p$. We demonstrate that the former disadvantage of the benevolent lending is greater than the latter advantage of it under a range of parameterization conditions.

In brief, this chapter proposes a new game-theoretic framework for analysing the strategic interaction between lending institutions and producers, and demonstrates the possibility of paradoxical reactions (which we call the 'paradox of benevolence'). In the process it draws attention to how we may want to reorganize international lending, paying particular attention to the market structure that the recipient country confronts, so as to ensure that the benefits reach their intended target.

17.2 THE FACTUAL CONTEXT

In this section we present a number of real-world contexts to which our theoretical framework is applicable, and explore the policy implications of our analysis, though the latter is picked up once again in Section 17.6.

One country lending money to another or giving aid with an eye on enhancing its own exports is not unusual at all. Many industrialized countries give loans to developing countries with the explicit requirement that the latter then use these to buy goods from the former (Eaton 1989; Fleisig and Hill 1984). Virtually all Organization for Economic Co-operation and Development (OECD) countries have special provisions for supplying export credit. This is money given to other nations specifically for those nations to buy goods from the donor nation. Moreover, importantly, a lot of this credit is given at concessionary rates, and, in particular, at lower rates than market interest rates. This is done, ostensibly, to help the recipient nations. Sweden, for instance, has the Swedish Export Credit Corporation or AB Svensk Exportkredit (SEK). This was established in 1962 'for the purpose of financing exports of Swedish capital goods and services on commercial terms' (OECD 2001, Sweden p. 3). Up to 1978 SEK used to grant credit on strictly commercial terms. Since then there has been a programme of subsidized lending. Subsidies are funded from Sweden's Development Aid Budget. As

OECD (2001, Sweden p. 10) notes, 'Concessionary credits are mainly tied to Swedish exports', though this is not necessarily so.

In United States, the Trade and Development Agency (TDA), formerly known as the Trade and Development Programme (TDP), has two objectives—to give subsidized credit to help developing and middle-income countries and to promote the export of goods and services to those countries. In the United States, tied aid has legal authorization because the Trade and Development Act, 1983, in particular, its sections 644 and 645, explicitly authorize the Eximbank and United States Agency for International Development (USAID) to provide tied aid and credit to other nations.

These are just two among many examples found in OECD (2001). There is reason to believe that the subsidized international credit sector, which aims to promote export and help the recipient country, is substantial. As Fleisig and Hill (1984, pp. 322–323) noted, 'Outstanding direct subsidized and export credits of the major lending countries (Canada, Italy, Japan, the United Kingdom and the United States) amounted to $55 billion at the end of 1978. These lenders offered substantial subsidies, charging interest rates between 7 and 8%, at the same time that private lenders charged between 5 and 15%'.

Under the requirement that the export credit should be used to import goods from the donor nation, the loan-recipient countries may be forced to choose a seller from a limited number of potential sellers. That is, the provision of export credit with such a requirement could end up creating or at least bolstering the sellers' market power. The paradoxical result of our model suggests that some of these recipient countries may have been better off if they were exposed to the private credit market with its nonconcessionary lending.

There are accounts galore of countries that have received subsidized international credit but have adamantly remained basket cases. There are a number of reasons for this. The money may have been dissipated in consumption and not invested diligently; there may have been corruption and leakage at the level of the government. But, in addition, our model suggests that there may be another previously unexplored reason why the beneficiaries may not have done well. This is to do with an unholy alliance between subsidized credit and the market structure of firms and banks that confront the borrowing country. The money may have leaked out to international producers with market power. One implication of our model is that, when an export credit is offered to a country at a *concessionary* rate, it should be ensured that the recipient country uses the credit to import goods from competitive markets.

Our model also yields important implications for the organization of international lending by multilateral organizations, such as the World Bank and the IMF, that give subsidized credit. The IMF, for instance, provides finance to its member countries under different types of credit arrangements ('facilities'). These include regular facilities at market-related interest rates and a concessionary facility for low-income countries (the poverty reduction and growth facility, PRGF). PRGF arrangements cover a three-year period, with repayments over 5.5–10 years at an interest rate of 0.5%.[2] See IMF (2001a) for further details on types of fund arrangements offered by the IMF.

A number of IMF-supported programmes (in particular, practically all concessionary financing arrangements) have included a variety of structural conditionalities. Concerning trade-related conditionality, the IMF often requires trade liberalization measured by the trade restrictiveness index that combines the average level of tariff protection as well as the coverage of nontariff barriers (IMF 2001b). Our model indicates that the IMF should also keep an eye on the structure of the markets from which the borrower countries import goods. In particular, if a borrower country imports goods from industries with substantial market power, it may be better off by having to borrow from a nonconcessionary facility rather than a concessionary facility, and so careful investigation is needed regarding the type of lending arrangement that is suitable.[3]

The model also highlights the crucial role of the *mechanism* through which the limited foreign exchange is released to the importers in the borrower country by the borrower government (or Central Bank). The chapter suggests that the rules for allocating the limited foreign reserves followed by the government can make a crucial difference in determining what effect international credit or aid has on the well-being of the recipient nation. Hence the model, despite its use of a rather stylized framework, depicts theoretically the general idea explored empirically by Burnside and Dollar (2000) on how the nature of governance in the borrowing nation can critically determine whether aid (or subsidized international lending) will work to its advantage or not.[4]

As a final point, we discuss an application of our theoretical framework to intracountry rural credit markets. In rural regions of developing countries, peasants often face short-term money shortage in the preharvest season. Hence, borrowing is widespread in such times with repayment occurring after the harvest when the peasant regains liquidity. According to a large-scale survey of contractual relationships in rural India (see Bardhan 1984, chapter 9, for details), landlords often lend money to their own share tenants, where the loans can be for many different purposes—consumption to tide over the lean season or production purpose loans. Interestingly, Bardhan reports that these loans can often be without interest. If such a peasant faces a monopolistic product market from which he buys the goods that he needs, then our theoretical framework suggests that the paradoxical result can occur.[5] That is, such a peasant could be better off if he were exposed to a profit-maximizing money lender rather than to a benevolent money lender. In this context, government subsidized credit to poor peasants may not be the panacea that it is often made out to be.

17.3 THE MODEL

In this model there is a developing country—henceforth *South*—and an industrialized country—henceforth *North*. These countries have their own currencies, but for all intercountry trade and exchange the only acceptable currency is the North's currency. This is the 'hard' currency. We shall refer to the South's currency as the 'soft' currency.

The South, in our model, has a shortage of 'hard currency'. This is so in the sense that if it could buy more hard currency at the going exchange rate, it would do so and use it to buy more foreign goods. The fact of a country facing a shortage of hard currency suggests some rigidity in the exchange rate. We treat the exchange rate as fixed and, without loss of generality, we treat it as fixed at 1. Although one reason for making this assumption is to make the model tractable, we also feel that this is not as strong an assumption as may appear at first sight. The fact that many Third World nations *do* face a shortage of hard currency suggests that exchange rates are at least partially rigid in reality. We suspect that there are innate factors in the structure of international economic relations which cause this. How else can one explain why, even after developing country governments go for a free float and allow the exchange rate to be market-driven, shortage of hard currency persists?

Another assumption in this chapter concerns the modelling of the government of a developing country. We treat the government not as a strategic agent, nimbly maximizing some payoff, but as a somewhat mechanical bureaucracy which has some rigid rules to which it adheres. In particular, we model licensed importers in the South to which the government (or the Central Bank) allocates its limited foreign exchange balance; the importers are given the right to buy goods abroad and sell them in the South. One reason why we do not treat the government as a strategic agent is for tractability; the model has a surfeit of strategic agents. However, we also believe that this description is fairly realistic in the case of many developing and transition economies. For instance, in the case of Pakistan and India, it fits reality quite well especially through the 1970s and 1980s.[6]

We shall in this chapter focus on one good which the South likes to consume but it does not produce. The good is in fact produced by a firm based in the North, which sells the good (not only in the North but also) in the South through the licensed importers. The Northern firm produces the good at a constant marginal cost c, faces no fixed cost, and chooses the price p at which it sells to the South. Though in our formal model we work with one such firm, our qualitative results would be unchanged under n oligopolistic firms.

On the demand side we assume, without loss of generality, that the South has one consumer, who is a price taker. Imagine first that the consumer has free access to the hard currency at the going exchange rate. In such a case let the consumer's inverse demand function for the good sold by the North be given by

$$p = a - bx, \tag{17.1}$$

where $a > c$, $b > 0$, p is the price of the product, and x is the amount demanded. This will be called the *unconstrained demand curve*. Without a shortage in hard currency and in the absence of licensed importers (i.e., assuming that the consumers buy directly from the Northern producers), standard monopoly analysis shows the equilibrium price and quantity to be

$$p^* = \frac{a + c}{2} \text{ and } x^* = \frac{a - c}{2b}.$$

This point is illustrated in Figure 17.1 by the point E^*. Note that the total amount of hard currency needed to buy the equilibrium amount of the good is given by $p^*x^* = (a^2 - c^2)/4b$.

We shall from here on consider the case in which the South's foreign exchange reserve R, though positive, is insufficient for this point E^* to be attained. In other words, we are making the following assumption.

ASSUMPTION 1: $0 < R < (a^2 - c^2)/4b$.

That is, the shortage of hard currency is such that the Northern firm cannot fully capture the monopoly rent associated with the unconstrained demand curve.

It is being assumed here that, what the South suffers from is not a problem of insolvency but illiquidity. In other words, it expects to have adequate access to foreign exchange in the future. The simplest way to make this formal is to suppose that the South's currency becomes convertible in the future. So in the future its demand is not constrained by its foreign exchange reserves. We will assume that this foreign exchange constrained position lasts for one period (which can of course be very long) and it is this one period that our model studies.

So the Southern government has a reserve of R units of hard currency. How does the government use this? We will assume that the government sets a quota for each of the m (2) importers. That is, each importer is given the right to acquire foreign exchange up to this quota limit by giving up an equivalent amount of soft currency. With this foreign exchange the importers buy goods from the North which they then sell to the Southern consumers. We shall, for simplicity, assume that all importers are treated identically, and so each importer has access to R/m units of the hard currency. It will be assumed that the importers take the international price of the product as given and constitute a Bertrand oligopoly in the domestic market.

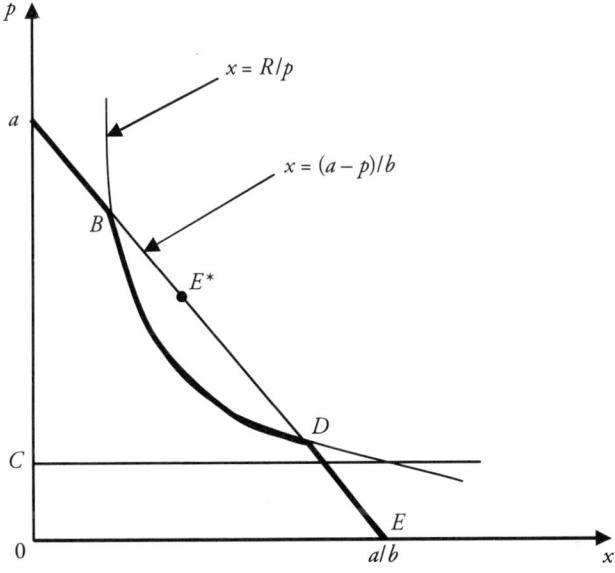

Figure 17.1: Constrained Demand Curve

It will be shown later in Section 17.5 that, for the purpose of our analysis, such a model works the same way as an alternative model in which the Southern government gives consumers direct access to a fixed amount of foreign exchange. Given this mathematical equivalence, in what follows we proceed with our analysis by supposing that the Southern government announces that the consumer can acquire up to R units of hard currency. In other words, the amount of foreign good, x, that the consumer buys must satisfy

$$x \leq R/p. \qquad (17.2)$$

Keeping in mind that Eq. (17.1) implies that the demand function (with no foreign exchange constraint) is given by $x = (a - p)/b$, and combining this with (17.2) we see that the *actual* demand function of the South is given by

$$x = \min\left\{\frac{a-p}{b}, \frac{R}{P}\right\}. \qquad (17.3)$$

This is demonstrated by the thick line in Figure 17.1.[7]

We now incorporate international lending into our model; we will consider the following two cases:

CASE I. There is a nonprofit 'international organization' that lends hard currency credit to the South at a subsidized interest rate.

CASE II. There is a profit-maximizing international bank (based in the North) that gives hard-currency credit to the South.

We shall, throughout, assume, without loss of generality, that the interest rate prevailing in the North is zero. The Southern consumer and government do not have direct access to the Northern credit market, but the international organization and the Northern bank have access to it. So for these latter agents the opportunity (interest) cost of lending money to the South is zero. Given our focus on illiquidity (rather than insolvency) problems faced by the South, we assume that the South never defaults.

The analysis of Case I is straightforward. Let us suppose that the international organization lends to the South at the opportunity cost interest, that is, an interest rate of zero. Once South has access to such credit, the foreign exchange constraint of R becomes immaterial. South's demand for the product is given by Eq. (17.1) and the equilibrium price and quantity are given by p^* and x^*, which are represented by point E^* in Figure 17.1.

Case II is the interesting case, and what we go on to show, later, is that the Southern country may be better off in this case than under Case I. But first we need to depict the equilibrium that will arise in Case II.

Since the central issue in the analysis of Case II is the strategic interaction between the firm and the bank, we derive the reaction functions (more precisely '*implicit* reaction functions') of the firm and the bank and then characterize Nash equilibria. Let us start with the firm. Consider first the case where $R = 0$, that is, for whatever the South buys from the North it has to first borrow money from the bank.

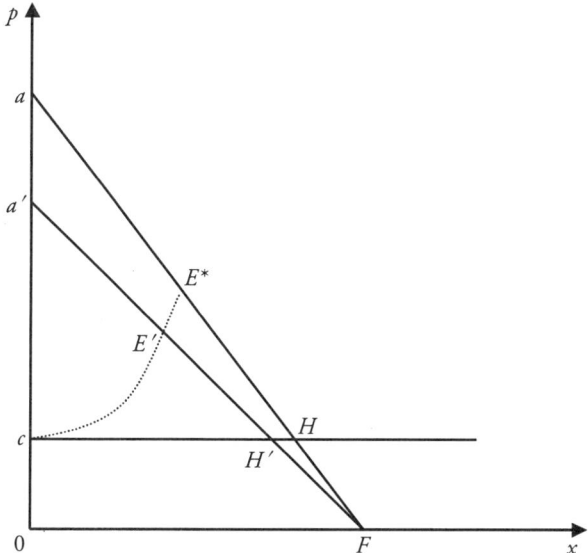

Figure 17.2: Firms Best-Response Curve

In Figure 17.2, aF is the South's unconstrained demand curve [given by Eq. (17.l)]. Suppose the bank charges an interest rate of i. Then if the firm charges a price of p, the effective price to the Southern consumer is $(1 + i)p$. Hence the effective demand curve is given by the line $a'F$ where $Oa = (1 + i)Oa'$. Standard monopoly analysis implies that the firm's best response is to choose a price that is represented by the midpoint of line segment $a'H'$, shown by point E'. By considering different interest rates, i, and plotting the mid-point that represents the firm's best response for each i, we obtain the firm's best response curve. This is represented by the broken line $E^*E'c$. We call it the firm's 'implicit reaction function'.[8] The reader should also check that, if c were 0, the firm's implicit reaction function would be a vertical line from E^* down to the horizontal axis. The reason why we call this an 'implicit' reaction function is because, unlike in a conventional reaction function where the two variables chosen by the two players are represented on the two axis, here the interest rate i, chosen by the bank, is not represented on any axis, but is implicit in the effective demand curve.

Now let us bring in the fact that $R > 0$, as shown in Figure 17.3. If the interest rate, i, charged by the bank is such that the effective demand curve is $a'F$, then the actual demand curve (the one which takes into account the fact that up to R units, the South does not *need* to borrow money) is given by the thick line, going through points B and D. The firm's implicit reaction function is E^*K' and point B, where E^*K is a truncated segment of the $E^*E'C$ curve in Figure 17.2. To see this, gradually increase the value of i, starting from $i = 0$. The firm's best response is represented by point E^* when $i = 0$ and by point E' (see Figure 17.2) when i is positive but sufficiently small. Then, as i rises E' moves in the southwest direction. But before E' reaches point K in Figure 17.3, the firm's best response point will jump to point B. Let us denote by K' the point where the jump occurs. To see

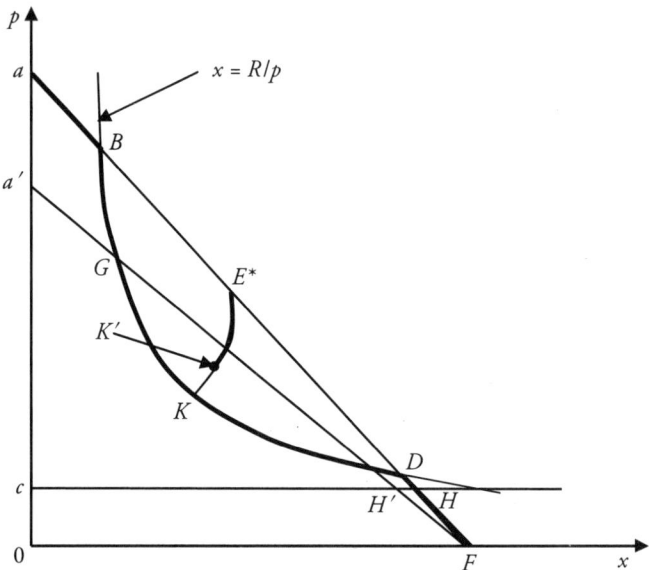

Figure 17.3: Best Response with Constraint

that this will happen, suppose that i is such that the line, $a'F$, passes through point K in figure 17.3. Clearly, the firm is strictly better off by choosing the price that corresponds to point B rather than point K. This is because at both prices revenue is the same and the total cost is smaller at point B. Hence, there exists point K', where the firm is indifferent between choosing point K' and point B.

Now we turn to the bank's reaction function. First suppose that the firm has fixed a price p such that $R/p \geq (a - p)/b$ holds. In this case, the South does not borrow hard currency because the consumer's demand given by the unconstrained demand curve (i.e., $p = a - bx$) is feasible without borrowing any hard currency. Then, any value of i is the bank's best response, because the bank cannot make any profits from lending to the South, for all $i \geq 0$.

Next suppose that the firm has fixed a price p such that $R/p < (a - p)/b$ holds. This condition means that, under the price, the consumer's demand given by the unconstrained demand curve is not feasible without borrowing hard currency because the Southern government has only R (>0) units of hard currency. Graphically, the price is strictly between the prices represented by points B and D in Figure 17.3. Given such price, the bank can make a profit from lending hard currency to the South, which is given by

$$\pi_B(i) \equiv i \left\{ p \left[\frac{a - p(1 + i)}{b} \right] - R \right\}$$

Graphically, the bank's profit is represented by area $QRST$ in Figure 17.4, where the firm has fixed a price at $p = p'$ and the bank has chosen i represented by $a'F$. Given p', the bank chooses i so that the area $QRST$ is maximized. The maximization implies that the bank chooses i such that point T in Figure 17.4 becomes the

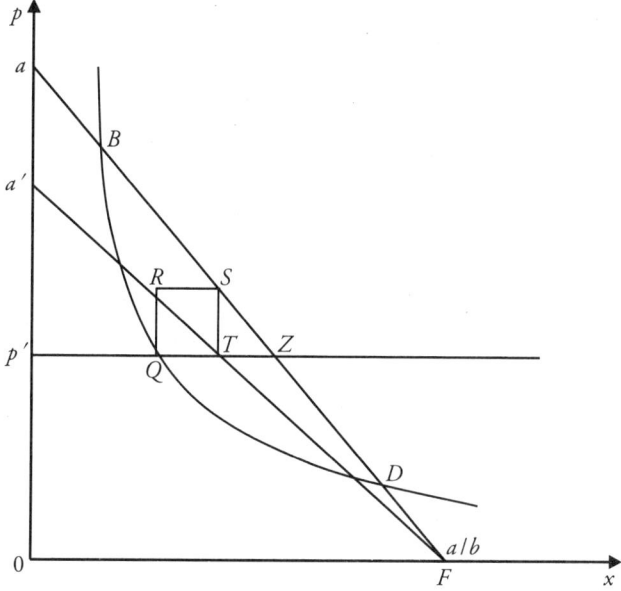

Figure 17.4: Bank's Profit

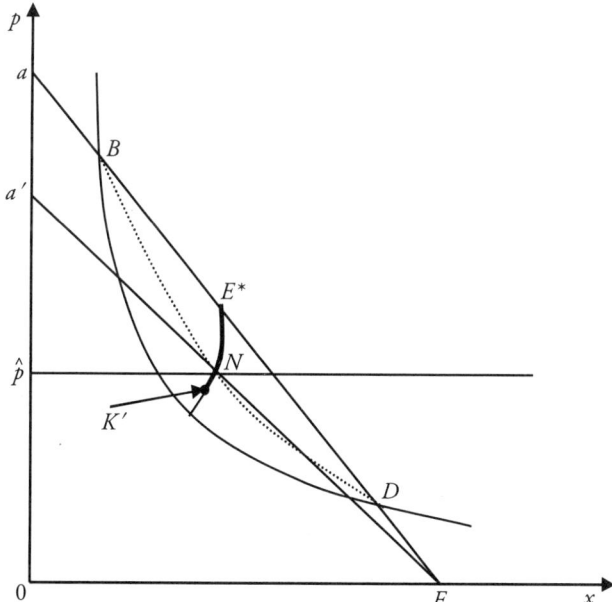

Figure 17.5: Bank's Implicit Reaction Function

mid-point of QZ. Then, for any given p', the bank's best response is to choose i such that corresponding $a'F$ line goes through the mid-point of QZ. Plotting such mid-points for different values of p', we obtain the broken line in Figure 17.5. We call it the bank's 'implicit reaction function'.

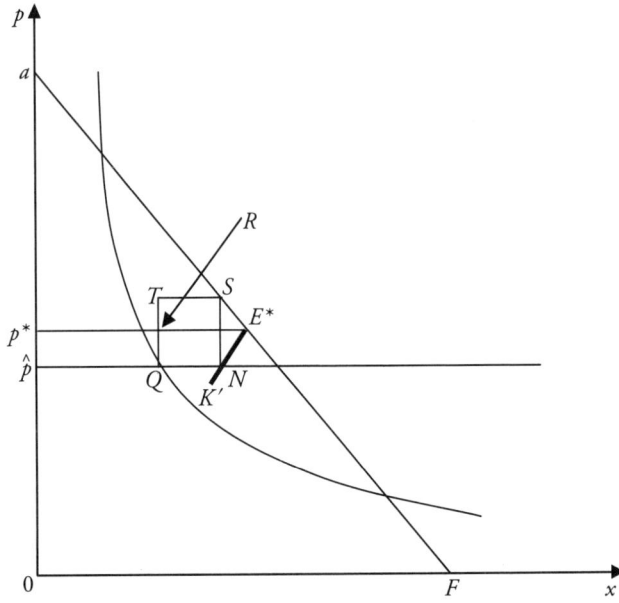

Figure 17.6: Welfare Aggregate

We are now ready to identify the Nash equilibria. Superimpose the firm's implicit reaction function (E^*K' in Figure 17.3) here. A Nash equilibrium is depicted by the point of intersection of the two reaction functions, shown here by point N, where the equilibrium price is given by $\hat{p}$ and the interest rate is the one implicit in the effective demand curve $a'F$. This is an equilibrium in which a positive amount is borrowed. We call this the N-equilibrium. Note that the N-equilibrium does not always exist because the broken line does not necessarily intersect with E^*K'. Note also that there exists another Nash equilibrium, where the firm chooses the price that corresponds to point B and the bank chooses a very high interest rate. This is an equilibrium in which no lending occurs.

17.4 THE PARADOX OF BENEVOLENCE

We now demonstrate that the paradox of benevolence can happen in the N-equilibrium. The aggregate welfare earned by the South in the N-equilibrium is shown in Figure 17.6 as the area $STQ\hat{p}a$.

Let us call this, in brief, W^π, where π is a reminder that this is the welfare of the South when the lender of credit is a profit maximizer. Let us denote South's aggregate welfare when the Northern lender is benevolent (and charges no interest) by W^b, where b is for benevolence. Our claim is that there are parameters of the model where

$$W^b < W^\pi.$$

We will say that the 'paradox of benevolence' occurs if this inequality is true.

To prove this we need to first depict W^b. Recall that when the South can freely borrow from a benevolent lender (Case I) equilibrium occurs at point E^* and the price of the Northern good is given by p^*. Hence W^b is the area of aE^*p^*. By examining Figure 17.6 it is clear that *a priori* we cannot say which is larger, W^b or W^n. Now, we are able to state the central result of the paper.

PROPOSITION 1 (*The paradox of benevolence*): *For any parameter values that satisfy Assumption 1, there exists a value $\tilde{c}$ (>0) such that, holding all parameter values except c fixed, the model exhibits the following property for all $c \in [0, \tilde{c}]$:*
   The N-equilibrium exists and the paradox of benevolence occurs in that equilibrium.

PROOF: See Appendix A.

To understand the logic behind the result, let us compare the Northern firm's profit-maximizing behaviour in Case I and Case II. Let $MR(p,x)$ denote the Northern firm's marginal revenue when it sells x units of the good to the South at the price of p. In Case I, the international organization lends to the South at the interest rate $i = 0$ and the Northern firm sells x^* units of the good at the price *of* p^* so that the marginal revenue becomes equal to the marginal cost, that is, $MR(p^*,x^*) = c$. In Case II, the profit-maximizing international bank charges $i > 0$. Given the shortage of hard currency in the South, the positive interest rate reduces the South's willingness of pay, which in turn reduces the Northern producer's marginal revenue. In order to sell x^* units to the South, the Northern producer can now charge only $[1/(1 + i)]p^*$, where $MR([1/(1 + i)]p^*, x^*) = [1/(1 + i)] MR(p^*,x^*)$.

   First consider the case where the marginal cost c is zero. Then, in Case I, under the Northern firm's optimal choice (p^*,x^*) its marginal revenue $MR(p^*,x^*)$ is zero. In Case II, although the positive interest rate reduces the South's willingness to pay, the Northern firm's marginal revenue when it sells x^* units is unaffected and still zero (i.e., $MR([1/(1 + i)]p^*, x^*) = [1/(1 + i)] MR(p^*,x^*) = 0$, if $MR(p^*,x^*) = 0$). Hence, the Northern firm's optimal quantity is x^* in Case II as well as in Case I. That is, the interest rate charged in Case II does not result in any additional quantity distortion. On the other hand, the Northern firm must reduce its price from p^* to $[1/(1 + i)]p^*$ to sell x^* units. And, given the positive amount of the South's foreign reserve $(R > 0)$, the South gets some benefit from the lower price charged by the Northern firm. The result is that the South is strictly better off in Case II (i.e., the paradox of benevolence occurs) when $c = 0$.

   Now let the marginal cost c be strictly positive, so that $MR(p^*,x^*) = c > 0$ holds under the Northern firm's optimal choice (p^*,x^*) in Case I. In Case II, the South's lower willingness to pay now implies that the Northern firm's marginal revenue when it sells x^* units is strictly below the marginal cost c (i.e., $MR([1/(1 + i)]p^*, x^*) = [1/(1 + i)] MR(p^*,x^*) < c$). This results in an additional quantity distortion; that is, the Northern firm's optimal quantity in Case II (denoted $\hat{x}$) is now strictly less than x^*. However, when the marginal cost c is small, the degree of this distortion is small. Then, this negative impact on the South's welfare is more than offset by the benefit of the lower price, and hence the South is strictly better off in Case II when c is small enough. This is what the proposition states.

   The result can also be understood graphically. First let $c = 0$. Then, as we have already seen, the firm's implicit reaction function is a vertical line from E^*. Hence,

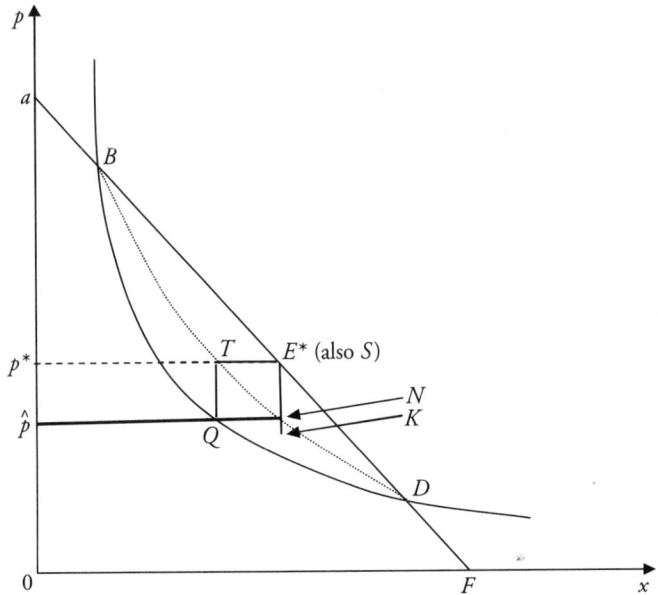

Figure 17.7: Nash Equilibrium with Zero Cost

as shown in Figure 17.7, the N-equilibrium point, N, is now vertically below E^*. Since the Northern firm's optimal quantity is x^* in Case II as well as in Case I, the South can capture area aE^*p^* (which is the South's consumer surplus in Case I) as a part of its consumer surplus in Case II. In addition, due to the lower price charged by the firm, the South also captures area $TQ\hat{p}p^*$ as its consumer surplus. The result is that the South's consumer surplus in Case II, represented by area $E^*TQ\hat{p}a$, is greater than its consumer surplus in Case I, represented by area E^*p^*a. Now let $c > 0$. Then, as shown in Figure 17.6, the N-equilibrium point, N, is not vertically below E^* any more, and the firm's optimal quantity is $\hat{x}$, where $\hat{x} < x^*$. This additional quantity distortion in Case II reduces the South's consumer surplus. However, if c is relatively small, area SE^*RT is smaller than area $RQ\hat{p}p^*$, which implies that the South is still better off in Case II than in Case I.

We conducted numerical simulations to compute the zones of paradox in the (R, c)-space, in particular, a space in which the horizontal axis represents R and the vertical axis represents c. For each value of R we have computed the maximum value of c, denoted c^{max}, such that the paradox occurs for all non-negative values of c less than c^{max}. The computation is made for the case where $a = 10$ and $b = 0.4$ or 0.5 and the results are displayed in Table 17.1. The table says that, for instance, with $b = 0.4$, we have $c^{max} = 2.23$, 6.06, or 3.00 when $R = 10$, 20, or 40, respectively. Namely, the paradox of benevolence occurs for all $c \in [0, 6.06]$ when $a = 10$, $b = 0.4$, and $R = 20$. Note that $c < a$ ($= 10$ in these examples) must hold for the Northern firm to sell a positive amount of goods to the South. The numerical examples therefore seem to indicate that the paradox of benevolence occurs in nontrivial ranges of parameter values.

It is interesting to note the table exhibits an 'inverted-U' shape in the (R, c)-space. That is, holding other parameter values fixed, the value of $c^{\max}$ is increasing in R when the value of R is relatively small, and it is decreasing in R when the value of R is relatively large. Although we have worked out a number of examples and identified this property in all of them, we have been unable to prove that this is the general property.

Table 17.1: Numerical Examples for the Paradox of Benevolence (the value of $c^{\max}$ when $a = 10$)

R	$b = 0.4$	$b = 0.5$
5	0.94	1.22
10	2.23	3.22
15	7.00	6.29
20	6.06	5.16
25	5.16	4.16
30	4.35	3.31
35	3.63	2.57
40	3.00	1.91
45	2.44	1.23
50	1.91	–
55	1.37	–
60	0.73	–
65	–	–

17.5 COMPETITION AMONG LICENSED IMPORTERS

In Section 17.3 we began with the realistic assumption that, in the South, the government gives some designated importers the right to acquire hard currency from the central bank in order to import goods for domestic sale. We then pointed out that, if these importers took the international price, p, of the good and the interest rate, i, as given and chose the domestic sale price (i.e., they played a Bertrand game), we could ignore these importers for the purpose of our analysis. Given this, we derived our result under the assumption that government allocated foreign exchange directly to the consumers rather than to the designated importers. In this section we show that we can indeed ignore the importers in order to derive our results.

As before, the Southern demand for the Northern good is given by

$$x = \frac{a - r}{b},$$

where r is the price that the consumers have to pay. There are now m identical importers. They can buy the good (subject to having the requisite foreign exchange) from a Northern producer at a price, p, chosen by the Northern producer. It is assumed that the Southern importers take this price as given. Each of these importers is given access to R/m units of foreign exchange by the Southern government. If they want more foreign exchange, they have to borrow this from a

Northern bank at an interest rate of i. Hence, if an importer wants to buy x units of this good from the North it has to incur a total cost, $TC(x)$, given by

$$TC(x) = \begin{cases} px & \text{if } px \le \dfrac{R}{m} \\[2ex] \dfrac{R}{m} + (1+i)p\left(x - \dfrac{R}{mp}\right) & \text{if } px > \dfrac{R}{m} \end{cases} \qquad (17.4)$$

Now, each of these m importers have to choose a price at which it offers to sell the product to the Southern consumers. If r_i denotes the price offered by importer i, then we may denote the strategy n-tuple of the m importers by $(r_1, \cdots, r_m)$. The profit earned by importer i may then be denoted by $\pi_i(r_1, \cdots, r_m)$.

Our aim is to characterize the Nash equilibrium (Bertrand equilibrium in this case) of this game. We will in particular be interested in the symmetric Nash equilibrium. In other words, we define r^* to be an 'equilibrium' if, for all $i=1, \cdots, m$, $\pi_i(r^*, \cdots, r^*) \ge \pi_i(r^*, \cdots, r_i, \cdots, r^*)$, for all r_i.

Fortunately, to characterize such an equilibrium we do not need to fully characterize the π_i function. We will here make the following reasonable assumptions. If every importer charges the same price r then each importer faces a demand of $(a - r)/bm$. If all importers, except importer i, charge r and importer i charges $r_i (\ne r)$ then the consumers respond as follows. If $r_i < r$, importer i faces a demand equal to $(a - r_i)/b$. All consumers who fail to buy from i, direct their demand at price r to the other importers. If $r_i > r$, all consumers go to importers other than i. Only those with unmet demand turn to i. These are fairly usual assumptions; a formal statement of these occur in Basu (1993).

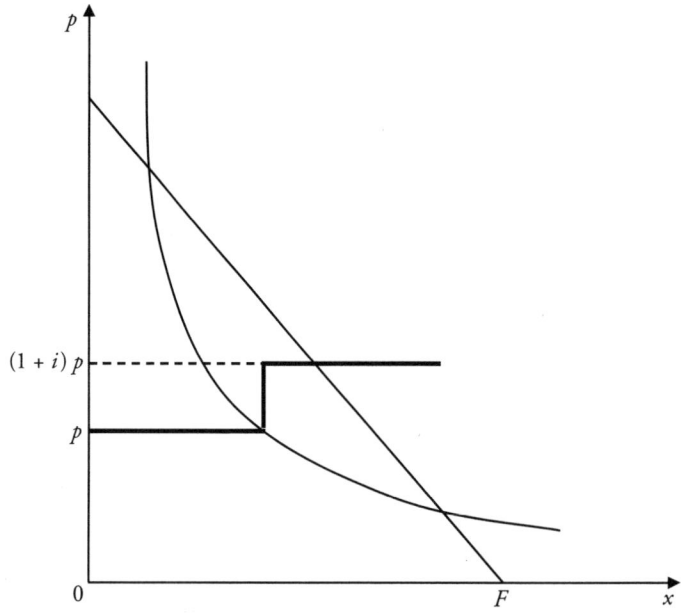

Figure 17.8: Marginal Cost Curve

Let us now suppose that the firm has fixed a price, p, such that $R/p < (a - p)/b$ holds. Also suppose that the bank has fixed an interest rate, i, such that $R/p < [a - p(1 + i)]/b$ holds. This condition means that, if government allocated foreign exchange directly to the consumers then the consumers' demand given by the unconstrained demand curve is not feasible without borrowing hard currency and so they borrow a positive amount of hard currency from the bank. Under such p and i, the horizontal summation of all importers' marginal cost functions [derived from Eq. (17.4)] is the thick line shown in Figure 17.8. It is easy to show that in this case $r^* = (1 + i)p$ is an equilibrium. That is, if each importer charges r^* then no one can do better by deviating. To see this note that when everybody charges $(1 + i)p$, the profit earned by each importer is given by iR/mp. Clearly by undercutting this price, an importer can only do worse. If, on the other hand, an importer charges $r_i > (1 + i)p$, no one will buy from him. Hence, his profit will drop to zero.

The analysis in the previous paragraph indicates that, for any p and i that satisfy the conditions described above, the profits of the firm and the bank are identical with or without the designated importers. Also, consumers face the same marginal price and demand the same amount of the good in the two cases. A similar equivalence can be shown for other combinations of p and i. Since we focus on the welfare consequences of the strategic interaction between the firm and the bank, this equivalence allows us to ignore the importers in our analysis.

17.6 POLICY IMPLICATIONS

The model and the results described in this chapter have important policy implications. First, it cautions aid donor agencies not to presume that subsidized credit, given to a developing country, necessarily benefits the recipient relative to the case in which credit is made available by a profit-maximizing bank or financial institution. At first sight it seems that the availability of subsidized credit cannot make the recipient nation worse off. However, we have shown that, depending on the structure of the import market, the advantages of subsidized credit may flow into the hands of corporations that sell goods to the recipient nations. In such a situation the donor agency has to think of ways, other than subsidized credit, for reaching benefit to nations. The classical literature on aid-tying used to be concerned with this question. What we have shown in this chapter, however, is that the flow-back of benefit to the North can occur even when aid is not tied, but depends on the market structure of imports and the strategic position of the donor.

In trying to reach out to poor nations, most international organizations use the method of lowering interest rates. The IMF uses this for the most indebted and poor nations, while combining the generous loan terms with 'conditionalities', which pertain to macroeconomic policies such as the need to keep the fiscal deficit under control and money supply growth in check. What this chapter alerts us to the fact that such policies may not be enough to plug the holes through which the benefits of cheap credit get frittered away. The 'market structure' of trade may be the main route through which the immiserization occurs. Hence, before lending at concessional rates, it is worth examining and advising recipient governments on the channels and structure of trade and methods of releasing limited foreign exchange reserves.

The model suggests (though we have not really gone into this) that there may be advantages to the South of giving the import rights to a single agent. This would empower the importers *vis-à-vis* the Northern manufacturer and may end up benefiting the Southern consumer. Second, the Southern government may stand to gain by being more pro-active in the foreign exchange market. Releasing the foreign exchange as quotas to different agents may not be a good idea.

Let us take up the first point first. In our model the Southern importers do poorly because they compete against one another both in the product market and the international credit market. If they could behave collusively, they could exercise market power. However, collusive behaviour is difficult to sustain on its own—a point made persuasively in the context of international borrowing by governments by Fernandez and Glazer (1990). However, in our model since the borrowers are agencies within a nation, the government can enable them to exercise market power. The system of 'canalized' imports used by some nations, for instance, India, could have potentially played this role. The practices, of canalized imports has been inefficient and bureaucratically cumbersome. Its potential has not been understood, let alone realized.

Let us now turn to the second subject of *how* to ration the limited foreign exchange reserve. The method analysed in this chapter—namely, one where the foreign exchange is rationed out to the importers—is not the only one. Governments could (and they often do) place quantity restrictions on the amount each importer may import. The analysis of this is not trivial since, while each importer will of course take the quantity ration as given, the government should be modelled as choosing that quantity ration, given which the total import *value* equals the amount of foreign exchange the government has (or wants to release). There can be other more sophisticated kinds of rationing, for instance one in which the amount of foreign exchange released to an importer depends on the terms of trade. Each such ration will change the market outcome and the total benefit generated to the South and may even avert the paradox of benevolence. In the future it will be worth examining formally the welfare effects of different systems of releasing limited foreign exchange. Southern governments could choose a system consciously to maximize the welfare of its consumers.

A.17 APPENDIX

PROOF OF THE PROPOSITION: We first analyse the firm's best response given i (≥ 0) chosen by the bank.

First consider i that satisfies

$$i \geq \frac{a^2 - 4bR}{4bR}.$$

(A1)

Under (A1), the South does not borrow any hard currency for any p chosen by the Northern firm. To see this, note that (A1) is equivalent to '$R \geq p[a - (1 + i)p]/b$ holds for all p'. Given such i, the firm chooses p such that the South spends R units of hard currency to purchase the good; namely it chooses p such that $p[(a - p)/b] = R$ holds. Hence, the firm's best response is given by

$$p = \frac{a + \sqrt{a^2 - 4bR}}{2} \equiv p^B.$$

(A2)

Any (p, i) such that $p = p^B$ and $i \geq (a^2 - 4bR)/4bR$ is a Nash equilibrium. Graphically, in this Nash equilibrium, the firm chooses the price that corresponds to point B in Figure 17.3 and the bank chooses high enough i so that $a'F$ does not intersect the rectangular hyperbola twice. In this equilibrium (we call it B-equilibrium), the bank's profit is zero and the firm's profit is given by

$$\frac{(p^B - c)(a - p^B)}{b} \equiv \pi^B.$$

(A3)

Next consider i that satisfies

$$i < \frac{a^2 - 4bR}{4bR}.$$

(A4)

Consider the monopolist who faces the demand curve given by $p = (a - bx)/(1 + i)$. It charges the price given by Eq. (A5) and the quantity demanded is given by Eq. (A6):

$$p = \frac{a + (1 + i)c}{2(1 + i)} \equiv \tilde{p},$$

(A5)

$$x = \frac{a - (1 + i)c}{2b} \equiv \tilde{x},$$

(A6)

Note that the Northern firm can earn π^B by choosing $p = p^B$ regardless of the value of i chosen by the bank. Then, given i, the firm chooses $\tilde{p}$ if and only if $\tilde{\pi} \equiv (\tilde{p} - c)\tilde{x} \geq \pi^B$. Hence the Northern firm's reaction function is given by

$$p(i) = \begin{cases} \dfrac{a + (1 + i)c}{2(1 + i)} \equiv \tilde{p} & \text{if } \tilde{\pi} \geq \pi^B \\[3mm] \dfrac{a + \sqrt{a^2 - 4bR}}{2} \equiv p^B & \text{otherwise.} \end{cases}$$

(A7)

Next we analyse the bank's best response given that the firm chooses p that satisfies

$$p[(a - p)/b] > R.$$

(A8)

Given such a price, the demand given by the unconstrained demand schedule (which is $p = a - bx$) is not feasible unless the bank sets $i = 0$. The bank chooses i that maximizes its profit given by

$$\Pi(i) \equiv i \left\{ p \left[\frac{a - p(1 + i)}{b} \right] - R \right\}.$$

(A9)

Note that, given (A8), the bank can choose $i > 0$ such that $\Pi(i) > 0$. The standard maximization exercise then implies that the bank's best response is given by

$$i(p) = \frac{pa - p^2 - bR}{2p^2}.$$ (A10)

Now we characterize a Nash equilibrium in which the bank lends a strictly positive amount of hard currency to the South. Inserting (A10) into (A5), we obtain

$$f(p) \equiv 2p^3 - cp^2 - (2bR + ac)p + bcR = 0.$$ (A11)

Note that $f(0) = bcR \geq 0$ and $f(c) = c^2(c - a) - bcR \leq 0$. This means that Eq. (A11) has exactly one root that is strictly greater than c. We denote the root by p^*. If there exists a Nash equilibrium in which the bank lends a strictly positive amount of hard currency to the South, such equilibrium is characterized by $(p, i) = (p^*, i(p^*))$. This constitutes a Nash equilibrium of the game if and only if $(p, i) = (p^*, i(p^*))$ satisfies $\tilde{\pi} \geq \pi^B$ and (A8), or equivalently if (A12) and (A13) hold:

$$\frac{[a - (1 + i(p^*))c]^2}{4b(1 + i(p^*))} \geq R - \frac{c(a - \sqrt{a^2 - 4bR})}{2b},$$ (A12)

$$p^*[(a - p^*)/b] > R.$$ (A13)

Note that p^* is continuous in c, which implies that $i(p^*)$ is also continuous in c.

Let $c = 0$. Then $f(p) = 2p^3 - 2bRp$, and so $p^* = \sqrt{bR}$ and $i(p^*) = (a\sqrt{bR} - 2bR)/2bR$. We find that, when $c = 0$, (A12) is equivalent to $a \geq 2\sqrt{bR}$ and (A13) is equivalent to $a > 2\sqrt{bR}$. Note that Assumption 1 implies $a > 2\sqrt{bR}$ holds when $c = 0$, and that both p^* and $i(p^*)$ are continuous in c. This implies that there exists $c^*(>0)$ such that both (A12) and (A13) hold for all $c \in [0, c^*]$.

Next, we assume $c \in [0, c^*]$, and let W^π denote South's aggregate welfare in the Nash equilibrium represented by $(p^*, i(p^*))$. As stated in the text, the social welfare is represented by the area $STQ\hat{p}a$ in Figure 17.6, which is given by

$$W^\pi = (1/2)[a - (1 + i(p^*))p^*]x^* + i(p^*)R,$$ (A14)

where $x^* \equiv (a - (1 + i(p^*))p^*)/b$. When $c = 0$, we have

$$W^\pi = \frac{a^2}{8b} + \frac{\sqrt{bR}(a - 2\sqrt{bR})}{2b} > W^b = \frac{a^2}{8b},$$ (A15)

where strict inequality holds because $a > 2\sqrt{bR}$ by Assumption 1. Note that p^*, $i(p^*)$, and x^* are all continuous in c. This implies that there exists $c^{**} > 0$ such that $W^\pi > W^b$ holds for all $c \in [0, c^{**}]$. Finally, let $\tilde{c} \equiv \text{Min}[c^*, c^{**}]$, and we obtain the desired result. Q.E.D.

NOTES

1. For works that either defend or debate this proposition, see Basu (1991), Bhagwati (1970), Darity and Horn (1988), Deshpande (1999), Gwyne (1983), Hyson and Strout (1968), Taylor (1985), and Winkler (1929).
2. PRGF was established in 1999. The predecessors of the PRGF had been the structural adjustment facility (SAP) and the enhanced structural adjustment facility (ESAF).
3. There is now a lot of evidence from cross-country studies on how trade liberalization

and greater openness in general leads to the growth of income (see, for instance, Ben-David 1993; Frankel and Romer 1999; Sachs and Warner 1995). What we show is that, if the trade liberalization and openness lead to a more competitive trade environment as one may expect, then this may also increase the efficacy of the concessionary facility offered by the IMF.

4. See also Collier (1997) and Hansen and Tarp (2001).
5. It is plausible that poor peasants often buy goods from sellers with substantial market power. Bardhan (1984) argues that highly personalized ties between transacting agents that are typically observed in isolated rural villages often result in monopolistic power. See Bhaduri (1983) for a similar argument regarding 'personalized rural market'.
6. Writing in the very early 1990s on Pakistan, Baysan (1992, p. 468) observed, 'Distinct from import bans and restrictions, value limits on individual licenses against cash for imports of machinery and millwork have been (and still are being) maintained .... These ceilings ... function as nontariff barriers ... and serve as a nonprice rationing mechanism for the allocation of foreign exchange'.
7. If we were thinking of this as an intracountry, credit market problem, we could think of consumers who have a 'true' demand curve (i.e., in the absence of any liquidity problems) given by Eq. (17.1) but have little liquid cash, maybe because this is the preharvest, lean season. If the liquid cash available with the consumer is given by R, then his effective demand function for the good in questions is given by Eq. (17.3).
8. The mathematical properties of this function are spelled out in Anant Basu, and Mukherji (1995).

REFERENCES

Anant, T. C. A., K. Basu, and B. Mukherji, 1995, 'A model of monopoly with strategic government intervention', *Journal of Public Economics*, 57, 25–43.
Bardhan, P., 1984, *Land, Labor, and Rural Poverty*, Columbia University Press, New York.
Basu, K., 1991, 'The international debt problem, credit rationing, and loan pushing: Theory and evidence', *Princeton Studies in International Finance*, No. 70.
——— 1993, *Lectures in Industrial Organization Theory*, Blackwell Publishers, Oxford.
Bayson, T., 1992, 'Trade Policies in Pakistan', in D. Salvatore (ed.), *National Trade Policies, Handbook of Comparative Economic Policies*, Vol. II, Greenwood Press, New York.
Ben-David, D., 1993, 'Equalizing exchange: Trade liberalization and income convergence', *Quarterly Journal of Economics*, 108, 653–79.
Bhaduri, A., 1983, *The Economic Structure of Backward Agriculture*, Academic Press, London.
Bhagwati, J., 1970, 'The tying of aid', in J. Bhagwati and R. Eckaus (eds.), *Foreign Aid*, Penguin Books, Harmondsworth.
Burnside, C. and D. Dollar, 2000, 'Aid, policies, and growth', *American Economic Review*, 90, 847–68.
Collier, P., 1997, 'The failure of conditionality', in C. Gwin and J. M. Nelson (eds.), *Perspectives on Aid and Development*, Johns Hopkins University Press, Baltimore, MD.
Darity Jr., W. and B. Horn, 1988, *The Loan Pushers: The Role of the Commercial Banks in the International Debt Crisis*, Ballinger, Cambridge, MA.
Deshpande, A., 1999, 'Loan pushing and triadic relations', *Southern Economic Journal*, 65, 914–26.
Eaton, J., 1989, 'Foreign public capital flows', in H. Chenery and T. N. Srinivasan (eds.), *Handbook of Development Economics*, Vol. 2, North-Holland, Amsterdam.

Fernandez, R. and J. Glazer, 1990, 'The scope for collusive behavior among debtor countries', *Journal of Development Economics*, 32, 297–313.

Fleisig, H. and C. Hill, 1984, 'The benefits and costs of official export credit program', in R. E. Baldwin and A. O. Krueger (eds.), *The Structure and Evolution of Recent U.S. Trade Policy*, University of Chicago Press, Chicago.

Frankel, J. and D. Romer, 1999, 'Does trade cause growth?' *American Economic Review*, 89, 379–99.

Gwyne, S. C., 1983, 'Adventures in the loan trade', *Harper's Magazine*, September, 22–6.

Hansen, H. and F. Tarp, 2001, 'Aid and growth regressions', *Journal of Development Economics*, 64, 547–70.

Hyson, C. D. and A. M. Strout, 1968, 'Impact of foreign aid on U.S. exports', *Harvard Business Review*, 46, 63–71.

IMF, 2001a, 'Structural conditionality in fund-supported programs', http://www.imf.org/external/np/pdr/cond/2001/eng/struct/index.htm.

——— 2001b, 'Trade policy conditionality in fund-supported programs', http:// www.imf.org/external/np/pdr/cond/2001/eng/trade/index.htm.

OECD, 2001, *Export Credit Financing Systems*, OECD, Paris.

Sachs, J. D. and A. Warner, 1995, 'Economic reform and the process of global integration', *Brookings Papers on Economic Activity*, 1, 1–118.

Taylor, L., 1985, 'The theory and practice of developing country debt: an informal guide for the perplexed', *Journal of Development Planning*, 16, 195–277.

Winkler, M., 1929, *Investments of US Capital in Latin America*, World Peace Foundation Pamphlets, Boston.

18 The Retreat of Global Democracy

One concomitant of globalization and technological progress, that has either gone unnoticed or been hushed up by those who did notice it, is that it has a natural corrosive effect on global democracy. As a consequence of this phenomenon, even if individual countries become democratic, the aggregate of global democracy may well be on the wane. The purpose of this chapter is to advance and defend this hypothesis, comment on its consequences (global instability) and suggest anti-dotes (restructuring international organizations such as the World Trade Organization [WTO], the International Monetary Fund [IMF], and the World Bank).

Democracy entails many things—the existence of a variety of political and legislative institutions, avenues for citizens to participate in the formation of economic policies that affect their lives and, in the ultimate analysis, a certain mindset. Yet at the core of it and in its simplest form, democracy requires that (1) people have the right to choose those who rule them and (2) the vote of one person counts as much as another person's. Even this simple principle runs into paradoxes and puzzles, as Lewis Carroll, in his original incarnation as C. L. Dodgson, had discovered and written about. But the simplicity of these requirements has the advantage that we can easily check whether a society satisfies them.

Next, note that globalization, almost by definition, means that nations and people can exert a greater influence on other nations and the lives of citizens in other nations. Moreover, it is not true by definition but is a fact that this power of one nation to influence another is by no means symmetric. The United States, for instance, can cut off the trade lines of Cuba. It can do so not only by curtailing its own trade with Cuba but also by threatening punitive action against those who trade with or invest in Cuba. This is not just a hypothetical possibility—the Helms-Burton Act in the US Congress is testimony to how it can actually happen. Cuba, on the other hand, can do little to hurt the American economy or polity. Likewise, China can do things to Taiwan, which Taiwan can in no way reciprocate.

The author is grateful to Shanti Prasad for editorial suggestions.
First published in *Indicators*, Vol. 1, (2002), 7787.

As the world shrinks and powerful governments develop a variety of instruments and ways to influence the lives of citizens in other nations, it is no longer enough for people to be able to choose the leaders of their own nations. Since democracy requires the ability to choose the leaders who have influence over one's life, citizens in a globalizing world such as today's, especially those of poor, weak nations, need to be able to vote in the elections of the rich and powerful nations. Since such transnational voting does not happen (and even its hypothetical suggestion sounds absurd to us), globalization is bound to cause a diminution of global democracy. This is the 'basic proposition' of this chapter.

18.1 GLOBALIZATION AND INTERFERENCE

The big and the powerful have always considered it natural to encroach on the sovereignty of others. This attitude is best exemplified by the story, no doubt apocryphal, that used to make the rounds in India, of the Indian diplomat in Moscow showing a map of South Asia to Stalin. 'India is a very big country', Stalin observed, and then, pointing to Sri Lanka, 'What is the name of this little Indian island?' 'This is not an Indian island, sir', the diplomat responded. 'It is a sovereign nation'. 'Why?' Stalin asked.

Fortunately, in today's world, to have influence in the affairs of another nation, it is no longer necessary to occupy the other nation's land or even to go to war with it. Moreover, even when there is war, unlike those of yesteryears, these are less battles of territory than acts of reprisal or punitive action to make nations conform to certain kinds of behaviour. And, given the march of technology, the stronger nations are able to take this action with very little direct confrontation and loss of life. A simple bit of statistics captures this changing nature of war. If we take the ratio of the number of dead civilians to the size of military casualties in armed conflicts, we find that it has risen almost relentlessly, from less than 1:1 in the first decade of the twentieth century to more than 5:1 in the 1990s. At least in part, this increase reflects how powerful nations can take action against others with minimal military casualty in their own ranks.

More important is the fact that military action, even of this arm's length kind, is now often unnecessary. Thanks to globalization, there is a variety of instruments that nations can use to influence outcomes elsewhere. Foremost among these is money. Thanks to the ease of instantaneous electronic links and the improving system of global guarantees, capital has flown across national boundaries like never before. It is true that in the heydays of imperialism, capital did go from one nation to another, but, almost invariably, it took the form of money moving between the territories of the imperialist nation and its colonies. In other words, the presence of the army in (or with direct control over) another territory was a prerequisite before money went there. That is no longer the case. With a gentle tap on a mouse, one can today move funds to lands with which one may have no tangible contact. And capital has flown to distant lands at unbelievable rates. In 1969 the World Bank, for instance, lent US $1.8 billion. By 1999 this amount had grown to US $32.5 billion. Private-sector capital flow has grown even faster, and by 1999 World Bank lending had become a miniscule 2 per cent of the total private-sector lending to developing nations. A rapid withdrawal of such capital can have devastating effects on the debtor nations, as we saw in 1997 when the Asian superperforming economies succumbed to financial crisis.

Like capital, international trade (after a decline in the years between the two world wars) has risen steadily. These global linkages have fuelled unprecedented growth rates of national incomes (during the 1990s China grew at around 8 per cent a year and India 6.5 per cent), but they have also created new vulnerabilities. Governments and international organizations can now use the threat of disrupting these flows (or the lure of releasing greater flows of money or goods) to enforce conformity to certain kinds of behaviour. And such threats have been used. International organizations have given money while insisting that the developing countries fulfil certain conditions, many of which have had nothing to do with ensuring repayment. These conditions have, at times, even been contradictory, such as requiring the debtor nation to practice democracy and to privatize certain key sectors, unmindful of the fact that this move was often against the collective wishes of the people. Some of these conditions have been blatantly in the interest of the donor nation. In 1998, during the Asian crisis, the rescue package put together with money from several industrialized nations, most prominently Japan and the United States, had clauses that required Korea to lift bans on imports of certain Japanese products (which Japan had long been trying to sell to Korea) and to open up its banking sector to foreign banks (an item that had long been on America's bilateral agenda with Korea). These were such surprising clauses that even a cautious magazine such as the *Economist* commented on their obviously donor-motivated raison d'etre. Some of these demands may well be good for the borrower, but that is not the issue here. From the point of view of assessing global democracy, what is relevant is that people of the weaker nation have very little say in the imposition of these policies.

It is once again these same features of globalization that have made it possible for some nations to use sanctions to bring other nations into compliance. Even terrorist groups have tried to harness the use of modern technology to wield influence in distant terrain. In addition, nations have sought to leverage the sanctions by threatening action not just against the nation that it seeks to punish but against other nations that do not join in the punishment. The classic example of this is the Helms-Burton Act in the United Staes, which seeks to take punitive action against companies and governments that trade and invest in Cuba. Clearly this practice will have a profound effect on the lives of the Cubans (and some on the nations, such as Italy and Canada, that do business with Cuba). Yet they have little say in the matter, because they have no say in the choice of the US president. There is enough evidence to believe that Clinton was a reluctant signatory to this contentious act, which has been challenged by some European nations and Canada. But he realized that signing this would shore up the conservative side of his image, and, at the same time, Cubans and Canadians having no say in his being president of the United States meant that he had no substantial voter constituency that would have responded negatively to his signing this act.

Given that the benefits of democracy are ample, as modern research has shown, this erosion of global democracy must have negative fallout. Indeed, it is arguable that the rise in global unrest and instabilities is a manifestation of this retreat of democracy, and the inchoate demands of the protestors in the streets of Seattle and Washington in 2001 may be founded in an intuitive but ill-articulated perception of this erosion of democracy. This condition may explain why these protests have attracted a disproportionate number of anarchists

18.2 DOLLARIZATION AND DEMOCRACY

The lack of global democracy is also holding back some important changes that are needed for a more efficient functioning of the world economy. One consequence of the freer flow of capital from one country to another, and which has received less than adequate attention, is that it has led to an intertwining of different markets. Thus a fall in the Thai housing market can cause a collapse of the Thai baht in a way that could not have happened before. Likewise a fall in the Indian rupee can today cause a meltdown of the Indian stock market in a way that was inconceivable even ten years ago.

The reason for this vulnerability is the large presence of overseas investors in any nation. Suppose you are a New Yorker who wants to buy shares, perhaps through some foreign institutional investor or some mutual fund, in the Mumbai stock market. To do that, your dollars will first have to be converted to rupees and then used to buy the shares. Your aim, like that of virtually all overseas investors, is not to hold rupees but to make some money and, eventually, convert back to dollars (basically, any globally accepted currency) to spend on clothes, housing, and so on in the United States. Now suppose the Indian exchange rate begins to fall. As a foreign investor, you will have good reason to sell the Indian stocks and take your money out of India because, even if the stock prices remain unchanged, your earnings in dollars will be smaller if you leave your money in India with the rupee falling. So while a fall in the exchange rate with no decline in stock prices gives the Indian investor no reason, *ceteris paribus*, to take money out of the stock market, the foreign investors will have good reason to get out of India. But if there are a sufficient number of foreign investors and they all begin to sell their stocks, the stock prices will begin to decline, and then there will be reason enough for Indian investors to sell the stocks as well. So now the stock market will be brought down as well.

Likewise, if the Thai housing market begins to collapse and it hurts the profitability of Thai companies, the collapse may cause the stock prices to fall. If Thailand had no foreign investors, it would be the end of the matter. But if there are foreign investors, they will, after selling their stocks, change their bahts for dollars since they had gone into Thailand originally for the Thai stock market. Hence, now the exchange rate will start collapsing.

These linkages between domestic markets in a developing country and the exchange rate are new and have played a larger role in the rapid spread of the 1997 East Asian crisis from one country to another and from one market to another than has been recognized by economists. But governments and citizens have been aware of this at the level of intuitive perception, which has given rise to demands for currency unions and dollarization. Indeed there are gains to be had from groups of countries coming under single currencies and, ultimately, converging to a one-currency world, something that Stanley Jevons had recommended, prematurely, in 1878.

The main advantage of dollarization for a developing country is that it will delink the various domestic markets. The housing market suffering losses will be less likely to disrupt international trade, for instance. To dollarize, somewhat surprisingly, is the equivalent of compartmentalizing the bellows of a hovercraft, so that one puncture does not bring down the whole craft.

The main disadvantage of dollarization is the loss of autonomy. By coming under the control of the US Federal Reserve Board, Argentina lost control over its own monetary, and to a certain extent even fiscal, policy. Even if Argentina had gone for dollarization, on the ground that under the former Currency Board system (which locked the peso unalterably to the dollar) it is already virtually dollarized, most other nations will consider the cost of coming under another's central bank control too big a loss of autonomy to contemplate. The only way the advantages of common currencies will be feasible is if we can think of central banks, which are answerable to all the nations that use the common currency. The European central bank does have this feature of multicountry democracy, which is the reason the euro is expected to be a net gain for all the nations that share it.

Unfortunately, global democracy is so underdeveloped that currency unions for countries that need them most—namely, the developing nations—remain a faint hope. Not only can we not think of a global democratic government (an idea that Bertrand Russell had campaigned for) with a global central bank, the main international financial institutions that we have, such as the IMF and the World Bank, remain largely answerable to the industrialized nations. Even when such organizations work for the poor countries, it is the *industrialized nations'* perception of the poor nations' well-being that is catered to.

18.3 DEMOCRATIC GLOBAL INSTITUTIONS

What can be done about the erosion of global democracy? Since in the social sciences political correctness demands optimism, it may sound strange to respond with 'not very much'. But that would be true, at least, in the next one or two decades. Utopian schemes such as a global government or a global bank that is answerable to all nations in the world are a distant dream. The process of globalization will course on, and intercountry democracy will continue to get bruised. It will be some time before this situation can bring us to discussing global governance and banking. In the meantime, what is open to us are small measures, namely, ones of strengthening the democratic structure of global institutions such as the World Bank, the IMF, and the WTO. These are small but extremely important measures, for they can contribute to global stability, and moreover, even as an end in themselves, they are morally desirable. The lesson, therefore, is ironically quite different from what the protestors in Seattle and Washington want, as do some archconservative groups in industrialized nations, to wit, the dismantling of these organizations. On the contrary, we need to restructure these organizations and recognize that they have an especially important role to play today. The process can, of course, be hijacked so that these institutions become batons in the hands of the powerful nations. On the other hand, it is naïve to believe that removing these institutions will defang the powerful. We need these mitigating institutions, but those committed to global democracy will have to be ever vigilant. There is enough evidence that powerful politicians in powerful nations like to think of international institutions as valuable only to the extent that they can use them to their own advantage.

On 20 January 2000, Senator Jesse Helms (R, NC), arguably the most important congressional voice until recently in the United States, told the council members of the UN: 'If the United Nations respects the rights of the American people, *and serves them as an effective tool of diplomacy, it will earn and deserve their respect and support.* But a United Nations that seeks to impose its presumed authority on the American people without their consent begs for confrontation and—I want to be candid with you—eventual US withdrawal'. (My italics—but also my hunch that Helms would want it this way.) Helms had gone on to express distaste for 'supranational institutions', including the international criminal court, which was created last year. What is most damaging about this kind of remark is that any organization that meets with Jesse Helms' approval immediately becomes suspect from the point of view of the world as a whole. The last thing that a credible international organization will now want is Jesse Helms' blessing.

What we need to work on is to give nations, rich and poor, equal say, at least in international organizations that are supposed to play a mediating role in world economics or international relations. This horizontal equity is violated in most organizations through at least one of two routes. First, there is the open channel, which gives a larger share of votes to the nations contributing more to these organizations. This is certainly true of the IMF and the World Bank. The second route is through the lack of transparency of decision-making. One can see the importance of this for democracy by looking at policymaking within a country. If the process of decision-making is visible to all, it becomes difficult for any group or lobby to hijack the agenda. Big businesses and the military, which are usually close to government, are able to push through their interests much more in Pakistan than in India, and the reason is simply that government is more open to scrutiny in India. The same holds true for international organizations. Big and powerful nations, by virtue of contributing senior personnel and money to these organizations, have much greater access to them. So if decisions occur behind opaque walls, they are much more able to divert the agenda to suit their own interests. Take the case of the WTO. While it does follow the important principle of one-country–one-vote, it is widely perceived as a preserve of powerful and rich nations. This perception is the result of what some analysts call the 'green room' effect, that is, what goes on behind the scenes. It is the green room where the agenda gets set on what is to appear on the table for all member nations to discuss and vote on, and a lot of the end results get determined at that stage. If the WTO is to become a more democratic institution, it must not allow its green room to be hijacked by a few.

This problem is nowhere more obvious than in the drafting of international labour standards, which are ostensibly being designed in the interests of the workers of developing countries. But, ironically, the biggest opposition to such standards has come from the poor countries, and not just from their governments but also from the trade unions and grass-root workers. The apprehension in the Third World is justified. The form that these standards are tending to take—and the increasing talk of using trade sanctions to impose these standards—is close to what protectionist lobbies in industrial nations seek. This is not surprising, given the greater access that the lobbies of rich countries have in the corridors of power in international organizations.

Many people seem astonished by this criticism of global institutions. The fact that it is seen as outrageous just to *question* the practice of richer nations (who contribute more funds) of exercising more voting power in these organizations simply shows how far away we still are from *global* democracy. It does not at all seem outrageous that Bill Gates does not have multiple votes in the US elections on the ground that he contributes more to government coffers. In fact, the suggestion that he could have more votes sounds outrageous because democracy within a nation is a much more settled idea. But it is time to give serious thought to how we can give more equal voting power to different nations, irrespective of their wealth. Money itself gives a lot of advantage. One of the basic tenets of democracy is that we should not compound this advantage by giving the rich extra voting power.

In the same Security Council address mentioned above, Jesse Helms complained that, while the UN 'lives and breathes on the hard-earned money of the American taxpayers', UN officials had the audacity to declare that 'countries like Fiji and Bangladesh are carrying America's burden in peacekeeping'. Plainly, he was piqued not because some individual UN official had said that might not be accurate, but because it was minor nations like 'Fiji' and 'Bangladesh' that were being given such importance. Clearly, the fact of each country's having equal say is as yet unacceptable even as an idea.

But, fortunately, opinions change. Multinationals nowadays talk in terms of environmental responsibility and the importance of respecting labour standards, even when those ideas imply having to take a cut in profits. This attitude seems to violate age-old beliefs and also the textbook description of multinational corporations. When the one-person–one-vote idea first came about, the rich feudal landlords must have been shocked and cried foul at this blatant injustice and the chaos in the process of decision-making that it would cause. But no longer does this democratic principle *within a nation* seem strange.

Now, with the call for restructuring international organizations ringing out from the streets of Seattle and Washington, and also congressional committees such as the recent one headed by Allan Meltzer, this is a good time for us to think through some of these issues, not just from the point of view of economic efficiency and greater cost effectiveness but from the point of view of representation of the poor. For the sake of global stability, economic efficiency, and also morality in international relations, we must try to impart a greater democratic structure to our international organizations. This may not be in the individual interest of every state, especially the big and the powerful, but is certainly in the interest of us all, collectively.

Index